140/-
AT 169891
126/-

STRENGTH AND FAILURE OF VISCO-ELASTIC MATERIALS

BY

G. M. BARTENEV AND YU. S. ZUYEV

Translated by

F. F. and P. JARAY

PERGAMON PRESS

OXFORD · LONDON · EDINBURGH · NEW YORK
TORONTO · SYDNEY · PARIS · BRAUNSCHWEIG

Pergamon Press Ltd., Headington Hill Hall, Oxford
4 & 5 Fitzroy Square, London W.1
Pergamon Press (Scotland) Ltd., 2 & 3 Teviot Place, Edinburgh 1
Pergamon Press Inc., 44–01 21st Street, Long Island City, New York 11101
Pergamon of Canada Ltd., 207 Queen's Quay West, Toronto. 1.
Pergamon Press (Aust.) Pty. Ltd., Rushcutters Bay, Sydney, N.S.W.
Pergamon Press S.A.R.L., 24 rue des Écoles, Paris 5e
Vieweg & Sohn GmbH, Burgplatz 1, Braunschweig

First English edition 1968

This is a translation of Прочность и Разрушение Высокоэластических Материалов (Prochnost' i Razrusheniye Vysokoelasticheskikh Materialov) published by Izdatel'stvo "Khimiya" Moscow 1964.

Library of Congress Catalog Card No. 67-23704

08 003110 2

CONTENTS

TRANSLATORS' PREFACE

IN THE translation of technical literature it is most important to maintain, as far as possible, the exact meaning of every word. We have, therefore, sometimes chosen to render some word verbatim if it appeared to us more descriptive, even if a different word might be in more common usage in English. Occasionally, in order to avoid ambiguities, we have put one or the other in parenthesis.

For instance:

The original title of the book is: *Prochnost' i Razrusheniye Vysoko-elasticheskikh Materialov*. The literal translation of the word "vysoko" is "highly". In English, however, the word "highly" in this context does not describe a material with a nearly ideal elasticity, but implies a degree of super elasticity. The physical states of the materials described in this book, however, are often clearly defined in scientific English as "visco-elastic", though they may refer to the low-temperature, high-elastic end of the visco-elastic range. Professor Bartenev has agreed that the translator's rendering of the term as "visco-elastic" is appropriate, but in general the words "visco-" and "high-elastic" should be taken as ambivalent.

The Russian word *deformatsia* would in English be most commonly rendered as "strain". We have generally translated it as "deformation" which appeared to us clearer and have used the word "strain" more in the mathematical sense (σ/E).

For the Russian word *dolgovechnost'* the most commonly used terms in English are: "time to failure, life, duration of loading, duration under load, breaking time", etc. In conformity with the Russian original we have used the term "durability" which is correct, descriptive and covers all the many instances and connections in which it is uniformly employed in the book.

The word *fluktuatsia* we translated sometimes with "fluctuation", sometimes with "oscillation", depending on the context.

In agreement with the diagrams we have sometimes used the words "temperature of flow" rather than "melting temperature" or "melting point".

Further, throughout the book, we have used the literal transcription of the Russian nomenclature of their synthetic rubbers which are as follows:

SKBN means butadiene-acrilonitrile rubbers.
SKN–18 contains 18% nitrile.
SKN–26 contains 26% nitrile.
SKN–40 contains 40% nitrile.
SKS–30–SKB–30 means butadiene-styrene rubbers.
SKS–30 contains 30% styrene.
NK—natural rubber, etc.
Nairit polyisoprene rubber—similar to Neoprene.
But have used Nylon for Kapron throughout, etc.

Acknowledgement and our thanks are due to Professor Bartenev himself for his helpful discussions during the translation and to Dr. L. R. G. Treloar of the Department of Polymer and Fibre Science, Manchester College of Science and Technology, whose advice, particularly with the first chapters, has been invaluable.

F. F. and P. JARAY

AUTHORS' PREFACE

THIS book on strength and failure of visco-elastic materials is a monograph, describing the present state of the problems and summarizing the results of the numerous investigations into the strength, durability† and mechanism of failure of these materials.

Under visco-elastic materials we understand linear and three-dimensional polymers or materials based on them which possess visco-elasticity and flexibility over a wide temperature range, including low temperatures. In foreign literature the equivalent term "elastometric" has in recent times frequently been employed; this, in our opinion, is a less felicitous term. The most typical representatives of visco-elastic materials are vulcanized and unvulcanized rubbers and also other linear amorphous and partially crystalline polymers with a low glass-transition temperature.

In order to give an up-to-date presentation of the problems of strength as a whole, a considerable place has been allotted in the monograph to information about the theory of strength and to the mechanism of failure and deformations in solids, including solid polymers. Statistical theories of strength, which at the present time play an ever increasing role in calculations of the strength of materials and structures, are examined in a special chapter. The basic material of the book is devoted to the visco-elastic polymers which, as distinguished from the classic solids, possess a clearly expressed specific strength, connected, partly, with the entropic character of their deformations and their capacity for orientation.

† *Translators' note*. The most commonly used terms in English are "time to failure, life, duration of loading, duration under load, breaking time", etc. In conformity with the Russian original we are using "durability" which is correct, descriptive and covers all the many instances and connections in which it is uniformly employed in the book. (See Translators' Preface.)

The greatest attention is given to the physical and physico-chemical aspects of strength problems of visco-elastic materials as this approach gives the best explanation of the mechanism of failure and, consequently, provides a basis for the selection of ways of strengthening materials and the creation of sound methods for testing them. The kinetic character of the process of failure under the effect of stress and thermal movement (fluctuation or oscillation mechanism of failure) is especially emphasized, and also the correlation of durability at static and dynamic conditions of deformation. It must be emphasized that the material, presented in connection with these questions about vulcanized and unvulcanized elastomers and rubbers has a general application for all polymers which are in a visco-elastic state.

In connection with the fact that there exists only unco-ordinated literature on the tearing of rubbers, and that there is a lack of clarity regarding the physical meaning of the characteristic energy of tearing, a special chapter on the theory of the tearing of rubbers has been included in the book.

A considerable part of the monograph is devoted to the questions of failure (cracking) and durability of visco-elastic materials in various chemically and physically aggressive environments. The consideration of the effect of environment on the strength of polymers is indispensable, as under normal conditions traces of chemically active impurities in the atmosphere have a considerable influence on those properties, as well as in connection with the expansion of the field of utilization of polymeric materials in various aggressive media.

Failure of the materials under the action of mechanical stresses takes place by the breakage of Van der Waals forces or of chemical bonds, and may therefore be regarded in general as a consequence of the overcoming of the interaction between the particles of the body. This process can take place not only under the action of mechanical stress, but also under the influence of other factors (heat, solvents, chemical agents) which lead to the fact that the laws are the same both for static fatigue and corrosion failure. This similarity between processes of corrosion failure and static fatigue is by no means limited to elastomers but is inherent in all materials and is from this point of view of more general interest.

Chapters 1–9 of the monograph have been written by Doctor of Chemistry, Professor G. M. Bartenev (W. J. Lenin Laboratory for

Physics at the M.S.P.I.); a great part of Chapter 1 was written by him jointly with candidate of physico-mathematical science, I. V. Rasymovska. Chapters 10–14 were written by Candidate of physical science, Y. S. Zuyev (Science Research Institute of Rubber Industry).

G. M. BARTENEV

Y. S. ZUYEV

CHAPTER 1

THE STRENGTH OF SOLIDS

THE mechanism and the laws governing the failure of brittle solids[1-8] and of solid polymers have features in common, especially at low temperatures. The statistical theory of strength, developed for brittle bodies, is applicable to all materials. In connection with this the nature and the fundamental laws governing the strength of solids, and in particular of solid polymers, are briefly examined in this chapter.

In crystalline and amorphous solids two fundamental types of fracture can be observed, namely brittle and ductile failure. Brittle fractures are those in which the applied stress does not cause any noticeable residual deformation connected with viscous flow, in the case of amorphous materials—or ductile deformation, in the case of crystalline materials. In a brittle failure the section of the specimen is the same before and after fracture, in contrast to the contraction of the specimen, to a greater or lesser extent, in the region of the fracture in a ductile failure.

The strength of a brittle material is characterized by one maximum value corresponding to the transition from elastic deformation to failure. The strength of a plastic or ductile material is characterized by two maximum values, the first corresponding to the transition from elastic deformation to a ductile one (sometimes the material, or a part of it, does at this moment not yet lose its load-bearing properties), the other to the transition from ductile deformation to fracture. Therefore the strength, in the widest sense of the word, is defined as the capacity of the material to resist not only failure but also ductile deformation.

Ductile deformation is observed in those cases where the yield point of the solid is lower than its brittle strength. The relation between these values changes with a change in the test conditions. The transition from a brittle failure to a ductile one with a rise in temperature is illustrated by the diagram (Fig. 1) suggested by Joffé. The temperature dependence of the yield point is steeper than that of the brittle strength. The two curves intersect at some point. This point corresponds to the temperature T_{br} of brittle

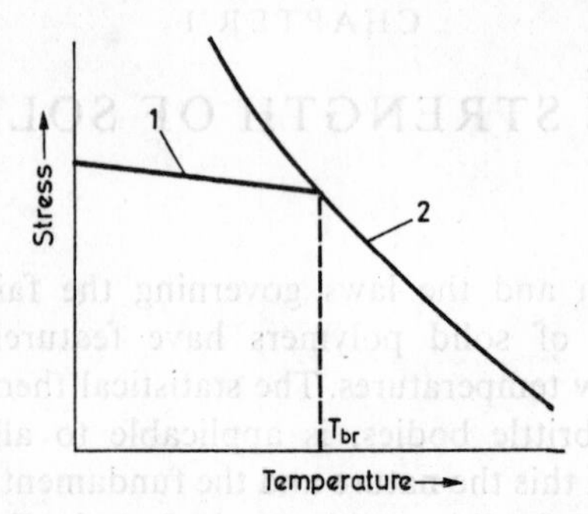

FIG. 1. Transition from brittle to ductile fracture with increasing temperature (diagram after Yoffe). 1 = brittle strength, 2 = yield point.

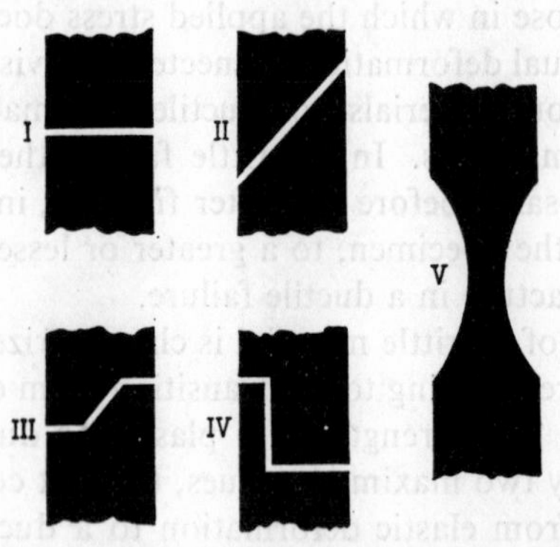

FIG. 2. Elementary types of failure under tension. I. Tear in plane normal to greatest stress. II. Failure in plane of greatest shear stress. III. Tear and shear. IV. Tear and shear in the direction of tension. V. Change of shape through the forming of a "neck" (deformation by shear at transition beyond yield point).

fracture. At higher temperatures failure occurs by ductile breakdown while at lower temperatures brittle fracture is observed.

Brittle fracture is the more dangerous one, therefore it was subjected to thorough systematical investigation, both theoretical and experimental. Brittle fracture of materials arises either as a result of tensile stress (parting) or as a result of shearing (sliding), the kind of failure being determined by the structure of the solid body and also by the type of stress. Under tension—the most dangerous kind of stress—the failure arises mainly through tearing while under compression it occurs mainly by shearing. In complicated cases of failure the fracture is a combination of local tearing and shearing along the planes of the easiest failure. Elementary types of failure which are met with in the tensile testing of solids are illustrated in Fig. 2.

1.1. The Theoretical Strength of Solids

Repeated attempts have been made to calculate the strength of solids, starting from the forces of interaction of their constituent particles (molecules, atoms, ions). The structure of the material is regarded as ideal, not impaired by any imperfections, defects or damage (ideal monocrystal).

The calculated strength, in contrast to that found from experimental data, is called the theoretical strength. The theoretical strength depends on the nature of the force of interaction between the particles (ionic, covalent, metallic bonds and others) and on the structure of the material. An exact calculation of the value of the theoretical strength is an extremely complicated task. Therefore more or less rigorous calculations have been carried out up to the present only for rock salt, the monocrystal of which is a three-dimensional lattice of ions of Na^+ and Cl^- between which Coulomb forces of attraction act.

The theoretical strength of rock salt has been calculated by Born[9] for the case of a uniform three-dimensional tension, and by Zwicky[10] for uniaxial tension. The results of these calculations, as also of subsequent calculations of theoretical strength, are, strictly speaking, correct only at a temperature of absolute zero.

For a homogeneous deformation the type of dependence of each atom's or ion's lattice energy potential on the distance between the particles in the direction of the tension is expressed by the formula of Mie

$$U = -\frac{A_m}{x^m} + \frac{B_n}{x^n}$$

where U is the energy potential of interaction, calculated for a single particle; x is the distance between particles in the direction of tension; A_m and B_n are constants characterizing structure and type of stress (for rock salt $m = 1, n = 9$).

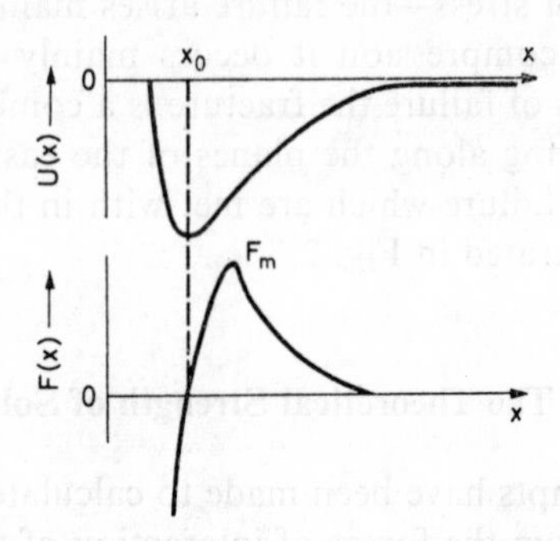

FIG. 3. The dependence of the energy potential U and absolute value of the quasi-elastic force F on the mean distance in the ionic crystal lattice.

The energy potential is expressed by a curve, as shown on Fig. 3. The quasi-elastic force F, acting on a single particle, is $-\partial U/\partial x$, and the external force (equal but opposite in sign) is $\partial U/\partial x$. With increase of the distance x the energy potential U approaches a limit, usually zero, and the absolute value of the quasi-elastic force F passes through a maximum.

If the external force acting on the particle is smaller than the maximum quasi-elastic force F_m, only a stretching of bonds takes place without their rupture; if it is greater than F_m, the material breaks up into separate particles (in the case of three-dimensional stress); or separates along particular ionic or atomic planes, perpendicular to the direction of stress (in the case of uniaxial tension). The value of the maximum quasi-elastic force F_m, multiplied by the number of atoms or ions, situated on a single plane of the solid body in the unstressed condition at right angles to the direction of tension, is equal to the theoretical strength.

The theoretical strength of rock salt, according to the calculations of Born on three-dimensional stress, is approximately 170 kg/mm^2; according to the calculations of Zwicky, for uniaxial, stress, 200 kg/mm^2. De Boer[11] has refined Zwicky's calculation, assuming Van der Waals forces of interaction between particles and, depending on the method of calculation, obtained two different values: 270 kg/mm² and 400 kg/mm².

Strict calculations of the theoretical strength by the methods of Born and Zwicky are impossible in practice for the majority of solids (and for amorphous solids also in principle). In connection with this Polanyi, Orowan and others have suggested semi-empirical methods for the calculation of the theoretical strength. Orowan[12] calculated the maximum quasi-elastic force on the separation of two single atomic planes of a solid body. Since for the majority of solids the energy potential of interaction of particles is unknown, Orowan employed the following approximate equation, expressing the dependence of the absolute value of the quasi-elastic force F on distance x between particles

$$F = F_m \sin\left[\frac{\pi}{a_0}(x-x_0)\right]$$

where a_0 is twice the distance from the position of equilibrium to the position corresponding to the maximum of the quasi-elastic force (Fig. 4).

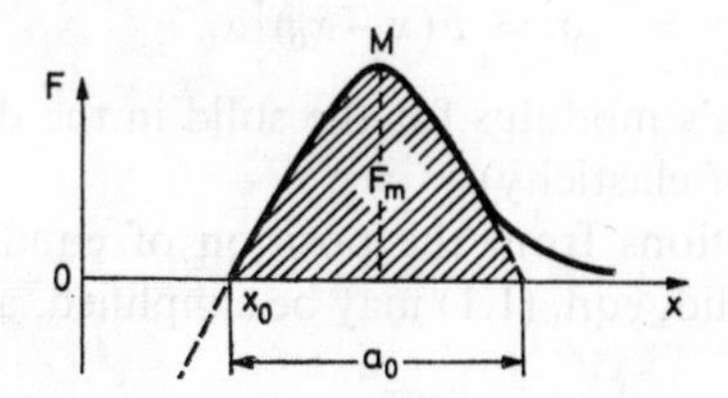

FIG. 4. Dependence of the quasi-elastic force on the interparticle distance (Orowan).

x_0 is the distance between two particles at equilibrium.

If two single planes move away from each other sufficiently slowly (quasi-statistically), the quasi-elastic force is exactly equal to the external force, and the stress σ is equal to

$$\sigma = NF$$

where N is the number of particles, on a single atomic plane; F is the absolute value of the quasi-elastic force acting on a single particle.

From this we have

$$\sigma = \sigma_m \sin\left[\frac{\pi}{a_0}(x-x_0)\right] \tag{1.1}$$

where $\sigma_m = NF_m$ according to Orowan is theoretical strength.

The work done by the external force along the path a_0 (see Fig. 4) equals

$$\int_{x_0}^{x_0+a_0} \sigma_m = \sin\left[\frac{\pi}{a_0}(x-x_0)\right] dx = \frac{2a_0}{\pi}\sigma_m$$

Orowan assumes that this work is transformed into the energy potential of the newly formed free surfaces with an area of 2 cm², therefore

$$\frac{2a_0}{\pi}\sigma_m = 2\alpha_{\text{surf}}. \tag{1.2}$$

where α_{surf} is the free surface energy of the solid body which is considered as known.

The constant a_0 in eqn. (1.1) is defined by Orowan according to Hooke's law, i.e.

$$\sigma = E(x-x_0)|x_0$$

where E is Young's modulus for the solid in the direction of the stress (modulus of elasticity).

At small deviations from the position of equilibrium, where Hooke's law is valid, eqn. (1.1) may be simplified, and becomes

$$\sigma \approx \sigma_m \frac{\pi}{a_0}(x-x_0)$$

From this it is easy to define the constant a_0:

$$a_0 = \frac{\sigma_m \pi x_0}{E} \tag{1.3}$$

Substitution of this expression in eqn. (1.2) leads to the formula of Orowan, which is

$$\sigma_m = \sqrt{\frac{\alpha_{\text{surf}}E}{x_0}} \tag{1.4}$$

For the calculation of the theoretical strength the simplified formula derived from eqn. (1.3) is often used. If one takes into account that the maximum quasi-elastic strength is reached at 10–20% elongation of bonds, then $a_0/2$ is equal to $0{\cdot}1$–$0{\cdot}2x_0$; consequently, $\sigma_m = \chi E$ where $\chi = 0{\cdot}07$–$0{\cdot}13$ or on the average

$$\sigma_m \approx 0{\cdot}1E \tag{1.5}$$

Previously to Orowan a cruder calculation had been made by Polanji[14] who assumed the quasi-elastic force to be constant from x to $x_0 + a_0$, and thereafter zero.

The application of Orowan's equation (1.4) to polymers is difficult as there are no data about their free surface energy, especially in the orientated state for which the calculations of the theoretical strength are also of interest.

Kobeko[15] employed a different method of calculation of the theoretical strength of polymers, using the Morse formula

$$U = -De^{-B_0(x-x_0)}(2 - e^{-B_0(x-x_0)})$$

where D is the energy of dissociation per single particle.

$$B_0 = \sqrt{\omega/D}\cdot\frac{1}{A_0};$$

ω is the frequency of oscillation of particles; A_0 is the amplitude.

The maximum value of the quasi-elastic force F_m found from the Morse formula equals $B_0D/2$; this leads to the theoretical strength $\sigma_m = NB_0D/2$, where N is the number of particles (in orientated polymers the number of chains) per unit of a cross-sectional area of the unstressed specimen. The energy of dissociation is calculated from the heat of combustion, the constant B_0 is calculated from optical data.

From the Morse formula the modulus of elasticity can also be calculated, i.e. $E = 2NB_0^2Dx_0$. Therefore $\sigma_m = E/4B_0x_0$. According to the calculations of Kobeko for the various types of bonds $\sigma_m/E \approx 0{\cdot}1$ and eqn. (1.5) is again obtained.

In this way the approximate eqn. (1.5) is applicable for any type of bonds the different rigidity of which is characterized by the value of Young's modulus E.

The calculation of the theoretical strength of bodies between the particles on which act Van der Waals forces shows that for dipole forces the theoretical strength is approximately fifty times less than for chemical bonds. From the value of Young's modulus and the theoretical strength, calculated with eqn. (1.5) one can form an opinion concerning the type of bonds responsible for breakage. This is especially important for calculation of the strength of polymers.

In Table 1 are listed approximate values of theoretical and practical (measured) strengths of various materials.

TABLE 1

Theoretical and Technical Strength of Various Materials

Material	Young's modulus E (kg/mm²)	Theoretical strength (kg/mm²)	Technical strength (kg/mm²)	Type of chemical bond
Steel	2×10^4	2000	100–250	Metallic
Rock salt	4×10^3	400	0·44	Ionic
Glass	$6–8\times10^3$	600–800	8–12	Semi-polar
Poly (methyl-methacrylate)	400–600	40–60	10	Covalent
Nylon fibres	450	45	70–80	Covalent
Acetate fibres	600	60	13–17	Covalent
Asbestos fibres	2×10^4	2000	190–300	Semi-polar
Glass fibres	$8–10\times10^3$	800–1000	200–300	Semi-polar
Flax	$1{\cdot}1\times10^4$	1100	80–140	Covalent
Filiform iron crystal ("whisker")	2×10^4	2000	1400	Metallic

For Nylon fibres the theoretical strength was found to be less than the technical strength. This physically unwarranted result indicates that the formulae for the calculation of the theoretical strengths of solids must be used with great caution when dealing with polymers. This, obviously, has its reason in the fact that the moduli of elasticity of solid polymers are basically determined by intermolecular forces (the modulus of elastic tension of the individual polymer chain is one or two orders larger), while the strength is determined by chemical bonds.

Not long ago eqn. (1.5) was much more precisely defined by Demishev and Rasumovskaya.[16] They showed that the coefficient

χ depends on the exponents m and n (in the formula of Mie), in the following way:

$$\chi = \frac{1}{m+1}\left(\frac{m+1}{n+1}\right)^{k'}$$

where $k' = (n+1)(n-m)$, $n > m$ for all materials.

For different solids the values of χ are different, and lie within the limits 0·03–0·15. Demishev[17] suggested a method of assigning values to m and n from data from acoustic and dilatometric measurements.

1.2. Actual Strength of Solids

The theoretical strength is usually greater by two to three orders of magnitude than the actual observed (technical) strength of materials. The fundamental reasons for this phenomenon were first explained by Griffith.[1]

According to Griffith there are in any material, especially in its surface layers, microcracks of various dimensions and orientations. Under the action of the applied stress there appears on the edges of the microcracks an additional or overstress P which can exceed by many times the average stress σ in the cross section of the specimen. If the value of the increased stress at the apex of the most dangerous crack is equal to the theoretical strength σ_m, there occurs a catastrophic propagation of the crack (with a speed near the speed of sound) and the specimen breaks in two. The applied stress at this stage corresponds to the so-called maximum technical strength σ_k. For an overstress smaller than the theoretical strength, when σ is less than σ_k, the crack does not grow, according to the Griffith hypothesis.

The coefficient of the stress concentration at the apex of the microcracks equals $\beta = P/\sigma$. It depends on the form and dimensions of the crack and on its orientation relative to the direction of the tension. Therefore the greatest technical strength is not a constant of the material. It changes from specimen to specimen, as the different specimens have different critical defects.

Griffith calculated the greatest technical strength (the critical stress σ_k) from the following condition: the crack grows only when the decrease of elastic energy in the specimen in the process of crack growth (owing to the unloading of the material around the increasing crack) is equal to or greater than the increase of the energy potential due to the formation of the new fracture surface.†

The change of the elastic energy ΔW in the specimen which has the shape of a thin strip during the formation of a crack with the length c, orientated perpendicularly to the direction of the extension equals

$$\Delta W = -\frac{\pi\mu c^2\sigma^2}{4E}\,\delta \tag{1.6}$$

where μ is Poisson's constant, c is the length of the crack, E is Young's modulus, and δ is the thickness of the strip.

It was assumed that the crack has the form of a very elongated elipse (Fig. 5). The change of the elastic energy caused by the

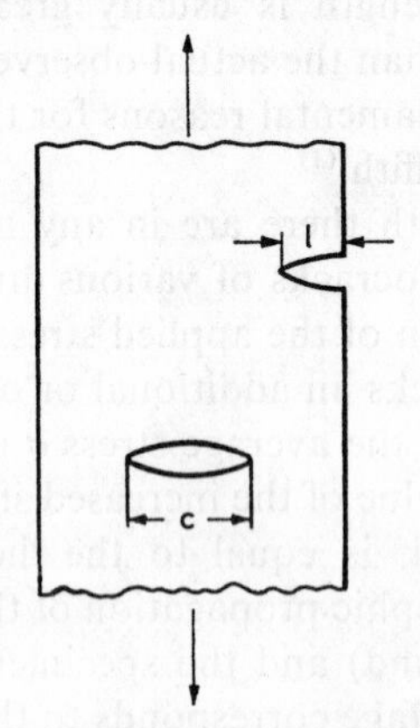

FIG. 5. Elliptical crack (Griffith).

presence of the crack is negative, as the presence of the crack leads to the unloading of the material around it, and to a decrease of the elastic energy of the specimen.

The surface energy of a crack of length c is equal to $2\delta c\alpha_{surf}$ (where α_{surf} is the free surface energy of the solid body.‡ The total

† It is assumed that the initial length of the crack is much less than the cross-section of the specimen and that the stress σ' in the section which contains the crack is practically coincident with the stress σ, calculated on the nominal area of cross-section of the specimen.

‡ Griffith understood by α_{surf} the surface tension which is not strictly correct, although in fact no considerable error arises, as the free surface energy and the surface tension are numerically similar.

energy of the specimen containing the crack equals

$$W = W_0 - \frac{\pi\mu c^2\sigma^2}{4E}\,\delta + 2\delta c\alpha_{surf} \tag{1.7}$$

where W_0 is the elastic energy of the specimen without crack; the second term of the right-hand side of the equation expresses the decrease of the elastic energy of the specimen caused by the crack; the third term represents the increase of energy of the specimen as a result of the formation of new surfaces.

The decrease of the elastic energy with increase of the length of the crack by a small amount dc is, in accordance with eqn. (1.6), equal to $\pi\mu c\sigma^2\delta dc/2E$. At the same time the surface energy increases to $2\delta\alpha_{surf}/dc$. According to Griffith the equality of these changes of energy (which is equivalent to the condition $\partial W/\partial c = 0$) is the condition for failure. It follows that the greatest practical strength of the strip containing an internal crack of length c equals

$$\sigma_K = \frac{2}{\sqrt{\pi}}\sqrt{\frac{\alpha_{surf}E}{\mu c}}$$

provided that the length of the crack c is much smaller than the width of the strip.

Subsequently, several investigators[18] have introduced a mathematical refinement into Griffith's theory which leads to an equation in which Poisson's ratio μ is missing:

$$\sigma_K = \frac{2}{\sqrt{\pi}}\sqrt{\frac{\alpha_{surf}E}{c}} \tag{1.8}$$

This equation is employed in the case of planar stress.

If microcracks (Fig. 5) are present on the edge with a length l equal to half the length of an internal crack, the greatest practical strength equals

$$\sigma_K = \sqrt{\frac{2}{\pi}}\sqrt{\frac{\alpha_{surf}E}{l}} \tag{1.9}$$

The correctness of the correlation $\sigma_K\sqrt{l}$ = const. was confirmed by Griffith in experiments with glass.

It follows from eqn. (1.9) that the strength decreases in environments which lower the free surface energy. This was confirmed by Obreimov's[19] experiments. Using Griffith's formula, Berdennikov[20] determined the free surface energy of glass in vacuum and in water under the critical stress σ_K.

Equations (1.8) and (1.9) have been derived for cracks which are perpendicular to the direction of the tension (angle $\theta = 90°$). If the crack lies at an angle to the direction of the tension different from 90°, then, according to Kontorova,[21] the quantity $\sigma_K \cdot \sin^2\theta$ should replace σ_K in these formulas.

Pines,[22] unlike Griffith who examined a thin strip with a crack, investigated the strength of a rod and obtained the following formula:

$$\sigma_k = \frac{3}{2}\sqrt{\frac{\alpha_{\text{surf}} E}{S/L}}$$

where α_{surf} is the free surface energy; S is the area of the crack; L is the characteristic linear dimensions of the specimen, for instance, the radius of the rod.

Sack[23] considering a massive specimen with an elliptical crack obtained a result which did not differ materially from Griffith's result.

Griffith's ideas were further developed in a number of works[24–26] and others, in which the problem of strength is considered from the point of view of the theory of elasticity (see also ref. 18).

The value of Griffith's theory lies in the clear formulation of the position; the practical strength as distinguished from the theoretical strength depends to a great extent on the imperfections and defects in the solid body. From this follows, in particular, that the strength in a series of identical specimens must vary from specimen to specimen and depends on the character of the most critical defect in a given specimen (see Chapter 5).

1.3. Development of the Concept of Defects and Cracks in Solids

Frenkel',[27] Elliott,[28] Barenblatt[18] and others examined the conditions under which Griffith's crack will grow or close up. On Fig. 6 are shown curves, corresponding to eqn. (1.7), for the dependence of the energy of the specimen on the length of the crack, for two values of the stress: σ_1 (curve 1) and $\sigma_2 > \sigma_1$ (curve 2). The condition of failure according to Griffith is $\partial w/\partial c = 0$, that is the maximum of the curve, corresponding to an unstable equilibrium at some critical length of the cracks (c_1 and c_2 on Fig. 6). If the length of the crack is greater than the critical length its further growth leads to a decrease of the total energy of the specimen, and the crack must grow spontaneously. If the length of the crack is less than the critical length, its decrease leads to a decrease of the total energy, and the crack must close up spontaneously. Therefore a crack of critical length is unstable. It can be seen from Fig. 6 that under a high stress the critical length of the crack is smaller than under a low stress ($c_2 < c_1$). The relation between the stress and the critical length of the crack c_k follows from eqn. (1.8): $\sigma \sqrt{c_K} =$ const.

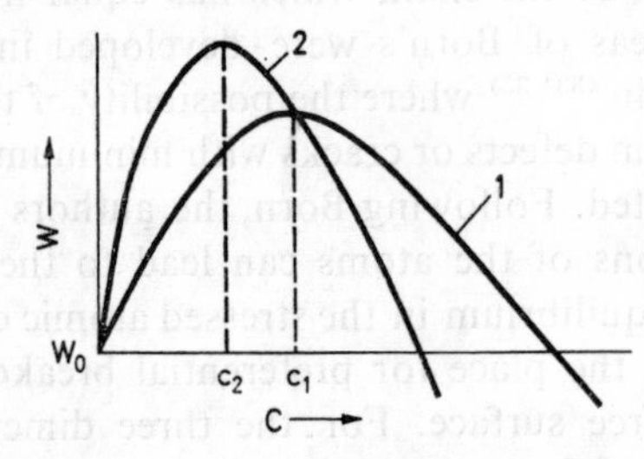

FIG. 6. Change of energy W of a specimen containing the crack with the change of the initial length of crack C.

However, in Griffith's work it is assumed that under the stress $\sigma < \sigma_K$ the length of the crack remains unchanged. In reality a crack in an ideal brittle body must completely close up under these circumstances; it was Elliott who first drew attention to this, and it was especially emphasized by Frenkel.

We must bear in mind that Griffith considered the problem at a temperature of absolute zero (for further details see refs. 18–29) and to speak of a closing-up of the crack makes sense only at temperatures above absolute zero, if one takes into account the molecular mechanism of the growth and closing of the cracks, and also the role of the thermal motion in that mechanism at $\sigma < \sigma_K$ (see § 1.11).

Smekal[5] developed a more general conception of defects as places where under the action of the tension the initiation of microcracks is possible (Griffith's initial microcrack could more correctly be called a defect). A crack, initiated by a defect, differs from the defect itself. When the stress is relaxed, according to

Rebinder's conception, the crack can close up, right up to the defect from which it had started.[30-32] This invalidates Frenkel's and Elliott's[27, 28] criticism of Griffith's theory.

The question whether there exist in solid bodies any cracks which are in equilibrium has been examined in a number of works, starting from the structure of solids. Born[9] showed that in a stressed chain of atoms, apart from the state of equilibrium in which all interatomic distances are equal, there exist other states of equilibrium in which the distance between any two atoms is greater than the remaining interatomic distances, i.e. some defects of structure are formed. The energy potential of such a chain is smaller than that of the chain which has equal interatomic distances. These ideas of Born's were developed in the works of Orlov and Plishkin,[33-35] where the possibility of the existence of similar equilibrium defects or cracks with minimum energy potential was investigated. Following Born, the authors think that the thermal oscillations of the atoms can lead to the formation of such a defect-in-equilibrium in the stressed atomic chain, and that this may become the place for preferential breakdown with the formation of a free surface. For the three dimensional lattice (within the limits of the authors' model) the formation of a crack-in-equilibrium is not possible. These questions have, however, so far not yet been fully examined.

In order to understand the nature of the strength of solids it is important to know what the initial defects are in the original unstressed material. These can be either microscopic cracks, originating (especially on the surface, the most vulnerable place of the specimen) as the result of thermal, mechanical or other influences; or defects and imperfections of structure. Cracks originate at inclusions or other inhomogeneities which possess mechanical properties which differ from those of the basic material in modulus of elasticity, or the yield point.[4] With metal the role of defects is taken by regions of imperfect contact between the grains.[22] With monocrystals[36] it is the weakened places which are the origin of plastic displacements to the surface. Also places of concentration of residual stresses, of which there are always some in the material, can be "defects", etc. According to Volkov[37] there exists even in a polycrystal with an ideal structure of the separate grains an uneven distribution of stresses, and this lowers the strength of the particular regions of the structure.

Recently dislocation theories of defects have been developed. According to those theories the brittle failure of metallic monocrystals is always preceded by a local plastic deformation in the course of which dislocation microheterogeneities are formed; these act as stress concentrators and therefore cause the initiation and development of cracks. Mott[38] and Stroh[39] maintain that through such stress concentrators an accumulation of dislocations takes place, the movements of which are decelerated by any obstacle in the slip planes; various dislocation models have been suggested by Fujita, Cottrell, and others.[40–42]

In the works of Lichtman and Shchukin[43] ideas were expressed about the mechanism of brittle failure for the general case of dislocation microheterogeneities, and the combined role of shear and normal stresses was elucidated. Two basic stages of metal failure were distinguished: (a) the origin and gradual growth of microcracks in equilibrium under the influence of shear stresses in places with high concentrations of stress, caused by the heterogeneities of the plastic deformation, and (b) the loss of "equilibrium" cracks, and their rapid distribution over the whole section of the monocrystal under the influence of the normal stress. Both processes are facilitated by a lowering of the free surface energy as a result of the penetration of adsorption-active particles on various defects of the structure inside the crystal. The development of these ideas leads to the theoretical basis of what had been observed experimentally: the constancy of the product of normal and shear stresses for the brittle failure of differently orientated monocrystals which allows to introduce the value of this product as a measure of the strength of monocrystals.

The scheme presented for failure is extremely general[44] and applicable to many particular models. As far as the dislocation mechanism of the formation of microcracks in polymers is concerned, this is possible in crystalline polymers but in amorphous polymers dislocation obviously does not exist. We shall return to the question of defects arising otherwise than from dislocations in Chapter 5, where the statistical theory of strength of solids and polymers will be investigated.

Direct structural and optical methods of investigation of the submicroscopical cracks which appear in the deforming specimen in the early stages of failure in polymers have been employed in recent times.

Zhurkov and his collaborators[45, 46] employed optical methods (change of the refractive index with transmitted light, change of angular dependence of the polarization of scattered light, change of transparency) for the investigation of submicroscopic cracks. These investigations made it possible to establish that the reduced transparency of the deformed specimens is caused by the formation of heterogeneities in them (microregions with an index of refraction different from that in the rest of the material), with dimensions of the order of 100 Å. From the comparison of the experimental indices with the calculated ones one can draw the conclusion that the submicroscopic cracks lie in a plane perpendicular to the direction of tension and have a shape similar to a disc. For the determination of the dimensions and forms of the heterogeneities the method of small-angle diffraction by X-rays has also been used. All these methods have proved effective for the study of defects (cracks) under small stresses.

The shape of the microcracks, especially near its apex where the break of bonds takes place, is important for the calculation of strength.

The origin and the shape of a crack depends first of all on whether we are dealing with brittle or with ductile materials. Let us examine the models of cracks for the ideal brittle body, shown in Fig. 7. Griffith examined microcracks of elliptical shape. Rebin-

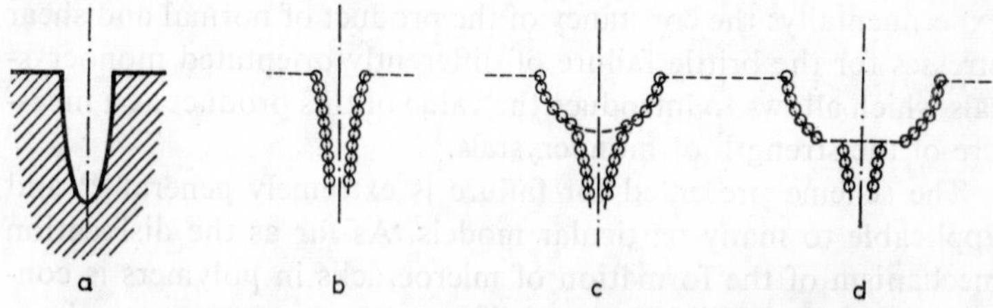

FIG. 7. Models of a crack (a, b, c are cracks in the ideal brittle material). (a) According to Griffith (within the limits of the theory of elasticity). (b) According to Rebinder.[30] (c) According to the proposals of Bartenev, Rasumovskaya, Rebinder.[47] (d) Crack in non-ideal brittle and plastic material.

der[30] was the first to introduce the conception of a crack as a cavity with asymptotically converging walls, where the free surface energy gradually changes from α_{surf} to 0. This makes the possibility of a closing of the crack after removal of the load understandable. Such a model of a crack has also been suggested subsequently in other works.[27, 28]

The model of a crack (see Fig. 7b), suggested not long ago,[47] is a generalization of Rebinder's model. The breakage of bonds takes place on the clearly defined boundary between free surface and solid centre (dotted line). A gradual slow increase of the interparticle distances takes place in the material near the tip of the crack. In the moment of the breaking of bonds the quasi-elastic force reaches its maximum value, and the interparticle distance changes rapidly. As a result the crack advances by an amount corresponding to one interparticle distance.

At the tip of the crack appear local (residual) deformations in the majority of actual solids† as the high stress concentrations exceed the yield point. In these non-ideally brittle materials, and even more in ductile ones, the atoms under the influence of the external stress already take part in ductile microdeformation displacement before and after the break. As a result the stress concentration in the roots of the cracks is reduced and their configuration changes (see Fig. 7). The growth of the cracks is irreversible: on removal of the load the permanent deformation at the apex of the crack does not disappear and the bonds cannot be re-established. This is one of the reasons why in many "Griffith" solids the cracks do not close up.

1.4. Critical Consideration of the Theory of Griffiths and of the Calculation of Theoretical Strength of Orowan

The majority of critical remarks about Griffith's theory were made by Griffith himself (limitation of application of the theory of elasticity, decrease of elastic moduli with increase of stress, influence of other cracks in the specimen on the one considered, etc.). However, these limitations do not apply to the ideal brittle body investigated by Griffith. Nevertheless, in principle there are two shortcomings in Griffith's theory which he himself does not point out.

† Apart from glasses and other brittle bodies at extremely low temperatures.

Firstly, the mechanism of failure has, according to Griffith, to be applied to the limiting case, i. e. to the athermic process ot failure, where thermal oscillations are absent (temperatures near absolute zero or critical speeds of failure). Griffith's theory has already more than once undergone criticism for not taking into accounf the time factor. But at absolute zero Griffith's theory is to a certain degree justified, as the time dependence of strength is absent at very low temperatures.

The second and fundamental shortcoming in Griffith's theory consists in the following[29]. Griffith based his definition of the critical stress on the condition that the change of the elastic energy δW during the growth of the crack in a brittle material equals the increase of the surface energy $d\varepsilon$ as a result of the formation of new free surfaces. This condition applies, however, only in the state of equilibrium when the speed of the crack propagation is zero. During the growth of the crack with a definite speed, as distinguished from zero, there takes place, as a result of a break of the interatomic bonds, a dissipation of elastic energy which ultimately is converted into heat. Consequently, corresponding to the laws of conservation of energy:

$$-\delta W = d\varepsilon + \delta Q$$

where δQ is the dissipated energy.

If $\delta Q = 0$, this means that the crack does not grow. It is probable, therefore, that Griffith's theory is in need of further refinement.

Orowan also does not take into account the process of dissipation of energy by bond breakage. This process is not connected with the usual mechanical losses in the mass of the material (for instance in the case of macroscopic inelasticity, ductility, etc.), as a brittle body is considered ideal if it shows in its various sections the property of elasticity right up to failure but is capable of dissipating the energy of elastic waves.† The process of dissipation of energy during the breakage of bonds can be understood by starting from the following argument.[29] A quasi-elastic process of extension of bonds is possible up to the point M (see

†In this definition the ideal brittle body differs in our opinion from the ideal elastic one which does not possess an inner friction nor the capacity of dispersing the energy of elastic waves.

Fig. 4), after which bond breakage takes place; this is not a quasi-elastic process but is spontaneous and proceeds with the speed of thermal oscillations.

Orowan's presentations of the mechanism of failure are also unclear. In his theory he did not consider cracks (at least not clearly) and assumed that the break takes place simultaneously over the whole area of the cross-section. However, the fact that he introduces new free surfaces into the energy calculation indicates that he is talking about the process of failure of an actual material which contains defects, as in a defect-free material failure is equivalent to dissociation, and no free surfaces are formed.

Orowan's calculation is a typical example of an identification of two different concepts: that of the theoretical strength and of the critical stress. The theoretical strength is the maximum quasi-elastic force in the ideal defect-free lattice (calculations of Born, Zwicky, Kobeko). So far as there are no defects in an ideal material, no particular possible surface of fracture is to be preferred and a free fracture surface is not formed. Therefore the failure of the ideal solid body at absolute zero is brought about by its disintegration into separate atoms (molecules), or along atomic planes. Such a process differs from the actual process of failure of solids, in which two or several new fracture surfaces are formed, dividing the solid body into macroscopic parts.

At absolute zero failure begins when the maximum quasi-elastic strength at the apex of the crack is reached. This value is called the critical overstress P_K.[29] In contrast to the theoretical strength, which is a constant of the material, (for a given type of stress) the critical overstress can vary somewhat from crack to crack.

Evaluating the calculation[29] of the dissipation of elastic energy by bond breakage at the apex of the crack shows that the theoretical strength σ_m is, according to Orowan, approximately lower than P_K by a factor of $1/1{\cdot}5$.

At absolute zero the crack can grow at a finite rate only under the condition that the overstress at its apex exceeds P_K. This is explained by the fact that part of the elastic energy, resulting from the unloading of the material surrounding the crack apex, turns into kinetic energy of the receding walls of the crack and adjoining parts in the solid body. Therefore the crack growing with a speed different from zero has at its apex the value of additional stress $\sigma'_m > P_K$. For infinitely slow growth of the crack $\sigma'_m = P_K$.

This has been confirmed by the results of Shand[48] who, starting from the experimental data and formulas of Weiber for the stress-concentration coefficient, calculated for glasses an additional stress at the apex of the crack σ'_m, at which the failure of the glass takes place at a high rate. This value is different for various glasses but exceeds in all cases the theoretical strength, calculated according to Orowan's formula, by a factor of 2–3.

Mott[49] and Bateson,[50] in their calculation of the speed of growth of a Griffith's crack in a brittle body, took into account the kinetic energy of the receding walls of the crack which is determined by the form of the relation with the rate of its growth. Bateson's formula for the speed of the growth of the crack is

$$v = \frac{V}{\sqrt{2(1+\mu)}}\left[1-\left(\frac{\sigma_m}{\sigma'_m}\right)^2\right]^{1/2}$$

where V is the speed of the transverse elastic waves in the solid body; μ is Poisson's ratio; σ_m is the theoretical strength; σ'_m is the additional stress at the apex of the crack which increases during the process of its growth.

At $\sigma'_m = \sigma_m$ the crack will not grow, but at $\sigma'_m > \sigma_m$ the speed of the growth will increase, reaching the maximum (critical) speed v_K at $\sigma'_m \to \infty$. Calculations based on the above formula lead to values of v_K which are extremely similar to those which were experimentally observed. For a more precise specification of Bateson's formula one has to substitute P_K for σ_m.

1.5. Mechanical Losses in the Failure of Solids

In the previous section we investigated the ideal brittle body and the dissipation of energy of the elastic waves which appear in it during crack propagation. Apart from this, in actual brittle bodies one can also observe losses of energy which are connected with ductile deformations (and in the case of polymers also with low elastic ones) in places of overstress, especially at the tips of microcracks. Thus three basic types of loss can be observed on failure of solids:

(1) Losses resulting from the damping of atomic vibrations after the breakage of bonds.

(2) Losses during the separation of the walls of the crack and unloading of regions of the specimen surrounding the crack.

(3) Losses due to the viscous, plastic (or low-elastic) flow in places of overstress.

The last two types of loss, which are observed at impact (by Holzmüller and others[51-53]) and at slow failure (by Roesler, Benbow, Berry and Svensson [54-56]), are usually mentioned in the literature. Specially interesting are the experiments on controlled slow failure, similar to the "equilibrium" process. These experiments showed that the energy of failure, that is the decrease of the elastic energy during the process of failure, depends basically not on the increase of the surface energy but on the mechanical losses.

The energy of failure is usually formally related to unit area of the fracture surface. Consequently

$$-\delta W = d\varepsilon + \delta Q = 2T_{\mathrm{ch.e.}}\, \delta\, dc$$

where $T_{\mathrm{ch.e.}}$ is the characteristic energy for fracture (in erg/cm^2); the remaining symbols are defined as in § 1.2.

Comparing this expression with eqn. (1.7) it can be seen that the expression $T_{ch.e.}$ should replace the free surface energy α_{surf} when dealing with real specimens. Therefore Griffith's equation (1.8) must be changed to read

$$\sigma_K = \frac{2}{\sqrt{\pi}} \sqrt{\frac{T_{\mathrm{ch.e.}} E}{c}} \tag{1.10}$$

Right away we notice that $T_{\mathrm{ch.e.}}$ is not a constant of the material as, for instance, α_{surf} but depends on the form and dimensions of the specimen. This is evident if one considers the second type of mechanical losses which depend clearly on scale and shape factors.

According to various authors[56] the value of energy failure $T_{\mathrm{ch.e.}}$ at 20°C for poly(methylmethacrylate) is approximately 5×10^5 erg/cm^2, and for the polystyrene 9×10^5 erg/cm^2, which exceeds the free surface energy almost by four orders of magnitude. Besides this, the existence of maxima in the curves for the temperature dependence of the fracture energy were discovered by Borgwardt[53] for impact and by Svensson[56] for slow failure.

These maxima are connected with the mechanical losses which are observed during the transition of polymers from the glassy to the visco-elastic state. This is confirmed by the displacement of these and other maxima in the direction of higher temperature as the rate of fracture is increased.

1.6. Failure of Solids Under Tensile Stresses below the Critical Stress

In the study of the strength of solids it has long since been noticed that their failure takes place both under a stress σ which is less than the critical stress σ_K, as well as for excess stresses at the apex of the crack $\sigma_m < P_K$. In engineering this phenomenon is called fatigue and is observed at temperatures which are far above absolute zero.

Under the influence of static loads fatigue appears less well defined than under varying or dynamic loads. Therefore one distinguishes between static and dynamic fatigue.†

Static fatigue appears in solids in the form of time dependence of strength. Recently it has been established that the time dependence of the strength of solids which relates the static stress applied to the specimen with the time to failure (durability) is not a result of the influence of some factors accompanying the failure but is determined by the nature of the process of failure itself.

The study of the surface relief of the breaks in solids, including solid polymers (crystalline and amorphous), and also the observation of the growth of cracks in the stressed material by microscopic and other methods leads to the conclusion that in all solids the cracks grow under tensile stresses which are considerably below the usually observed ultimate strength. Müller,[6] apparently, was the first to discover that with glass two stages of break can be observed. The first stage is connected with the slow growth of the primary crack which leads to the formation of a "mirror" fracture surface; the second one is connected with growth of the primary and secondary cracks with nearly the speed of sound, with the formation of rough zones. In the first stage the

† When testing metals one understands by their fatigue the premature failure which can be observed under cyclic loading. This is connected with the fact that at ambient temperature static fatigue of metals is hardly noticeable.

rate of growth of the crack depends on the stress (Fig. 8), the temperature and the environment of the specimen. At the temperature of liquid air there is in fact no mirror part on the fracture surface, the break takes on immediately a catastrophic character and no time dependence of the strength can be observed.

On the basis of investigations of the failure of different materials under various conditions three basic types of failure in solids can be distinguished.

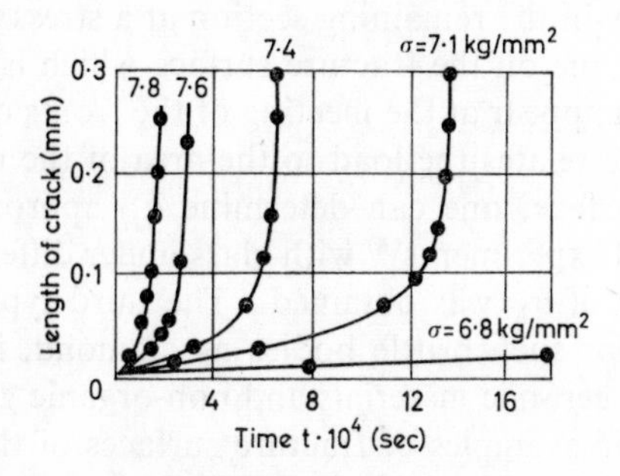

FIG. 8. Propagation of cracks in acetate cellulose under various loads.[60]

TYPE I. Defects in the mass and on the surface are equally critical or they appear simultaneously in the process of deformation. Then a large number of microcracks develop simultaneously throughout the specimen, and they subsequently coalesce to form one main crack. Under such circumstances the whole fracture surface of the specimen is rough (Fig. 9). This is how polycrystals[57] fail; the microcracks appear in crystallites as a result of the ductile deformation and on the weakened grain boundaries. In monocrystals a great number of pre-failure cracks appear owing to local ductile deformations in various regions of the mass.

TYPE II. Surface defects are more critical than internal ones, and the individual surface defects are approximately equally severe. In that case, under defined conditions the cracks grow with a common front towards the centre of the specimen.[58, 59] This is how non-brittle solid polymers fail under small stresses and long test times.

TYPE III. On the surface or in the mass of the brittle material there may be a defect with a certain amount of criticality from which the primary crack grows. As the primary crack grows the stress σ' in the still unbroken section of the specimen becomes much greater than the initial nominal stress σ, calculated on the whole cross-section of the specimen; thence, the growth of the crack accelerates. When the increasing stress σ' becomes equal to, and then exceeds the critical stress σ_K, a transition from the first stage of breakdown to the second takes place, and the primary crack grows with a speed near to the critical speed v_K. Secondary cracks begin to grow quickly in the remaining section at a stress $\sigma \geqslant \sigma_K$. They form a rough zone on the fracture surface which is covered with slip lines which appear at the meeting of the fronts of the growing cracks.(4) If one relates the load to the area of the rough zone of the fracture surface, one can determine σ_K approximately. (Indeed, in several experiments(6) with glass under different stresses σ the same value of σ_K was obtained.) The third type of failure is characteristic for such brittle bodies as diamond, ionic crystals, brittle plastics, ceramic materials and non-organic glass.

Characteristic examples of fracture surfaces of these materials are shown on Fig. 10 and 11. The mirror zone and the gradual transition to the rough one can be seen distinctly.

One has to note that the character of failure in brittle bodies is in reality more complicated than the two-stage process described above, as the transition from the first zone of the fracture surface to the second one often occurs without the sharp dividing boundary, as can be seen on the photographs.

The growth of the cracks can be observed by standard methods,(48, 60–65) for instance by the method of microcinephotography.(60) While studying the speed of growth of the crack, Shand(48) observed in the first stage velocities of the order of 10^{-4}–10^{-5} m/sec and less, and in the second one, rapid critical speeds of the order of 2000 m/sec. Critical velocities of the growth of cracks in brittle materials are one-half to one-third of the velocity of elastic waves in those materials. For instance, in silica the speed of the transverse elastic waves is 3510 m/sec, the speed of the longitudinal ones is 5365 m/sec, and the critical speed of growth of cracks (according to Shardin's measurements) is 2155 m/sec.

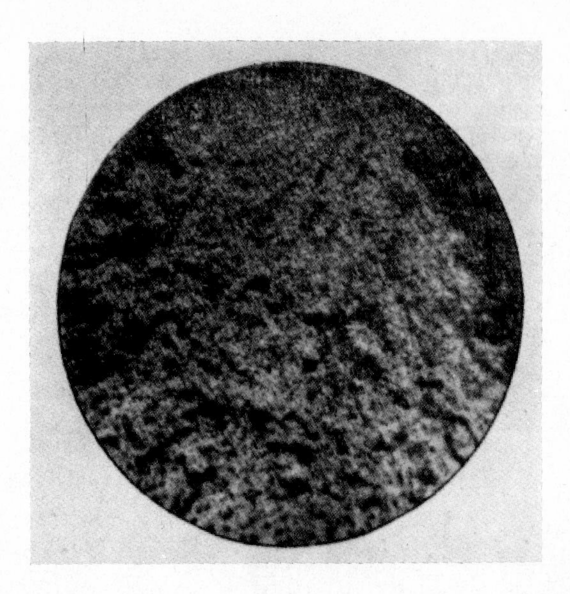

FIG. 9. Brittle fracture of steel under unidirectional tension at 196°C.

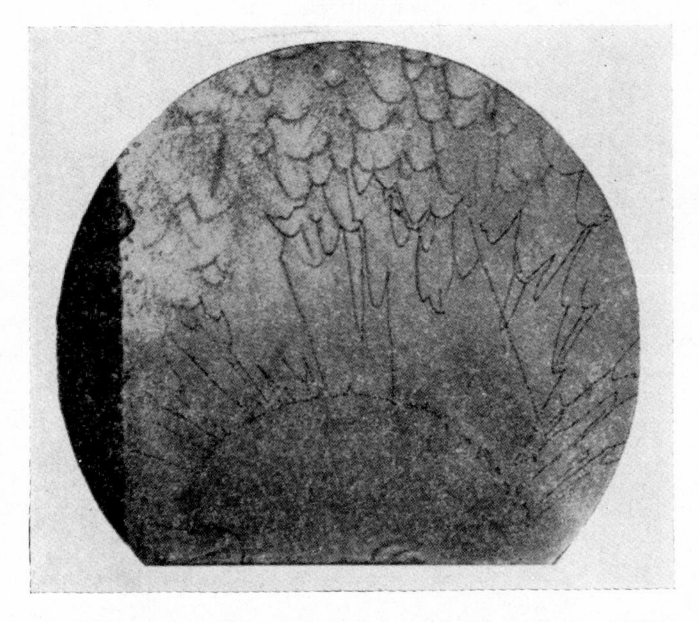

FIG. 10. Surface of break in poly(methylmethacrylate) at 20°C and stress near critical stress. (Enlargement × 200.)

FIG. 11. Surface of break of in-organic glass as a result of slow failure (enlarged × 5).

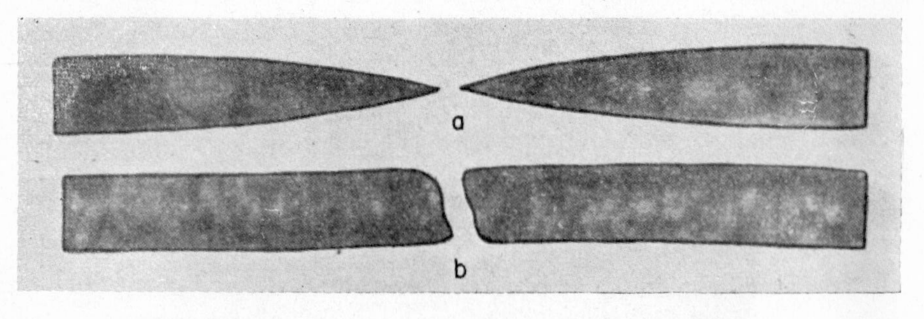

FIG. 19. Breaks of zinc monocrystals.[85] (a) Plastic break in inactive environment. (b) Brittle break in surface-active environment.

1.7. Types and Rates of Loading

In practice one meets an unlimited number of the most varied time rates of loading or of deformation of materials but all of them can in a larger or smaller degree be related to one of five general types of test. Two of them are static ones: test under constant load (or stress), and test at constant strain the remaining three are dynamic ones: test with constant speed of loading (or deformation), periodic or cyclic loading, and impact deformation.

The strength and durability of the materials depends essentially on the rate of loading and type of stressed state (see § 1.14). Various rates of loading can be compared under any one type of stressed state, for instance under tension.

The durability of the specimen can differ sharply on changing from one static state to another. Thus, for instance, under a given tensile stress $\sigma =$ const., the elastic energy of the specimen increases during the growth of the crack owing to the work done by the external force. Under these circumstances the elastic energy reaches the crack with the speed of elastic waves in the solid. At the moment when the speed of the growth of the crack approaches the speed of elastic waves the acceleration of growth of the crack ceases, and the speed of growth reaches its maximum (critical) value. Consequently, at the rate $\sigma =$ const., the initiated process of failure accelerates, and the stress σ' in the still unbroken part of the specimen increases continuously as the crack grows.

Matters are different, though, if the test is carried out at constant strain. As the crack grows the load on the specimen diminishes and the stored elastic energy can decrease so much that the speed of growth of the crack will begin to decrease, or become zero. The shorter the specimen, the more likely this is to happen, whereas with a long specimen the stored elastic energy is sufficient to ensure its use for the failure of the specimen. Under this condition (constant strain) the stress σ' may increase if the specimen is long, but at a slower rate than under the condition $\sigma =$ const., but it may also decrease if the specimen is short (about this aspect of the scale effect of strength see Chapter 5). Consequently, the second type of loading is less dangerous than the first one as it does not, under certain conditions, lead to catastrophic failures.

In the following paragraphs of this chapter we shall chiefly investigate the condition σ = const. as the theories of time dependence of strength were developed in conformity with this simplest of all types of loading. All other types of test are more complicated owing to the fact that the stress changes with time. Nevertheless it is in principle possible, if the time dependence under the condition σ = const. is known, to calculate the durability (time to break) under other types of test. Some methods of calculation will be discussed in Chapters 7 and 8.

1.8. Laws Governing the Time and Temperature Dependence of the Strength of Solids

The dependence of strength on time was first clearly established for silicate glass,[7, 48, 66–69] and in many works of foreign scientists this phenomenon was explained by the chemical or surface-active effect of atmospheric humidity.[67] In Fig. 12

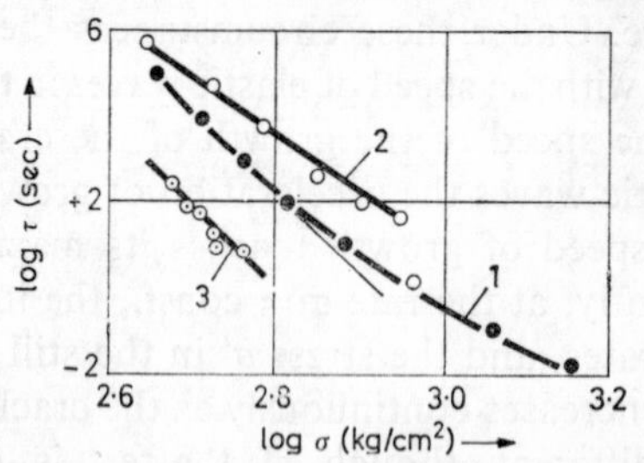

FIG. 12. Time dependence of strength of silicate glass in air. 1. Data.[69] 2. Data,[7] 3. Data.[67]

Preston's data on experiments with glass rods are shown and also the results of an experiment on the curvature of a strip of glass according to data by Holland and Turner and by one of the authors of this book.

For the time dependence of the strength of silicate glasses several empirical formulas have been suggested, of which Holland's and Turner's formula has become the most widely used:

$$\tau = B\sigma^{-b} \tag{1.11}$$

where τ is durability (time to failure), σ the applied tensile stress, B and b are constants.

At the same time, starting from experimental data, the first attempts were made to work out a theory of the time dependence of strength which was based on some ideas about the molecular mechanism of failure.(67, 70)

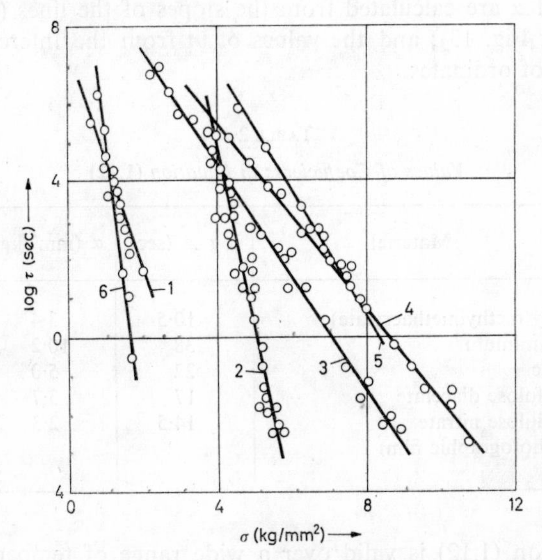

FIG. 13. Time dependence of strength of solids.(8) 1. Poly(vinylchloride). 2. Aluminium. 3. Poly(methylmethacrylate). 4. Zinc. 5. Celluloid. 6. Silver chloride.

A systematical study of the time dependence of the strength of solids and its connection with the mechanism of failure was undertaken by Zhurkov and his co-workers.(71–78) They studied the time dependence of strength for unidirectional tension on materials of various composition with widely differing mechanical properties. It was found that the time dependence of strength is characteristic for all the materials investigated (Fig. 13) and the relation between durability and stress for metals, plastics and polymer fibres† is expressed in the exponential law:

$$\tau = Ae^{-\alpha\sigma} \tag{1.12}$$

where A and α are constants; σ is the applied tensile stress.

† For elastomeric polymers the time dependence of strength is the same (see Chapter 6) as for silicate glasses in air (Holland's and Turner's formula).

This kind of dependence has also been observed earlier for several other materials.[79]

In Table 2 the values of the constants A and α in eqn. (1.12) are shown for five materials at room temperature;[71-73] the values of α are calculated from the slopes of the lines (see, for instance, Fig. 13); and the values of A from the intercepts on the axis of ordinates.

TABLE 2

Values of Coefficients in Equation (1.12)

Material	log A (sec)	α (mm²/kg)
Poly(methylmethacrylate)	10·5	3·4
Aluminium	38	10·2
Zinc	22	5·0
Cellulose diacetate	17	3·7
Cellulose nitrate (photographic film)	14·5	2·3

Equation (1.12) is valid over a wide range of temperatures; the coefficients A and α change with the temperature, but the durability lines radiate fanwise from a single point corresponding to $\sigma \approx \sigma_K$ (Fig. 14 and 15).

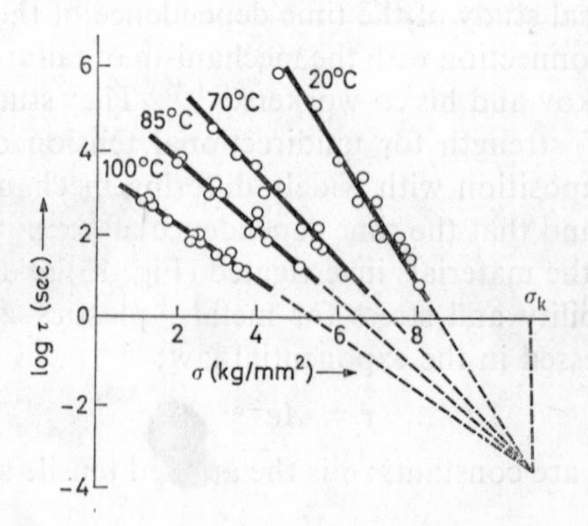

FIG. 14. Time dependence of the strength of celluloid at various temperatures.[71]

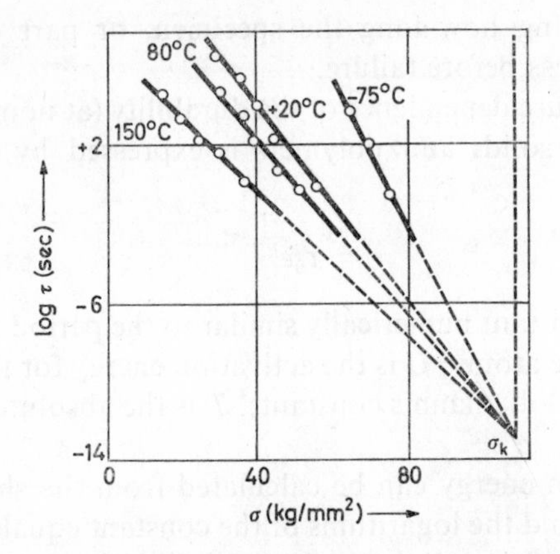

FIG. 15. Time dependence of strength of viscose fibre at various temperatures.(75)

The lower the temperature, the weaker the time dependence of the strength as shown by the change of slope of the lines. At relatively low temperatures and high rates of strain the fracture of the solid approximates to critical fracture. This means that at low temperatures the loading time has, practically, no effect on the magnitude of the breaking stress or, in other words, breakage will not take place if $\sigma < \sigma_K$ however long the material may be in the stressed state. This justifies the introduction of the concept of a "limiting strength" or the technical concept of a "time-independent strength". For inorganic glasses† and metals with a high melting point normal temperatures can be considered as low. For plastics the temperatures at which there is practically no time dependence, are -200°C and less.

The limiting strength as a constant of the material can have a definite physical meaning at any temperature if we regard it as the maximum strength, equal to σ_K, obtained at high strain rates.(67) At relatively low temperatures when the strength is independent of time the limiting strength, approximately equal to σ_K, is practically synonymous with the simple mechanical strength of the solid. Under all other test conditions one must not speak of strength

† In vacuum where the time dependence of the strength of glass is not distorted by the surface-active influence of humidity.

without indicating how long the specimen, or part of it, was subjected to stress before failure.

The temperature dependence of the durability (at nominal stress $\sigma =$ const.) for solids and polymers is expressed by the equation

$$\tau = \tau_0 e^{\frac{U}{kT}}$$

where τ_0 is a constant numerically similar to the period of thermal oscillation of the atoms; U is the activation energy for the process of failure; k is Boltzmann's constant;† T is the absolute temperature.

The activation energy can be calculated from the slope of the lines (Fig. 16), and the logarithms of the constant equals the value of the intercept of these lines on the axis of ordinates.

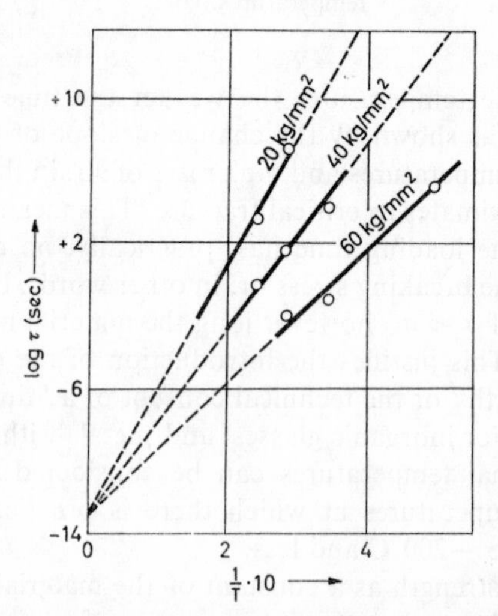

FIG. 16. Temperature dependence of durability of viscose fibre[75] under various stresses.

The activation energy U depends, for many solids, on the stress σ, and decreases according to the formula (Fig. 17):

$$U = U_0 - \gamma\sigma$$

† This constant in various units equals: $1{\cdot}37\times10^{-16}$ erg/deg C; $3{\cdot}28\times\times10^{-24}$ kcal/deg C; $1{\cdot}40\times10^{-22}$ kg cm/deg C; $1{\cdot}37\times10^{-21}$ n cm/deg C.

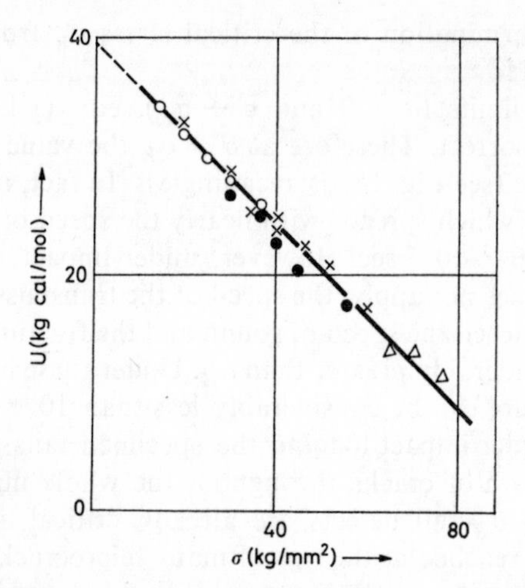

FIG. 17. Dependence of activation energy of the break of viscose fibre on stress[75] under various temperatures. ○ – at 150°; × – at 80°C; ● at 20°C; △ – at −75°C.

where U_0 is the activation energy of the elementary act of failure in the absence of stress and is similar in value to the sublimation energy for metals and to the energy of chemical bonds for polymers; γ is a coefficient which depends on the nature of the structure of the material.

On the basis of eqn. (1.12) it is easy to obtain the equation for durability which expresses the temperature–time dependence of the strength of solids:

$$\tau = \tau_0 e^{\frac{U_0 - \gamma\sigma}{kT}} \tag{1.13}$$

Equation (1.13) can be written as follows:

$$\log \tau = \log \tau_0 + \frac{1}{2{\cdot}3kT}(U_0 - \gamma\sigma)$$

From this it follows that the constants A and α of eqn. (1.12) are given by $A = \tau_0 e^{U_0/kT}$ and $\alpha = \gamma/kT$. Under a stress $\sigma = \sigma_K = U_0/\gamma$ the durability is the same for all temperatures ($\tau = \tau_0$). The values σ_K and τ_0 correspond to the coordinates of the point from which the family of lines representing the time dependence of strength originate (see Fig. 15). Thence follows the possibility of a different determination of the critical stress σ_K from eqn. (1.8); see also § 1.11.

Near the limits ($\sigma \to 0$ and $\sigma \to \sigma_K$), eqn. (1.13) is, strictly speaking, incorrect. Therefore at $\sigma = \sigma_K$ the value of durability $\tau = 10^{-12}$ sec (see Fig. 15) is meaningless. In fact, in a strip 1 cm wide a crack which spreads with nearly the speed of sound transverses it in 10^{-5}–10^{-6} sec. However, under impact loading when eqn. (1.13) does not apply, the speed of the transmission of elastic energy can exceed the speed of sound and the fracture stresses can become considerably greater than σ_K. Under these conditions the time for failure can be considerably less than 10^{-6} sec. This implies that under impact loading the specimen fails with a simultaneous growth of cracks throughout the whole mass, as under a stress $\sigma > \sigma_K$, all defects are already critical and a critical overstress is reached at the tips of many microcracks.

The validity of eqn (1.13) was observed over a wide range of time: from 10^3 sec up to 10^7 sec. This universal validity leads to the conclusion that for all solids the nature of the failure process is the same. Tests carried out in high vacuum[76] showed that though the environment has an influence on the time dependence of strength, yet it is not its fundamental cause.

It is stress and temperature which are the basic factors which cause the failure of solids: the stress decreases the activation energy U, and the thermal motion leads to the breakage of bonds; the probability of this depends on the value of the ratio U/kT. A surface active environment influences the values of the constants U_0 and γ, used in eqn. (1.13). In some cases strong surface-active environments play just as great a role as temperature.[80]

The constants which characterize the temperature–time dependence of strength of a number of materials[81] are shown in Table 3.

In Table 3 the value τ_0 is not shown, as this constant is of the same order for all substances and is approximately 10^{-12} sec.†

Few investigations have been made into the durability under complex stress conditions. Fundamentally, the question arises which stress one should consider in the formula for the speed of crack propagation and of the duration of the process of failure under a complex stress. The investigations of Solobiyrev,[82]

† According to the data of Fig. 14 τ_0 has another numerical value. However, all further experiments have shown that for solids the value of $\tau_0 = 10^{-12}$–10^{-13} sec.

Kurov and Stepanov[83] showed that the durability of metals under torsion can be expressed by eqn. (1.13). It is, however, less than the durability under tension.

TABLE 3

Values U_0 and γ for Several Materials†

Materials	U_0 (kcal/mol)	Energy of chemical bonds (for polymers) or energy of sublimation (for metals) (kcal/mol)	γ $\left(\frac{\text{kcal mm}^2}{\text{kg mol}}\right)$
Poly(methylmethacrylate)	54–55	52–53	3·18–3·37
Nylon fibre (orientated)	45	43	0·29–0·44
Zinc (polycrystalline)	25	27	1–3
Aluminium (polycrystalline)	54	55	4–9
Aluminium (monocrystal)	51	55	5·5
Silver	64	68	2·2
Platinum	120	127	1·5–8·3

This, in the authors' opinion, is due to the fact that the activation energy for the process of failure is less under torsion than under stress.

It must be noted that Zhurkov[84] proves on the basis of the same data[83] the independence of the activation energy from the type of stress.

† The numerical value γ is expressed in the table as kcal mm²/kg mol. In order to express the constant which in reality has the dimensions of mass, in mm³, one has to multiply the numerical value of γ in the table by $7{\cdot}1\times10^{-19}$ (taking into account that 1 kcal = $4{\cdot}27\times10^5$ kg mm and the Avogadro's number is $N_A = 6{\cdot}02\times10^{23}$).

1.9. Influence of Surface-active Environments on the Failure of Solids

The mechanical properties of solids, with other conditions equal, depend on their environment. The nature of this extremely general physical and chemical phenomenon was found by Rebinder,[30-32, 80, 85] and consists in the following: Surface-active environments lower the surface tension of the material without causing an irreversible change in the structure. The defects of the material serve as places of preferential adsorption of atoms or molecules of the active environment. They enter the defects by diffusion or migration.† The failure of solids is always accompanied by the development of new free surfaces. This process can be facilitated or accelerated if the free surface energy on the boundary of the solid body with the surrounding environment, i.e. the work of formation of new surfaces, is lowered in comparison with its value in vacuum. In the presence of a surface-active environment the initiation and development of plastic displacements and of incipient cracks is facilitated. On the microscale this means that the interaction with adsorption-active molecules (or atoms) helps the breakage and re-formation of the interatomic bonds in the solid material.

Rebinder and his co-workers[31] investigated the influence of typical organic surface-active environments (solutions of alcohols of high molecular weight, acids and their derivatives) on the processes of deformation of a number of poly- and monocrystalline metals; in those cases, even with a relatively small reduction of the free surface-energy as a result of adsorption, deformation is facilitated (the yield point and the factor of hardening of the metal are lowered).

Under varying loads active environments exert a considerable influence on the durability and strength. When the load is removed a so-called adsorption after-effect can be observed: the molecules of the environment hinder the crack from closing and are only gradually "squeezed out" of the crack. This facilitates failure under a new cycle of loading.

Further,[43, 44, 85] a basically new and extremely wide circle of various physico-chemical phenomena appear at the interaction of

† Under certain conditions the environment can cover the surface of the crack by any means of spreading.

solids with strongly adsorption-active environments. As far as metals are concerned such environments are melts of some other, lower melting metals. It has been established that under a very considerable reduction of the free surface energy the following effects are possible: (1) appearance of brittleness and decrease of strength right up to a complete loss of strength and ductility (in the limit to spontaneous disintegration to particles of colloidal dimensions), or (2) facilitation of deformation.

Under some conditions metallic melts can also have a strengthening effects as a result of a mass diffusion into the crystals (alloying).[80] The character of the adsorptive interaction and the degree of its effect depend on a number of physical and chemical factors—properties of the melts and their quantity, composition and structure of the solid body, time of contact with the melts, conditions of deformation (temperature, speed and type of stress).

On Fig. 18 can be seen that the stress–strain curve of a mono-

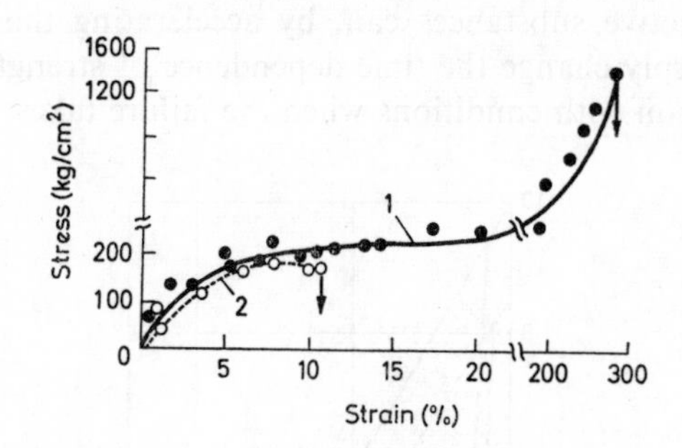

FIG. 18. Curves of deformation at 20°C of zinc (1) and zinc monocrystal amalgam[43] (2).

crystal of zinc in a non-active environment defines it as a typically ductile material (curve 1). However, the covering of the specimen with a very thin film of mercury leads to a sharp reduction of its strength (curve 2). As the solubility of mercury in zinc is extremely small the character of the ductile flow as a mass property of the material does not change but the brittle strength is sharply reduced: the break acquires a typically brittle character (Fig. 19, facing p. 25).

On Fig. 20 are shown the values of strength and deformation at break of the monocrystals of zinc in melts of tin and lead. Even small additions of surface-active tin (lead is a weak surface-active substance in relation to zinc) causes a sharp reduction in the strength and a transition from a ductile failure to a brittle one.

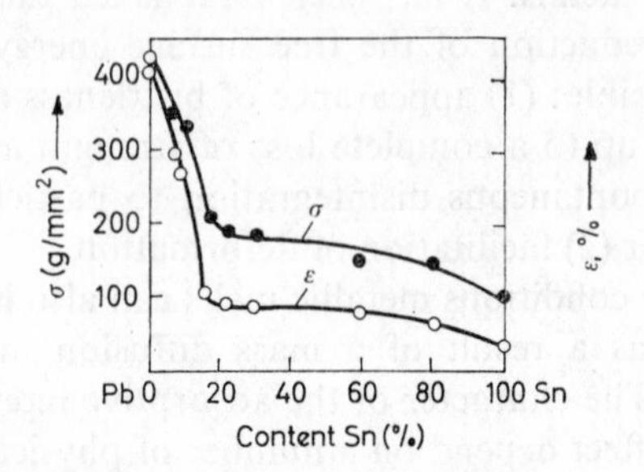

FIG. 20. Strength σ and elongation ε at break of zinc monocrystals depending on the concentration of tin in alloys with lead (alloy in contact with the specimen surface at 350°C).[85]

Surface-active substances can, by accelerating the growth of cracks, sharply change the time dependence of strength of solids in comparison with conditions when the failure takes place in an

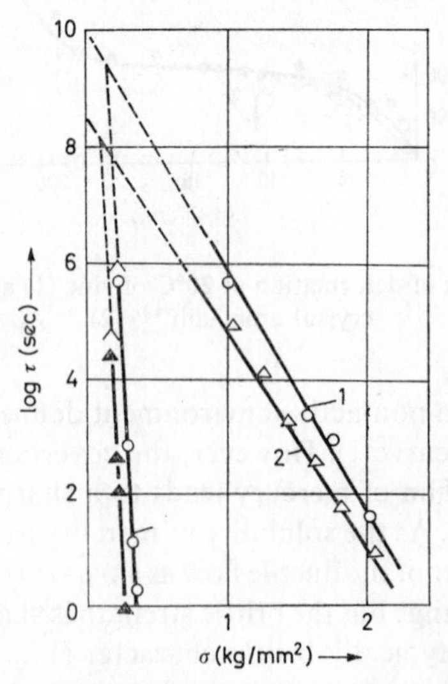

FIG. 21. Time dependence of the strength of zinc monocrystals.[80] 1, 2. Zinc at 20 and 50°C. 3, 4. Zinc amalgam at the same temperatures.

inactive environment. In conditions of a considerable reduction of the free surface energy the continuous time dependence of strength disappears: instead, a "threshold" of strength appears in the area of low stress; for stresses exceeding the threshold by a small amount, the specimen fails instantly whereas below this threshold the durability is practically infinite (Fig. 21). In this respect the effect of surface-active environments is analogous to the lowering of temperature, and so the failure ceases to be a thermally activated process.

The displacement of dislocations does not occur during the process of loading in brittle amorphous solids as it does in metals.

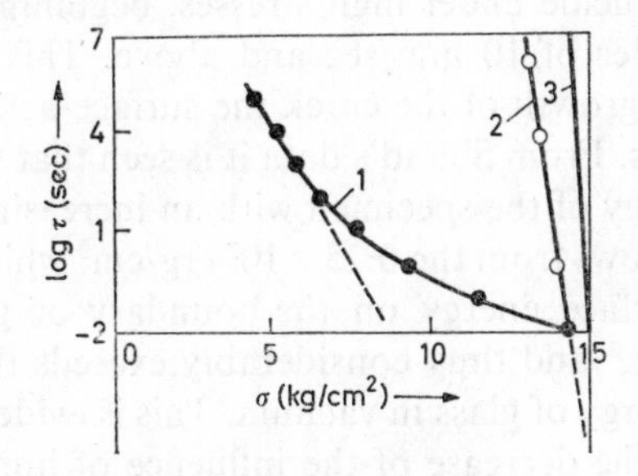

FIG. 22. Dependence of logarithm of durability of glass on the tensile stress. 1. In air.[66] 2. In vacuum.[68] 3. Theoretical dependence in vacuum.[93]

Therefore, also the mechanism of the effect of surface-active environments is simpler in those cases. Figure 22 shows the time dependence of strength of silicate glass in air (curve 1) and in vacuum (curves 2 and 3). In vacuum the time dependence of strength is a very steep function, and in air an exponential one. In contrast to glass the influence of humidity on the strength of solid polymers and metals is small. The time dependence of strength of these materials in vacuum and air is therefore the same.

Under small stresses the speed of crack propagation in glass is small, and the molecules of the environment have time to follow the growing crack and to influence the failure. As the speed of growth of the crack increases the molecules of the environment cannot reach its apex, and failure occurs as in the absence of surface-active environments. The greater the stress, the sooner the transition to rapid failure where the environment has no time to influence the process of failure. Consequently, under a stress near the critical one (σ_K) surface-active environments do not influence the strength of the solid. This is clearly seen from the data on Fig. 22. With the increase of stress the transition from the time dependence of strength in the environment (dotted straight line) to the time dependence of strength in vacuum (curve 2) takes place. The theory of this process has been examined by Bartenev and Rasumovskaya.[86]

Recently, Shand[87] showed that the speed of growth of cracks in inorganic glass in air and in vacuum differ greatly under small stresses but coincide under high stresses, beginning with a strain rate of the order of 10 mm/sec and above. This means that at high speeds of growth of the crack the surface-active influence of humidity ceases. From Shand's data it is seen that with tests in air the elastic energy of the specimen with an increasing rate of crack propagation grows from the $0{\cdot}25 \times 10^3$ erg/cm^2 which corresponds to the free surface energy on the boundary of glass–water, to $1{\cdot}5 \times 10^4$ erg/cm^2 and thus considerably exceeds the value of the free surface energy of glass in vacuum. This is evidently connected not only with the decrease of the influence of humidity but also at high rates of crack growth with the dissipation of a part of the elastic energy in the form of heat. In experiments in vacuum the decrease of elastic energy ($4{\cdot}10^3$ erg/cm^2) even at low rates of crack growth is greater by an order of magnitude than in experiments in air. This agrees with Obreimov's experiments and with Orowan's calculations of the splitting of mica where the free surface energy in water and in vacuum are respectively 375 and 4500 erg/cm^2.

1.10. Theoretical Presentation of the Time Dependence of the Strength of Various Solids

There exists no single mechanism of failure for the various solids. Dislocations and ductile deformations play a great role in the failure of crystalline bodies, and various kinds of defects and microcracks in that of brittle amorphous bodies.

Theories of time dependence of the strength of crystalline bodies have been suggested by Pines,[88] Orlov,[89] Vladimirov, and Gurevich[90] and others. For brittle materials there exist several theories which are similar to each other, for instance Gibbs and Cutler,[91] Stuart and Anderson,[92] and also the more general theory of Bartenev[93] which was published in 1955.

Some theories[89, 90] refer to solids which fail according to the first type (see p. 23) through the simultaneous formation on dislocations of a great number of microcracks which grow and thereupon merge into one main crack.

Orlov investigated the failure of polycrystalline bodies for which a connection has been observed[94–96] between the rate of creep $\dot{\varepsilon} = d\varepsilon/dt$ and the durability

$$\dot{\varepsilon}\tau = \text{const.},$$

where the constant is approximately 0·1.

From this correlation the author draws the conclusion that creep and incipient failure proceed simultaneously and are caused by the same mechanism—ductile deformation. The elementary act of failure precedes the elementary act of creep.†

Dislocations, slowed down by various obstacles create in many places of the polycrystalline specimen microcracks which thereupon grow, due to the migration of the dislocations to the surface of the microcracks. Each advance of the crack causes a redistribution of stresses which activates new dislocations. At each transition into another plane during the slip the crack partly closes and the process begins anew. The growing crack leaves a trace in the form of a little crack which alternates with sound material. These cracks do not disappear when the load is removed which also explains the irreversibility of the process of failure. When the section of the specimen has been sufficiently weakened, due to the presence of such cracks, it fails through the growing of the main crack.

Pines[88] considers that the crack in the solid body grows through an advance of "holes" or flaws towards the apex of the crack where the greatest stress gradients exist, and also through a

† At the same time the elementary plastic displacements which also lead to the origin and growth of microcracks appear as the cause of the process of creep.[80]

diffusion of atoms from the surface to the depth of the specimen at the flaws. The stress applied to the solid must lower the barrier potential of this process which is equal to the energy of activation of self-diffusion and this leads to a time dependence of strength, similar to eqn. (1.13).

However, the self-diffusion mechanism of the growth of the cracks calls forth a number of serious objections. Firstly, it is not common to all solids as it applies only to crystalline materials which contain flaws. Secondly, on the basis of this mechanism it is impossible to explain the rapid stage of failure which proceeds with the speed of sound; on the other hand, for the slow stages this mechanism is improbable on account of the small flaw concentration near the crack. Consequently, this mechanism does not have a general validity, although it can appear at high temperatures. It is possible that it takes place in some crystalline bodies but does not determine the process of failure. According to data(77) small impurities which considerably change the coefficient of self-diffusion do not influence the energy of activation of the failure process.

For the second type of failure (see p. 23) which is characteristic for non-brittle solid polymers and where from the internal surfaces grow simultaneously a great number of "silver cracks" the theory of time dependence of strength has been less fully developed. Some ideas about the character of time dependence of the strength of those materials are investigated in § 1.13 and the mechanism of strength is considered in detail in Chapter 3.

The theories of a number of authors(91–93) refer to brittle solids and partly to brittle solid polymers. The theory of time dependence of the strength of those materials which is investigated in detail in the following sections can at present be regarded as the most developed one although it is as yet far from complete.

1.11. Oscillation (Fluctuation) Theory of Strength of Brittle Bodies

The theory of time dependence of the strength of brittle bodies is based on the investigation of the kinetic growth of the cracks and starts from the two-stage process under uniaxial stress (see § 1.6).

The growth of each crack is regarded as the successive breakage of bonds at its apex under the action of the stress and the thermal oscillations of the atoms or molecules. Whilst the various kinetic theories on the growth of strength have common features, they have also differences which will be noted at the end of this section. We investigate now the theory of strength of brittle bodies in the absence of a surface-active environment† which was suggested by one of the authors[93] in 1955, and its further development.

In brittle materials the most critical defects are usually the micro- or submicrocracks which exist before the application of the load. The growth of one or several of the most critical

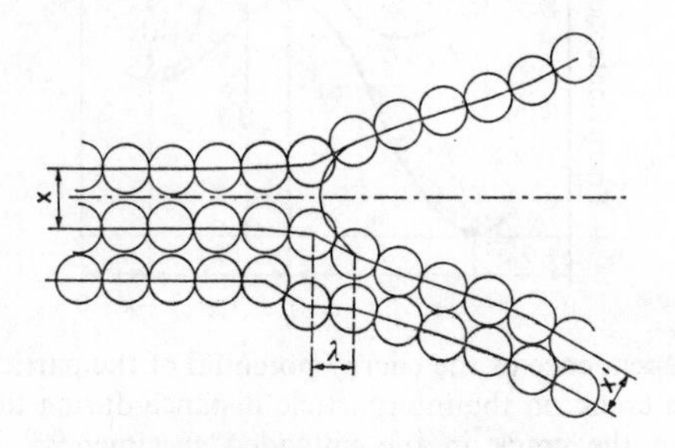

FIG. 23. Structure of crack in brittle material at its apex.

microcracks in the first stage of failure determines the durability of the specimen of the brittle material.

Let us examine the molecular mechanism of the processes of propagation and closure of a crack in a brittle body starting from the molecular model of the crack (Fig. 23). In the breakage of the bonds at the tip (independently of the type of bonds in the solid body) the barrier potential U is surmounted. In the reverse process (restoration of bonds between atoms, ions, molecules) the barrier potential U' is surmounted. Figure 24 shows the change of the potential energy with change of interparticle distance in the process of breakage and renewal of bonds. This diagram shows for the convenience of the following exposition, a crack in a loaded specimen ($\sigma = 0$).

† A theory which considers the influence of a surface-active environment on the strength of brittle materials has been presented in works of a number of authors.[47, 86]

The value of the potential energy of the particles in the volume (before the breakage of bonds) as a function of distance between the particles is determined by the interparticle distance x, and the particle on the free surface (after the breakage of bonds) by the interparticles distance x'; $x = \lambda_0$ is the interparticle distance in equilibrium in the mass; $x' = \lambda_0'$ in the surface layer (see Figs. 23 and 24). The left-hand minimum of the potential energy curve (see Fig. 24) corresponds to the equilibrium position of the particles

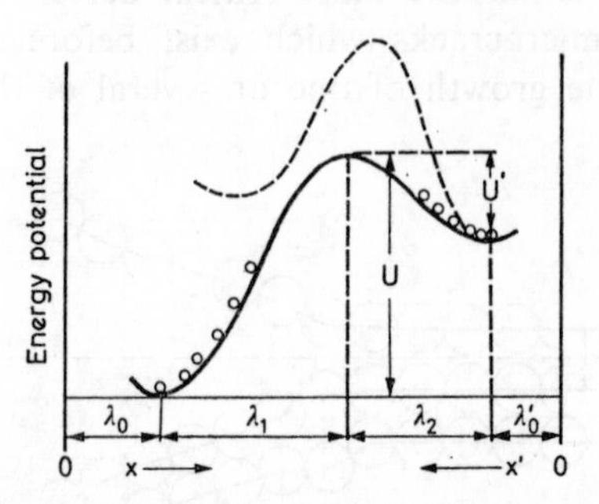

FIG. 24. Dependence of the energy potential of the particles at the apex of the crack on the interparticle distance during the closing of the crack in the unloaded specimen.[93]

in the mass remote from the crack; the right-hand one to the equilibrium position of the particles which emerge after the break onto the free surface. Since in the unstressed body the potential energy of the particles on the free surface is greater, the right-hand minimum is higher than the left-hand (see Fig. 24), and the barrier potential U, which equals the activation energy for the breakage of bonds, is greater than the barrier potential U'—the activation energy for the renewal of bonds. The difference $U-U'$ is the surface potential energy of the solid body attributable to two particles:

$$U-U' = 2\alpha'_{surf}\lambda_0^2$$

where α'_{surf} is the surface energy potential, erg/cm^2; λ_0 is the interparticle distance in equilibrium.

The value α'_{surf} is defined by the Gibbs–Helmholtz equation,

$$\alpha'_{surf} = \alpha_{surf} - T\,\partial\alpha_{surf}/\partial T$$

Here α_{surf} is the free surface energy of the solid body in the absence of a surface-active environment and represents a small adjustment. It may therefore be assumed that $\alpha'_{surf} \approx \alpha_{surf}$ so that α_{surf} in practice corresponds to the surface tension of the solid.

In the unstressed (unloaded) material the crack must gradually close right up to the defect (initial microcrack) from which it grew as the probability of the existence of particles in the left-hand minimum of the potential energy is greater than in the right-hand one.

Small stresses lead to a decrease of the asymmetry of the potential energy curve. At relatively high stresses the sign of the asymmetry changes as shown by the dotted curve (see Fig. 24), and the breakage of bonds becomes more probable.

Let us look first at the uniaxial extension of thin, strip-shaped specimens. The crack grows from the edge of the strip where the most critical defects are those deriving from the cutting of the specimen. The equation of durability of the thin strip does not differ in principle from the equations for specimens of other shapes.

A tensile stress σ applied to the object creates an overstress P at the apex of an edge crack with a length l growing in the direction perpendicular to the direction of the extension, given by

$$P = \beta\sigma' = \beta \frac{\sigma}{1-y} \tag{1.14}$$

where β is the coefficient of stress concentration.

The actual stress σ' in the remaining, and as yet unbroken section of the specimen is

$$\sigma' = \frac{\sigma}{1-y} \tag{1.15}$$

where $y = l/L$ is the relative length of the crack (L is the width of the strip).

It is supposed that critical failure (second stage) begins when the critical overstress P_K is reached at the root of the crack. This critical overstress is proportional to the absolute value of the derivative $\partial U/\partial x$ at the point of inflection of the rising branch of the potential energy curve (see Fig. 24) and is connected with the critical stress in the remaining section of the specimen as follows:

$$\sigma'_K = \frac{1}{\beta} P_K \tag{1.16}$$

The overstress P at the apex of the crack promotes the breaking of bonds and hinders their renewal. In order to allow for this it is necessary to subtract from the potential energy U the elementary work ωP which is performed by the applied local stress P along the path λ_1 (see Fig. 24) and to add to U' the work $\omega' P$ which is performed against the applied local stress along the path λ_2. Here $\omega = \lambda_1 \lambda \lambda_P$ and $\omega' = \lambda_2 \lambda \lambda_P$ are the elements of volume in which the breakage and renewal of the bonds takes place during thermal oscillations. λ_P is the elementary segment of the front (perimeter) of the crack and consists of one or several particles which are simultaneously involved in the oscillation;† λ is the elementary path of the order of the interatomic distance by which the segment of the crack's front advances at each oscillation (see Fig. 23); λ_1 and λ_2 have values which depend on the type of bond and on the structure of the solid.

The overstress P_0 at the apex of the crack at which the potential curve becomes symmetrical is called the non-critical local stress. Under less than critical stresses the crack closes, under greater ones it grows.

The stress σ_0 applied to the specimen which corresponds to the non-critical overstress P_0 at the apex of the crack is called the non-critical stress of the specimen

$$\sigma_0 = \sigma'_0(1-y_0) = \frac{P_0(1-y_0)}{\sigma}$$

where σ'_0 is the non-critical stress calculated on the area of the cross-section of the specimen including the area of the micro-cracks; y_0 is the relative length of the most severe initial crack in the specimen.

As $y_0 \ll 1$, $\sigma_0 \approx \sigma'_0$ and therefore $\sigma_0 \approx P_0/\beta$. Correspondingly the critical stress applied to the specimen $\sigma_K \approx \sigma'_K$ and therefore $\sigma_K \approx P_K/\beta$.

† In connection with this the activation energy of the break of bonds U refers in general to the group of bonds which enter into one kinetic unit.

The frequencies of oscillations which lead respectively to the growth and the closing of the cracks, taking into account that $P = \beta\sigma'$, equal

$$\nu_1 = \nu_0 e^{-\frac{U-\omega\beta\sigma'}{kT}}; \quad \nu_2 = \nu_0 e^{-\frac{U'+\omega'\beta\sigma'}{kT}} \tag{1.17}$$

where ν_0 is the frequency of the thermal oscillation of the group of particles which simultaneously take part in the breaking or renewal of bonds (for solid polymers $\nu_0 \approx 10^{12}$ sec^{-1}, for inorganic glasses, ionic crystals and other brittle bodies $\nu_0 \approx 10^{13}$ sec^{-1}; k is Boltzmann's constant; T is the absolute temperature.

The rates of growth and closing of the cracks by oscillations equal $v_1 = \lambda\nu_1$ and $v_2 = \lambda\nu_2$ respectively. Consequently, the actual speed of growth of the crack is equal to $v = v_1 - v_2$. Under the non-critical stress $\sigma = \sigma_0$ these speeds are equal ($v_1 = v_2$), and the crack neither grows nor closes (equilibrium state of crack). Under stresses which are not very near the non-critical one, when $\sigma > \sigma_0$ we have $v_1 \gg v_2$ and therefore the actual rate of growth of the crack equals†

$$v = \lambda\nu_0 e^{-\frac{U-\omega\beta\sigma'}{kT}} \tag{1.18}$$

This formula has been confirmed experimentally by observing the speed of crack propagation during the breaking of a brittle material (cellulose acetate) under high stress.[60] It was found that

$$v = v_0 e^{z\sigma'}$$

where, in agreement with eqn. (1.18) we have

$$v_0 = \lambda\nu_0 e^{-U/kT}, \quad z = \omega\beta/kT.$$

The critical value of the overstress $P_K = \beta\sigma'_K$ at the apex of the crack is determined by the condition $\omega\beta\sigma'_K = U$. Under these conditions the critical rate of crack growth for the transition to the rapid stage, according to eqn. (1.18) is $v_K = \lambda\nu_0$. For inorganic glass $v_K = 2000$ m/sec, for resin it is 400 m/sec[64] according to test data. By substituting the respective values ν_0: for glass, 10^{13} sec^{-1}; and for resin, 10^{12} sec^{-1} we obtain respectively $\lambda = 2$Å which is approximately equal to the mean distance be-

† Such a law of the speed of the crack's growth corresponds to the third hypothesis of failure by Alfrei.[97]

tween the atoms of silicon and oxygen in glass, and $\lambda = 4$Å which corresponds to the intermolecular distances in polymers.

Consequently, the stresses $\sigma_0 \approx \sigma_0'$ and $\sigma_K \approx \sigma_K'$ are the limits of the range of tensile stresses in which the initial speed of the crack's growth lies between the limits of 0 and v_K. At $\sigma \approx \sigma_0$ the crack does not grow, at $\sigma \approx \sigma_K$ the crack grows at the critical speed from the very beginning. Under all other stresses between σ_0 and σ_K the crack grows in the first stage of the break with an increasing speed but at the transition to the second stage of fracture it grows at the constant critical speed v_K. In the first stage of the fracture the speed of growth of the crack is expressed by the equation

$$v = \lambda \nu_0 \left(e^{-\frac{U - \omega\beta\sigma'}{kT}} - e^{-\frac{U' + \omega'\beta\sigma'}{kT}} \right)$$

This expression is used below to arrive at the expression of the durability of the specimen under a given nominal tensile stress σ; the stress σ' in the sections of the specimen remaining unbroken is related to the nominal stress σ at each given moment of time t by eqn. (1.15), where $y = y(t)$, and $\sigma' = \sigma'(t)$ but $\sigma = \text{const.}$ by definition.

The durability of the specimen under the action of the given stress in the interval $[\sigma_0, \sigma_K]$ is determined by summing the time of breakdown in the first and second stages:

$$\tau = \int_0^{l_K} \frac{dl}{v} + \tau_K \frac{L - l_K}{L} = L \int_0^{y_K} \frac{dy}{v} + \tau_K (1 - y_k) \tag{1.19}$$

where the integration begins from zero as the initial length l_0 the crack and, consequently, its relative length $y_0 = l_0/L$ is very small; l_K and y_K are the critical absolute and relative lengths of the cracks which correspond to the transition of the fracture to the rapid stage at $\sigma' = \sigma_K'$; τ_K is the time of failure of the specimen under suddenly applied critical stress when also secondary cracks simultaneously take part in the failure. The first term in eqn. (1.19) expresses the time of breakdown in the slow stage of failure, the second one in the rapid stage.

The result of the calculation of durability according to eqn. (1.19) for a solid polymer at 300°K is shown on Fig. 25 (curve 2). The calculation was based on the following numerical values:

$L = 1$ cm; $\nu = 10^{12}$ sec^{-1}; $\lambda_1 \approx 1$ Å; $\lambda = \lambda_2 = 15$ Å;† $\sigma_K = 12$ kg/mm^2; $\sigma_0 = 1$ kg/mm^2; $B = 10$ and the activation energy $U_0 = 45$ kcal/mol (definition of U_0 see below).

Excluding those stresses which are very similar to the non-critical ones and substituting σ' for σ in accordance with eqn.

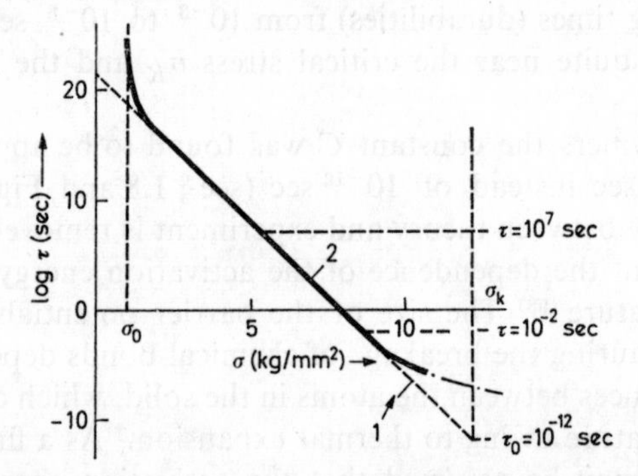

FIG. 25. Time dependence of strength of organic glass at 300°K, calculated: (1) according to eqn. (1.21); (2) according to eqn. (1.19).

(1.15), we obtain the following expression for durability (see more details in ref. 93):

$$\tau = \tau_K(1-y_K) + \frac{L\omega\beta\sigma}{\lambda\nu_0 kT} e^{\frac{U}{kT}} \left[\frac{kT}{\gamma\sigma} e^{-\frac{\gamma\sigma}{kT}} - \frac{kT}{\gamma\sigma_K} e^{\frac{\gamma\sigma_K}{kT}} + \right.$$

$$\left. + \mathrm{Ei}\left(-\frac{\gamma\sigma}{kT}\right) - \mathrm{Ei}\left(-\frac{\gamma\sigma_k}{xT}\right) \right]$$

where Ei $(z\sigma)$ is the integral exponent; $z = \gamma/kT$, $\gamma = \omega\beta$ but $\sigma_K = \sigma'_K(1-y_0)$. With $y_0 \approx 0$ we obtain $\sigma'_K \approx \sigma_K$. The value $\tau_K(1-y_K)$ can be disregarded since it is extremely small if stresses σ, very near to the critical stress, are not considered. The time dependence of strength

$$\tau = Ce^{\frac{U-\gamma\sigma}{kT}} \qquad (1.20)$$

† It was considered that the chemical bonds are responsible for the failure of polymers and that in a non-orientated material each third chain breaks, as on the average one of three chains is orientated in the direction of the tension. The distance between two neighbouring chains is taken as 5 Å.

(where $C = LkT/\lambda\nu_0\omega\beta\sigma$) is obtained by further simplifications which were found by Bartenev and correspond to eqn. (1.13) if one disregards the small dependence of the pre-exponential factor on temperature and stress.

The interpolation equation (1.20) is applicable over a sufficiently wide interval of durability τ, and covers the experimentally observed breaking times (durabilities) from 10^{-3} to 10^{-8} sec. It is not valid only quite near the critical stress σ_K and the non-critical stress σ_0.

For polymers the constant C was found to be approximately 10^{-5}–10^{-6} sec instead of 10^{-12} sec (see § 1.8 and Fig. 15). Such discrepancy between theory and experiment is removed if we take into account the dependence of the activation energy for failure on temperature.[98] The size of the barrier potentials which are overcome during the breakage of chemical bonds depends on the mean distances between the atoms in the solid, which change with the temperature owing to thermal expansion.† As a first approximation it may be assumed that the activation energy and the density of the solid decrease linearly with rise of temperature: $U = U_0 - \alpha T$, where U_0 is the activation energy for the failure process extrapolated to absolute zero, and α is a constant which depends on the type of bonds and on the structure of the material. Therefore eqn. (1.20) will have the form

$$\tau = \tau_0 e^{\frac{U_0 - \gamma\sigma}{kT}} \tag{1.21}$$

where τ_0 as distinguished from Zhurkov's[8] and Hsiao's[99] presentations is not the period of oscillation of the atoms but a complicated quantity as follows:

$$\tau_0 = \frac{LkTe^{-a/k}}{\lambda\nu_0\omega\beta\sigma}$$

Taking into account that $C \approx 10^{-5}$–10^{-6} sec for polymers, in agreement with experiment ($\tau_0 \approx 10^{-12}$ sec), a value of the constant $\alpha \approx 27$ kcal/mol degree, $Ke^{-\alpha/k} \approx 10^{-27}$–$10^{-28}$ kg cm degree was derived. Assuming further that $1/\lambda = n$ is the number of bonds per unit of the path of the crack (in cm^{-1}), one can

† With changing temperature, not only the mean distances between the particles change but also their relative positions and the structure of the material at the crack's apex. Consequently, the activation energy as well as the critical overstress at the apex of the crack must change.

finally write eqn. (1.21) as follows:

$$\tau = \frac{nLke^{-a/k}}{\nu_0} \cdot \frac{T}{\omega\beta\sigma} e^{\frac{U_0-\omega\beta\sigma}{kT}} \tag{1.22}$$

The obtained formulae are applicable to strip-shaped specimens, cut out of film or sheet material. Rods of radius R are the other shape of specimens frequently used. The same formula is employed for rods if we substitute in it πR for L.

The oscillation theory also applies to those cases where, in the first stage of failure in the section of the sample considered which contains the most critical defect or crack, several cracks grow simultaneously. In this case the number of cracks in the mass of the specimen must be small so that they do not interact at this stage and the microcracks outside the investigated section cannot change the conditions of stress in the section. In this case the parameter y is interpreted as the fraction of the failure area of the cross-section S at a given moment of time, i.e. for a sample in the form of a strip

$$y = \sum_{i=1}^{m} y_i$$

where m is the number of cracks in this section; y_i the relative area of the ith crack in the fracture section.

If we assume that the coefficient of stress concentration is the same for all cracks the expression for the durability will differ in this case from eqn. (1.22) only in the pre-exponential term by the factor $(m_1+2m_2)^{-1}$, where m_1 is the number of the surface cracks, m_2 the number of cracks in the mass, where $m_1+m_2 = m$. This difference is negligible.

The dimensions of the microcracks before the application of the load are not taken into account in this calculation. If y_0 is a fraction of the section which is occupied by the microcracks which exist before the application of the load the integration in eqn. (1.19) has to start not from 0 but from y_0.

As a result we obtain a more general formula for the durability

$$\tau = \frac{n(L/\varkappa)kTe^{-a/k}}{\nu_0\omega\beta(m_1+2m_2)\varkappa\sigma} e^{\frac{U_0-\omega\beta\varkappa\sigma}{kT}} \tag{1.23}$$

where $\varkappa = (1-y_0)^{-1}$. Introducing the product $\varkappa\sigma$ instead of σ means that at $y_0 \neq 0$ the actual initial stress in the section is $\varkappa\sigma$.

The scatter of the values for durability in tests on a series of identical specimens is explained by the various values of the coefficient $\varkappa$ for the different samples, which cannot be measured; therefore the value $\varkappa$ is estimated from the scatter of the test results.

Series of small and large test samples are characterized by a different mean value of the scale factor $\bar{\varkappa}$. Consequently, eqn. (1.23) implicitly takes into account the scale effect of the strength (see Chapter 5) if one understands under $\varkappa$ the value $\bar{\varkappa}$ and under τ the mean value of durability.†

Finally, one must note that the simplifications which lead to eqn. (1.20) and to the following ones are in essence equivalent to the assumption of irreversibility of the growth of cracks. For solids the irreversibility of the failure process can be observed under defined conditions. For instance, in solid polymers, at least at not very low temperatures, a low-elastic deformation can take place at the apex of the crack—because of the great over-stresses. This changes the configuration of the apex of the crack and, chiefly, hinders the closing of the crack on removal of the stress. For such polymers the rate of growth of cracks over a wide range of stress is expressed by eqn. (1.18) which, allowing for the change of the activation energy with temperature, takes the following form:

$$v = \lambda \nu_0 e^{a/k} \cdot e^{-\frac{U_0 - \omega\beta\sigma'}{kT}}$$

The substitution of this expression for v in eqn. (1.19) leads to eqns. (1.21)–(1.23).

It has to be emphasized that the basic physical precondition of the theories[91-93] is the oscillation mechanism of the breaking of bonds at the apex of the crack connected with the transition through a barrier potential (see Fig. 24). This model was proposed for the explanation of the failure of inorganic glasses at first by Gibbs and Cutler[91] and then by Stuart and Anderson[92] who introduced the energy barrier purely formally on the basis of Eyring's ideas on the transition through the barrier potential in the breakage of chemical bonds in molecules. The failure of bodies formed only by chemical bonds have therefore been investigated in those papers. It must be noted that Eyring's scheme was applied

† All empirical formulae including eqn. (1.13) are based on mean values of durability.

by them, strictly speaking, for individual molecules and not for solid bodies. In the above theory the barrier potential (see Fig. 24) has been introduced from general considerations for any type of bonds in solids.

It is clear that Stuart and Anderson also assume that failure in a solid body proceeds simultaneously along the whole section (the number of the breaking bonds is proportional to the number of bonds which exist in the cross-section of the specimen) which contradicts the actual mechanism of failure.

1.12. Non-critical Stress and Impact Failure

In the literature attention has been paid repeatedly to the existence of non-critical stress which was considered an engineering characteristic. The physical meaning of the non-critical stress which is introduced emerges from the oscillation theory. As shown on Fig. 25 (see p. 47), at small stresses the linear dependence

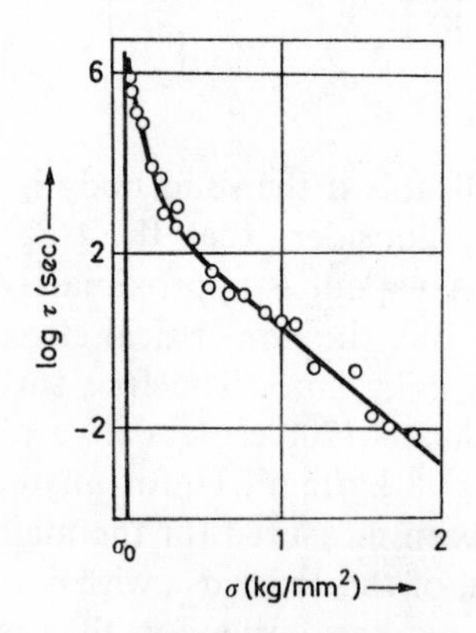

FIG. 26. Time dependence of strength of aluminium[72] at 500°C.

of log τ on σ breaks down and the curve bends upwards, approaching asymptotically a vertical line which corresponds to the non-critical stress $\sigma = \sigma_0$, or the axis of ordinates if $\sigma_0 \approx 0$.

Curves of durability were obtained for plastics and metals in a number of very prolonged experiments. Indeed, under small stresses a sharp rise of the curve of durability (Fig. 26) is shown which confirms the correctness of the conclusions of the theory. Some limit probably exists for solids below which—in the absence of reactive environments—no failure takes place. Holland and Turner[7], from an analysis of their data, come to the conclusion that the non-critical stress for inorganic glasses is about 30% of the technical strength.

The non-critical overstress P_0 at the apex of the crack (see § 1.11) can be approximately calculated from the formula

$$P_0 \approx \frac{2\alpha_{surf}}{(\lambda_1+\lambda_2)},$$

where α_{surf} is the free surface energy of the solid body, $\lambda_1+\lambda_2$ is the distance between the minima of the energy potential (see Fig. 24).

Taking into account that $P_c = \beta\sigma_0$, where β is the coefficient of stress concentration at the apex of the crack, and supposing $\lambda_1+\lambda_2 \approx 2\lambda_0$ (where λ_0 is the interparticle distance in equilibrium) we obtain the result that the non-critical stress of the specimen is approximately

$$\sigma_0 \approx \frac{\alpha_{surf}}{\beta\lambda_0}.$$

This formula is applicable if the solid body is in a surface-active environment. If one considers that the free surface energy of silicate glass in water vapour is approximately 300 erg/cm², and $\lambda \approx 1{\cdot}5$ Å and $\beta = 100$, the non-critical stress of glass in water or air must be $\sigma \approx 2$ kg/mm². The free surface energy at the glass–vacuum interface is 1200 erg/cm²; we obtain therefore for glass in vacuum $\sigma_0 \approx 8$ kg/mm². Unfortunately, the free surface energy has not yet been measured for the majority of solids.

Under high stress, of the order σ_K, with corresponding durabilities of $\tau < 10^{-3}$ sec the curve for the time dependence of strength will have a different shape, for instance as shown by the chain-dotted line on Fig. 25. This is explained by the fact that at very short times of failure the life of the sample becomes equal to the time of application of the load (impact experiments) and depends on the velocity of the elastic waves. In this region it will not be possible to achieve a static load $\sigma =$ const., and the strength depends to a large extent on the speed of loading. In impact loading experiments the process of failure acquires a basically different character, and the concept of the critical stress σ_K loses its meaning. This problem goes beyond the scope of this monograph.

The expounded theory of time dependence of strength must be regarded as an extremely approximate one. Equations (1.21)–(1.23) contain, for instance, the empirical constants a, β, U_0 and others. Further, the coefficient of stress concentration at the crack's apex can, strictly speaking, not be regarded as a constant, as it increases with the growth of the crack, reaches a maximum and then decreases.[100] The derivation of an equation for the breaking time, taking into account the variation of β presents considerable difficulty. The circumstances that the main contribution of β to the durability (breaking time) lasts only during the initial period of the growth of the crack, when the coefficient β has had no time to change materially, is some justification for the use of β = const. in the calculations.

1.13. Application of the Oscillation Theory of Strength to Solid Polymers

The investigations which have been carried out in recent years on the time dependence of the strength of solid polymers enable the physical significance of the constants U_0, ω and β, in eqns. (1.21)–(1.23) to be more precisely defined.

In Zhurkov's and Abasov's work it has been established that the "zero" activation energy U_0 corresponds to the energy of one chemical bond in the polymer.† Consequently, in a solid polymer the probability of a break during oscillation of more than one polymer chain is small. If this is so, the oscillating mass ω corresponds to a single breaking chain. The value U_0 was treated in those papers as the activation energy under the stress $\sigma = 0$, and no attention was paid to the temperature dependence of the activation energy. The reason for this was probably that, in so far as the activation energy U changes linearly with the temperature,

† Calculations of the energy of the chemical bond in solid polymers which were recently carried out by Gubanov and Chevychelov[101] showed that, for instance, the energy of the C–C bond in polymers agrees in practice with the activation energy U_0 which was found from the time dependence of strength.

this does not in itself involve a change in the form of eqn. (1.13). Strictly speaking, under "zero" activation energy U_0 one has to understand the activation energy at $\sigma \to 0$ and $T \to 0$ as this was mentioned repeatedly in § 1.11.

For a given value of the stress σ the actual force f which acts on the breaking polymer molecule will be the greater, the greater γ is. This follows directly from the oscillation theory as $\gamma = \omega\beta$ and the actual tensile force f which acts on the polymer chain ahead of the growing crack, is

$$f = \frac{\omega\beta\sigma'}{\lambda_1} = \frac{\gamma\sigma'}{\lambda_1}$$

where σ' is the mean stress in the remaining cross-section which contains the crack; it differs from the nominal stress σ in eqn. (1.15) by the factor $(1-y)^{-1}$; λ_1 is the value of the displacement of the ends of the chain at the place of fracture: $\omega = \lambda_1 S$ (here $S = \lambda \cdot \lambda_p$ is the elementary cross-section of one of the chains broken during the parting of the specimen).

There exists an important difference between the physical meaning of the constant γ in the papers of Zhurkov, Abasov and others[75-78, 81] and that of γ in the oscillation theory.

In the papers[75] γ is regarded as a structural constant which takes into account the heterogeneity of the stress distribution along the chains in the mass of the polymer. Its size is determined by the mechanism of stress redistribution amongst the chains. Thus the authors, following W. Kuhn and H. Kuhn,[102] F. Bueche[103] and others, suggest that the most highly stressed chains in the mass of the polymer break selectively. This mechanism of failure of polymers has repeatedly been criticized† and contradicts the observed mechanism of failure of polymers through the development of cracks and other heterogeneities in the material. Therefore

† The criticism of this mechanism of failure is given in Chapter 3. Recently Bueche's theory[103] of strength of polymers has been more precisely elaborated,[104] starting from the mechanism of break of the individual chains of the polymer without any development of cracks. The break of completely orientated linear polymers was investigated under the assumption that the polymer chains do not interact. This is far from the actual mechanism of break of polymers but can be regarded as a basis for approximate calculations of the theoretical strength of the completely orientated polymers whilst allowing for thermal oscillations.

the physical meaning of γ in the oscillation theory is also different and connected with the heterogeneity of the stress near cracks and other defects; the non-uniform distribution of stresses around the chains at a distance from the crack in the mass of the polymer is a secondary factor of failure.

Zhurkov and Abasov[81] define γ as the coefficient of overstress which indicates by how much the true local stress, under the action of which breakage of the polymer takes place is higher than the mean stress σ in the sample. The quantity γ does not appear as a non-dimensional coefficient but has the dimension of a volume. For polymers its value would be $\gamma \approx 10^{-18}$–10^{-19} mm^3 which is very much smaller than unity; this indicates that this definition is erroneous. The correct physical meaning of γ follows from the oscillation theory (see § 1.11), according to which, $\gamma = \omega\beta$ is the non-dimensional product of the coefficient of stress concentration and the oscillating mass.

The erroneous interpretation of the coefficient τ_0 in the equation of durability (1.13) as a period of oscillation of the atoms of the polymer chains was repeatedly noted in § 1.11.

The theory which was investigated in § 1.11 referred to the third type of failure which is observed in glasses and solid amorphous polymers at relatively low temperatures when these materials can be regarded as nearly ideally brittle ones. At higher temperatures, but not higher than the glass–transition temperature, a second type of failure can be observed in amorphous solid polymers when orientated "silver" cracks grow (see details in Chapter 3). As the planes of the silver cracks are strengthened by strands they themselves bear a part of the load and the growth of the silver crack takes place with a practically constant speed.

Although different mechanisms of failure can be observed in solid polymers at low and at high temperatures, the time dependence of their strength is expressed by the same eqn. (1.21).

Let us in connection with this briefly examine the process of failure of solid polymers when it is determined practically entirely by the growth of silver cracks.[105] The difference between this process and the brittle failure is: (1) the stress at the root of the silver crack does not depend on its dimensions; (2) not all chemical bonds but only a fraction break in the path of the silver crack, the remaining polymer material is drawn out into slip strands and this leads to an increase of the oscillating volume ω; (3) the coefficient β is smaller bacause the strands smooth out the stresses near the silver crack.

The growth of silver cracks is accompanied along their path by the breakage of some chemical bonds. The oscillation mechanism of bond breakage at the apex of the silver cracks is the same as for fracture cracks and, as shown experimentally, is characterized by the same value of activation energy. This means that the speed of growth of the silver cracks is in practice determined by the probability of chain breakage and not by the speed of the drawing out of strands. Therefore the speed of growth of the silver cracks is described by eqn. (1.18) in which σ' is substituted by σ. The time dependence of the strength of solid polymers is in that case expressed by an equation which agrees with eqn. (1.21) with the difference that the constants τ_0 and γ have different values.(105) However, the coefficient τ_0 differs in these two cases by less than one order of magnitude. In view of the lack of accuracy when determining τ_0 from experimental data ($\tau_0 = 10^{-12}$–10^{-13} sec) this difference is immaterial. It is further known that in some solid polymers the value of the coefficient γ changes with temperature: at high temperatures it is smaller than at low ones.(106) This is explained by the fact that at high temperatures a transition to another mechanism of failure takes place.

In all intermediate cases the times of growth of silver cracks and of fracture cracks are comparable and the picture of failure is a more complicated one. The main role is played by silver cracks at low stresses but by fracture cracks at high ones. Therefore in a graph showing log τ versus σ we see a curve instead of a straight line. The curvature depends on the relation between the values of γ for the growth of failure cracks and for the growth of silver cracks. For some polymers such a curve is clearly shown.(58)

A further refinement and examination of the oscillation theory of the strength of polymers is carried out in Chapter 4 which deals with the influence of molecular weight and polymer orientation on strength.

1.14. Influence of Type of Stress on the Strength of Solids

Griffith's formula and other formulae obtained from the physical theory of strength define the ultimate strength or the critical stress of solids under tension. For other types of stress analogous formulae based on the physical theory of strength are lacking. Nevertheless, the strength depends essentially on the type of stress.

Tension is the most critical type of stress. Experiments on materials are most frequently carried out under tension as this type of deformation can be achieved almost in a pure form (in contrast to compression, shear and torsion). Therefore the mechanical characteristics which are determined under tensile stress are the fundamental initial data for the calculation of the strength of components and structures.

The existence of correlations between values of strength under different types of stress is an object of the theory of elasticity. When testing the strength of materials methods are employed which allow the strength under one type of stress to be calculated from the strength under a different type. The methods of these calculations are based on various mechanical theories of strength[107] which do not especially take into consideration the physical and physico-chemical aspects of failure: the influence of time, of strain rates, scale factors, temperature, environment and others.

To the two basic forms of failure of solids — failure by tearing under the action of a normal stress and failure by shearing under the action of a shear stress — there are two corresponding basic theories of the strength of materials: the highest normal stress and the highest tangential stress. Apart from these most elementary cases others have also been suggested which have not proved themselves in practice. Two new relevant theories have been recently suggested: Davidenkov's and Friedman's unified theory of strength[57] and Volkov's statistical theory of strength.[37]

Davidenkov and Friedman combined the hypothesis of the highest tangential stress and of the highest extension. In this theory the relation of the highest tangential stress to the greatest applied tensile stress appears as a characteristic of the type of stress. In Volkov's theory which takes into account the microheterogeneity of actual material brittle failure appears in all possible types of stress as a result of the action of microscopic tensile stresses (even if the whole specimen is under compression). The conception that a limited deformation may be the cause of failure (hypothesis of limited deformations) is completely excluded in the statistical theory of strength; this fact is very important.

Mechanical theories of strength are natural supplements to physical theories, as the latter ones are usually developed for the simplest types of stress (tension, compression, flexure).

The most widespread method of testing the strength of materials under various types of stress is the testing of thin-walled tubes. In this connection Mustafin's and Sokolov's work[108] on the study of the strength of organic glass under the condition of planar stress is quite typical. Depending on the type of stress they distinguish between "soft" and "hard" loading. The numerical expression of the "softness" of the load is the relation $q = \tau_m/\sigma_1$ where τ_m is the maximum tangential and σ_1 the maximum normal stress at a given point (or local region). Examples of the most widely used types of stress and also the corresponding values of q are shown on Fig. 27. A tube stressed by internal pressure has a

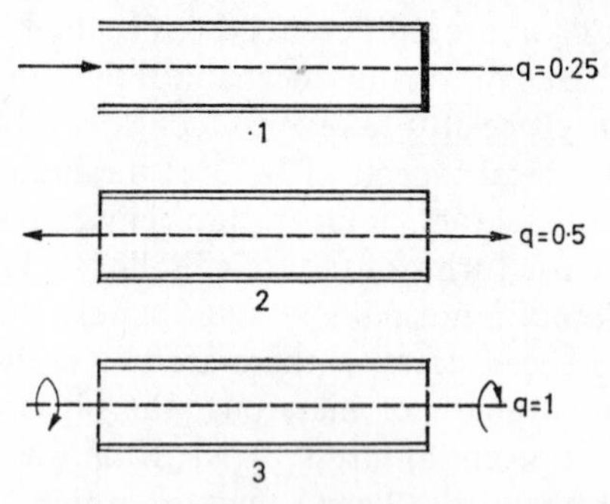

FIG. 27. The simplest types of planar stress in thin-walled tubes and corresponding to values q. 1. Inner pressure. 2. Axial tension. 3. Torsion.

considerable "hardness" of loading $q = 0{\cdot}25$; the combination of torsion with axial pressure ($q > 1$) is an extremely "soft" loading. They regarded the specimen as failed when a brittle fracture took place or when the form of the specimen changed as a result of ductile deformations and the specimen lost its diametrical or longitudinal stability. As the stress was planar it was characterized by two principal normal stresses σ_1 and σ_2

where in accordance with the theory of elasticity for any type of loading

$$q = \frac{\tau_m}{\sigma_1} = 0{\cdot}5\left(1 - \frac{\sigma_2}{\sigma_1}\right).$$

Mustafin and Sokolov established that for each temperature T there exists a maximum value σ_1 at some $q = q_0(T)$. For $q < q_0$ the failure is a brittle one, whilst for $q > q_0$ it is a ductile one (Fig. 28).

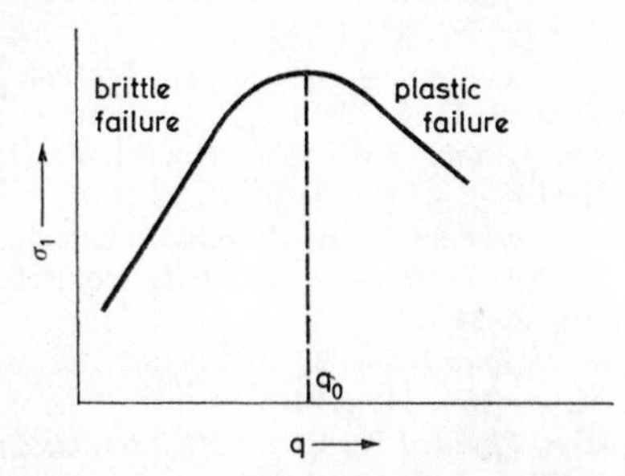

FIG. 28. Dependence of normal stress σ on the "softness" of load q at constant temperature.[108]

From Davidenkov's theory it follows that at $q = q_0$ the relation

$$\tau_m/\sigma_1 = R_t/R_\sigma$$

must be fulfilled at all temperatures where R_t is the resistance of the material to shear and R_σ the resistance to rupture, values which depend on the material and the temperature.

The most favourable stress (as brittle failure by tearing is the most critical one) is that which corresponds to the maximum value σ_1, i.e. $q = q_0$, where $q_0(T)$ is a function of temperature. If the value obtained is q_0, the other normal stress is given by $\sigma_2 = (1-2q_0)\sigma_1$.

According to Mustafin's and Sokolov's[108] data the most favourable stress for organic glass at a temperature of 45°C was found to be that for which $q_0 = 0{\cdot}5$, i.e. a unidirectional tension; at other temperatures the greatest strength was reached under other types of stress. If, however, the type of stress is given, the temperature at which the sample possesses its greatest strength is predetermined. Types of failure of inorganic glass under different stresses have been investigated in detail by Ponselet.[109]

Literature

1. A. Griffith, *Phil. Trans. Roy. Soc.* A **221,** 163 (1921).
2. A. F. Joffe, M. B. Kirpicheva, M. A. Levitskaya, *ZhRFKhO chast' fizicheskaya*, **56**, No. 5–6, 489 (1924).
3. A. P. Aleksandrov, S. N. Zhurkov, *Yavleniye khrupkogo razriyva*, Gostekhteoretizdat, 1933.
4. A. P. Aleksandrov, O. Khrupkosti, *Vestnik AN SSSR*, No. 7/8, 51 (1944).
5. A. Smekal, *J. Soc. Glass Technol.* **20,** 432 (1936); *Glastechn. Ber.* **23,** No. 3, 57 (1950); *Ergebn. exakt. Naturwiss.* **15,** 106 (1936).
6. K. Müller, *Z. Phys.* **69,** 431 (1931).
7. A. Holland, W. Turner, *J. Soc. Glass Technol.* **24,** 101 (1940); *J, Soc. Glass Technol.* **32,** 5 (1948).
8. S. N. Zhurkov, *Vestnik AN SSSR*, No. 11, 78 (1957); *Z. Physik. Chem.* **213,** 183 (1960).
9. M. Born, *Atomtheorie des Festen Zustandes*, Leipzig, 1923; M. Born, M. Geppert-Meyer, *Teoriya tverdogo tela*, Izdatinlit, 1938.
10. F. Zwicky, *Phys. Z.* **24,** 131 (1923).
11. J. De Boer, *Trans. Farad. Soc.* **32,** 10 (1936).
12. E. Orowan, *Nature* **154,** 341 (1944).
13. J. E. Stanworth, *Physical Properties of Glass*, Oxford, 1940.
14. M. Polanji, *Z. Phys.* **7,** 323 (1921).
15. P. P. Kobeko, *Amorfniye veshchestva*, Izd. AN SSSR, 1952.
16. G. K. Demishev, I. V. Rasumovskaya, *Steklo*, **15,** No. 1, 30 (1962).
17. G. K. Demishev, *Steklo*, **15,** No. 4, 1 (1962).
18. G. I. Barenblatt, *Zh. prikl. mekh. i tekhnich. fiziki*, No. 4, 3 (1961).
19. I. V. Obreimov, *Proc. Roy. Soc. (London)*, A **127,** 290 (1930).
20. V. P. Berdennikov, *ZhFKh*, **5,** 358 (1934).
21. T. A. Kontorova, *ZhTF*, **15,** 346 (1954).
22. B. Ya. Pines, *ZhTF*, **16,** 981 (1946).
23. R. A. Sack, *Proc. Phys. Soc.* **58,** 729 (1946).
24. B. A. Drozdovskii, Ya. B. Fridman, *Vliyaniye treshchin na mekhanicheskiye svoistva konstruktsionniykh stalei*, Metallurgizdat, 1960.
25. G. I. Barenblatt, *DAN SSSR*, **127,** 47 (1959).
26. V. V. Danasyuk, *Opredeleniye napriazenii i deformatsii vblizi mel' chaishei treshchiniy*, *Voprosiy mashinovedeniya i prochnosti v mashinostroyenii*, No. 6, Mashgiz, 1960, p. 114.
27. Ya. I. Frenkel', *ZhTF*, **22,** 1857 (1952).
28. H. A. Elliott, *Proc. Phys. Soc.* **59,** 208 (1947).
29. G. M. Bartenev, I. V. Rasumovskaya, *DAN SSSR*, **133**, 341 (1960).
30. P. A. Rebinder, *Fiziko-khimicheskiye issledovaniya protsessov deformatsii tverdiykh tel*, *Sbornik k XXX-letiyu Oktiabr'skoi revolyutsii Izd. AN SSSR*, (1947), p. 123; *ZhTF*, **2,** 726 (1932); *Z. Phys.* **72,** 191 (1931); *Izv. AN SSSR, Otd. matem. i est. nauk*, Ser. khim. **5,** 639 (1936).
31. B. I. Likhtman, P. A. Rebinder, G. V. Karpenko, *Vliyaniye poverkhnostnoaktivniykh sred na protsessiy deformatsii metallov*, Izd. AN SSSR, 1954.

32. M. S. Aslanova, P. A. Rebinder, *DAN SSSR*, **96**, 325 (1954).
33. A. N. Orlov, Yu. M. Plishkin, I. M. Shepeleva, *FMM*, **4**, 540, (1957).
34. A. N. Orlov, *FMM*, **7**, 481 (1959).
35. Yu. M. Plishkin, *FMM*, **9**, 178 (1960); *DAN SSSR*, **137**, 564 (1961); *Zhurn. prikl. mekh. i tekhnich. fiziki*, No. 2, 95 (1962).
36. A. V. Stepanov, *Sbornik k 70-letiyu akad. A. F. Joffe*, Izd. AN SSSR, 1950, p. 341.
37. S. D. Volkov, *Statisticheskaya teoriya prochnosti*, Mashiz, 1960.
38. N. I. Mott, *J. Phys. Soc. Japan*, **10**, 650 (1955); *Proc. Roy. Soc. (London)*, A **220**, 1 (1953).
39. A. N. Stroh, *Proc. Roy. Soc. (London)*, A **216**, 391 (1953); *Phil. Mag.* **46**, 968 (1955); *Phil. Mag.* **2**, 1 (1957).
40. F. E. Fujita, *J. Phys. Soc. Japan*, **11**, 1201 (1956).
41. A. Cottrell, *teoriya dislokatsii*, *Uspekhi fiziki metallov*, vol. I. Metallurgizdat, 1956; V. T. Reed, *Dislokatsiya v kristallakh*, Metallurgizdat, 1957.
42. V. L. Indenbom, A. N. Orlov, *UFN*, **76**, 557 (1962).
43. V. I. Lichtman, E. D. Shchukin, *UFN*, **66**, 213 (1958); *Uspekhi khimii*, **29**, No. 10, 1260 (1950); *DAN SSSR*, **124**, 307 (1959).
44. V. I. Lichtman, L. A Kochanova, L. S. Bryukhanova, *DAN SSSR*, **120**, 757 (1958).
45. S. N. Zhurkov, V. A. Marikhin, A. I. Slutsker, *FTT*, **1**, 1159 (1959).
46. S. N. Zhurkov, A. I. Slutsker, V. A. Marikhin, *FTT*, **1**, 1759 (1959).
47. G. M. Bartenev, I. V. Rasumovskaya, P. A. Rebinder, *Kolloid. zh.* **20**, 655 (1958).
48. E. Shand, *J. Am. Ceram. Soc.* **37**, 52, 559 (1954); **42**, 474 (1959).
49. N. F. Mott, *Engineering*, **165**, 16 (1948).
50. S. Bateson, *Phys. and Chem. Glasses*, **1**, 139 (1960).
51. W. Holzmüller, K. Altenburg, *Physik der Kunstsoffe*, Berlin, 1961, p. 402.
52. W. Holzmüller, P. Jung, *Plaste und Kautschuk*, **2**, 218 (1955).
53. A. Borgwardt, *Plaste und Kautschuk*, **6**, 68 (1959).
54. F. C. Roesler, *Proc. Phys. Soc.* B **69**, 981 (1956); J. J. Benbow, F. C. Roesler, *Proc. Phys. Soc.* B **70**, 201 (1957).
55. J. P. Berry, *SPE Trans.* **1**, 1 (1961); *J. Polymer Sci.* **50**, 107, 313 (1961); *J. Appl. Phys.* **34**, 62 (1963).
56. N. L. Svensson, *Proc. Phys. Soc*, **77**, 876 (1961).
57. Ya. B. Friedman, *Mekhanicheskiye svoistva metallov*, Oborongiz, 1952; Ya. B. Friedman, T. A. Gordeyeva, A. M. Zaitsev, *Stroyeniye i analiz izlomov metallov*, Mashgiz, 1960.
58. M. I. Bessonov, E. V. Kuvshinskii, *Plast. massiy*, No. 5, 57 (1961).
59. V. R. Regel', Yu. N. Nedoshivin, *ZhTF*, **23**, 1333 (1953).
60. S. N. Zhurkov, E. Ye. Tomashevskii, *ZhTF*, **27**, 1248 (1957).
61. R. N. Haward, *The Strength of Plastics and Glass*, London, 1949; A. M. Bueche, A. V. White, *J. Appl. Phys.* **27**, 980 (1956).
62. F. E. Barstow, H. E. Edgerton, *J. Am. Ceram. Soc.* **22**, 302 (1939).
63. H. M. Dimmick, I. M. McCormick, *J. Am. Ceram. Soc.* **34**, 240 (1951).

64. G. A. Kuz'min, V. P. Pukh, *Sb. "Nekotoriye problemiy prochnosti tverdogo tela"*, Izd. AN SSSR, 1959, p. 367.
65. H. Shardin, Conference on Fracture, Swampscott, Mass., U.S.A. Nat. Academy of Sci. Report, Paper No. 12, p. 10.
66. F. Preston, *J. Appl. Phys.* **13,** 623 (1942).
67. G. M. Bartenev, *Mekhanicheskiye svoistva i teplovaya obrabotka stekla*, Gosstroiizdat, 1960; *DAN SSSR*, **71,** 23 (1950); *ZhTF*, **21,** 579 (1951).
68. C. Gurney, S. Pearson, *Proc. Phys. Soc.* B **62,** 469 (1949).
69. T. Baker, F. Preston, *J. Appl. Phys.* **17,** 179 (1946).
70. T. A. Kontorova, *DAN SSSR*, **54,** 23 (1946).
71. S. N. Zhurkov, B. N. Narzullayev, *ZhTF*, **23,** 1677 (1953).
72. S. N. Zhurkov, T. P. Sanfirova, *DAN SSSR*, **101,** 237 (1955).
73. S. N. Zhurkov, E. Ye. Tomashevskii, *ZhTF*, **25**, 66 (1955).
74. S. N. Zhurkov, T. P. Sanfirova, *FTT*, **2,** 1034 (1960).
75. S. N. Zhurkov, S. A. Abasov, *Viysokomol. soyed.* **3,** 441, 450 (1961); *FTT*, **4,** 2184 (1962).
76. S. N. Zhurkov, B. Ya. Levin, E. Ye. Tomashevskii, *FTT*, **2,** 2066 (1960).
77. V. I. Betekhtin, S. N. Zhurkov, A. V. Savitskii, *FMM*, **10,** 453 (1960).
78. S. N. Zhurkov, B. Ya. Levin, T. P. Sanfirova, *FTT*, **2,** 1040 (1960).
79. W. Busse, E. Lessig, D. Loughborough, L. Larrick, *J. Appl. Phys.* **13**, 715 (1942).
80. V. I. Lichtman, Ye. D. Shchukin, P. A. Rebinder, *Fiziko-khimicheskaya mekhanika metallov*, Jzd. AN. SSSR, 1962.
81. S. N. Zhurkov, S. A. Abasov, *Viysokomol. soyed.* **4,** 1703 (1962).
82. V. P. Solobiyrev, *Izv. AN SSSR, OTN*, No. 4, 92 (1958).
83. I. Ye. Kurov, V. A. Stepanov, *FTT*, **4,** 191 (1962).
84. S. N. Zhurkov, *FTT*, **4,** 3353 (1962).
85. P. A. Rebinder, V. I. Lichtman, L. A. Kochanova, *DAN SSSR*, **111**, 1278 (1956); Ye. D. Shchukin, P. A. Rebinder, *Kolloid. zh.* **20,** 645 (1958).
86. G. M. Bartenev, I. V. Rasumovskaya, *DAN SSSR*, **150,** 784 (1963).
87. E. Shand, *J. Am. Ceram. Soc.* **44,** 21 (1961).
88. B. Ya. Pines, *ZhTF*, **25,** 1399 (1955); *FTT*, **1,** 265 (1959).
89. A. N. Orlov, *FTT*, **3,** 500 (1961).
90. I. V. Vladimirov, L. E. Gurevich, *FTT*, **2,** 1782 (1960).
91. P. Gibbs, J. B. Cutler, *J. Am. Ceram. Soc.* **34,** 200 (1951).
92. D. A. Stuart, O. L. Anderson, *J. Am. Ceram. Soc.* **36,** 416 (1953).
93. G. M. Bartenev, Vremenaya i temperaturnaya zavisimost' prochnosti tverdiykh tel, *Izv. AN SSSR, OTN*, No. 9, 53 (1955).
94. S. N. Zhurkov, T. P. Sanfirova, *ZhTF*, **28,** 1719 (1958).
95. P. Felthman, J. D. Meakin, *Acta metall.* **7,** 614 (1959).
96. B. Ya. Pines, A. F. Sirenko, *FMM*, **10,** 382 (1960).
97. T. Alfrei, *Mekhanicheskiye svoistva viysokopolimerov*, Izdatinlit, 1952, str. 505.
98. G. M. Bartenev, V. E. Gul', *ZhKhO im. D. I Mendeleyeva*, **6,** 394 (1961).

99. C. C. Hsiao, *Nature*, **186**, 535 (1960).
100. G. Neiber, *Kontsentratsiya napriazhenii*, Gostekhizdat, 1947.
101. A. I. Gubanov, A. D. Chevychelov, *FTT*, **5**, 91 (1963).
102. W. Kuhn, H. Kuhn, *Helv. chim. acta*, **29**, 1634 (1946); L. Treloar, *Fizika uprugosti kauchuka*, Izdatinlit, 1953, str. 180.
103. F. J. Bueche, *J. Appl. Phys.*, **28**, 784, (1957); **29**, 1231 (1958).
104. A. I. Gubanov, A. D. Chevychelov, *FTT*, **4**, 928 (1962); A. D. Chevychelov, *FTT*, **5**, 1394 (1963).
105. G. M. Bartenev, I. V. Rasumovskaya, *FTT*, **6**, 657 (1964).
106. Yu. M. Ivanov, *Zav. lab.* **27**, 455, (1961).
107. M. M. Filonenko-Borodich, *Mekhanicheskiye teorii prochnosti*, Izd. MGU, 1961.
108. Ch. G. Mustafin, B. P. Sokolov, Prochnost' organicheskogo stekla pri ploskonapriazhennom Sostoyanii, *Izv. AN SSSR, OTN*, No. 5, 179 (1959).
109. E. Ponselet, *Verres et refract.* **2**, 203 (1948); **3**, 149, 289 (1949). **4**, 156 (1950); **5**, 69 (1951); *Glass Ind.* **45**, 251 (1964).

CHAPTER 2

DEFORMATION AND STRENGTH OF POLYMERS

POLYMERS can be in the following fundamental structural states: two structural liquid (viscous-liquid and visco-elastic) and two solid ones (crystalline and amorphous). In both states the structure of the polymer depends on its thermal history. According to their structure polymeric materials can be divided in five basic groups: (1) amorphous, (2) amorphously orientated, (3) partially crystalline, (4) crystalline isotropic, and (5) crystalline orientated. To the first group belong the majority of rubbers and plastics, to the second and fifth the fibre- and film-forming materials, to the third crystalline rubbers, and so on.†

According to their chemical structure polymers are divided in linear, branched and spatial-structural or cross-linked ones and also in low- and high-molecular ones. The chain structure of high-molecular compounds of the same chemical composition can differ owing to stereo-isometry. The most important stereo regular polymers are the isotactic and the syndiotactic ones. Atactic (stereo non-regular) and isotactic polymers of the same chemical composition differ sharply in structure and properties. Atactic polymers, consisting of irregularly constructed chains, are amorphous and incapable of crystallizing even under tension. Isotactic polymers are usually in a crystalline condition, or crystallize easily under tension (natural rubber).

The strength of polymers is closely connected with their deformation properties which depend on the structure and physical composition. Under different physical conditions the polymer

† This classification has a conditional character, as a polymer can belong to the one or other group, depending on temperature and method of production; for instance amorphous polymers can be in the first or second group, crystalline polymers in the fourth or fifth, and so on.

shows different types of deformation and failure.[1] The relaxation theory of deformation of polymers was at first suggested by Kobeko and co-workers.[2] Later on Kargin and Slonimsky[3, 4] suggested a mathematical theory of visco-elastic polymers, starting from the general theory of Boltzmann–Walter and from ideas about the molecular structure of the polymers. They worked out a deformation theory for three types of linear amorphous polymers: glassy, visco-elastic and visco-fluid. These conditions (Fig. 29) are displayed under small stresses. Under great stresses

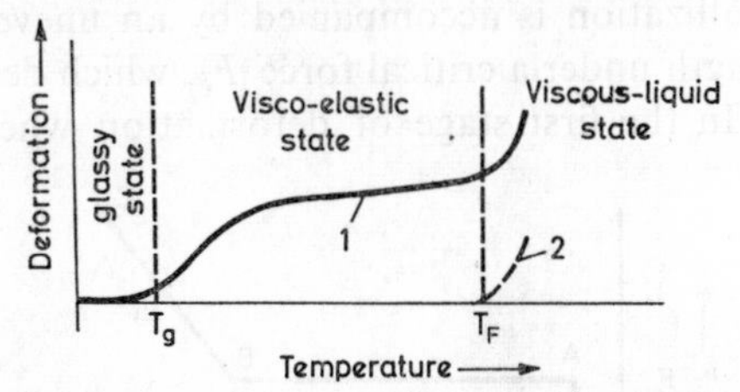

FIG. 29. Three physical states of amorphous linear polymers: T_g – Second order transition point; T_F – temperature of flow. 1. Thermomechanical curve. 2. Curve of remaining deformation.

the deformations of polymers acquire a number of new features which influence the strength and the mechanism of failure. These features show up extremely clearly under tension of solid polymers (crystalline and amorphous ones), when "cold flow" can be observed. In this way appears an orientated structure which greatly strengthens the material.

Kargin and Sogolova[5, 6] discovered the nature and the laws governing the deformation properties of crystalline polymers, the influence of orientation, structure and relaxation properties of their molecules on their strength. Kargin, Kozlov and co-workers[7] fully investigated the structure of polymer systems and showed in detail that in crystalline polymers polymorphism can be observed which has a considerable influence on mechanical and other properties.

Alexandrov and Lazurkin[8, 9] investigated the low-elastic deformation of solid amorphous polymers and suggested a relaxation theory of that phenomenon. Bartenev and co-workers[10] showed that at high temperatures amorphous polymers pass from a high elastic condition into a ductile one (analogous to the transition of simple solids from a brittle to a plastic condition).

2.1. Deformation and Strength of Crystalline Polymers

Polymers with a regular chain structure for which the crystalline state is characteristic differ in structure and deformation properties from ordinary crystalline bodies.[5, 6, 11–15] Only under small deformations do crystalline polymers behave like ordinary solid bodies. On the other hand, greatly deformed under tension, they undergo a structural transition from the initial crystalline phase to one which shows orientation along the direction of tension. Recrystallization is accompanied by an uneven change of the sample length under a critical force F_{cr} which depends on the temperature. In the first stage of deformation when $F < F_{cr}$ a

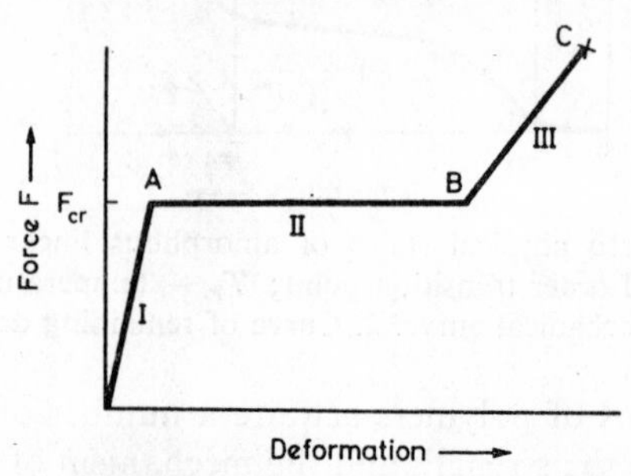

FIG. 30. Diagram of tensioning a crystalline polymer.

uniform small extension of the specimen (Fig. 30) arises; on reaching F_{cr} (point *A*) there appears suddenly in the sample a "neck" (Fig. 31) which under further stressing gradually spreads along the whole specimen. From some moment (point *B*) the sample stretches again, as a whole, right up to the break.

With decreasing molecular weight the strength of the crystalline polymer falls and the point of break *C* (see Fig. 30) is moved to the left along the curve of deformation. The strength can then become less than F_{cr} and then the polymer shows only brittle failure.

Great deformations of crystalline polymers are visco-elastic ones as they are connected with the change of the orientation of the polymer chain during the structural transition.

Kargin and Sogolova[5] established during experiments on fracture of polyamide and other crystalline polymers that they have in the non-orientated or transversally orientated state a dis-

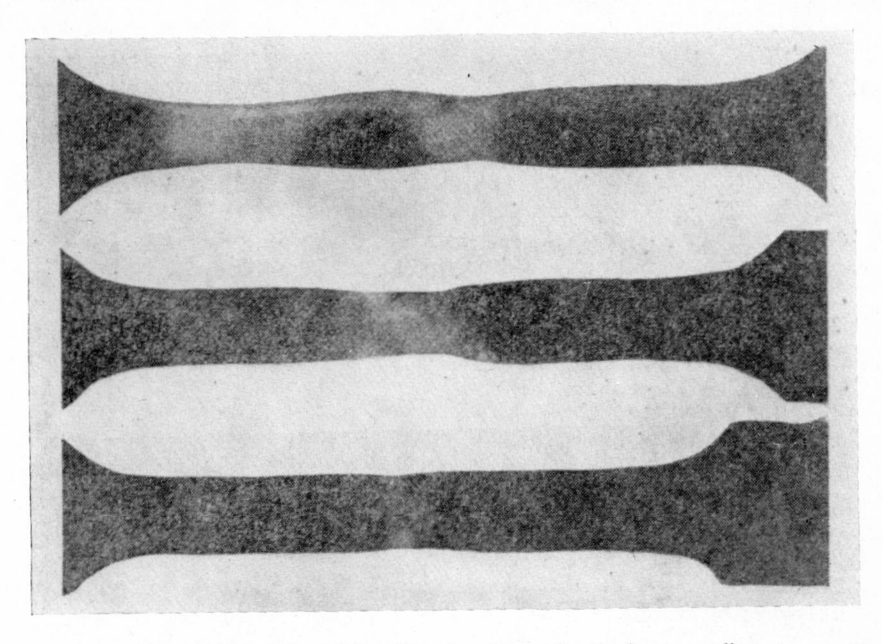

FIG. 31. Formation of "neck" on nylon under tension according to the data of J. Miklowitz.[13]

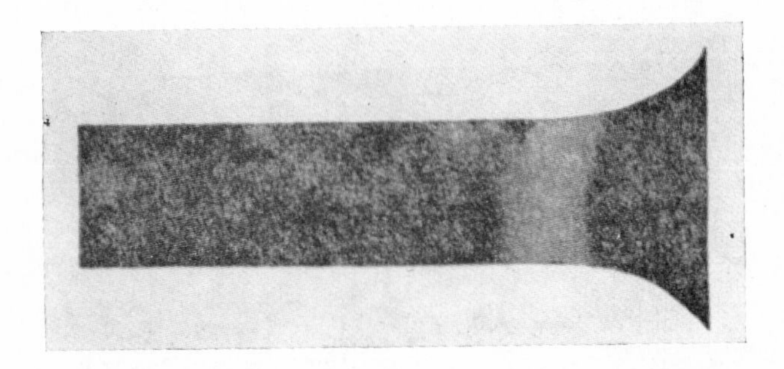

FIG. 36. Loaded sample of polyvinyl alcohol in polarized light (with crossed Nichols) before the formation of a "neck".

tinctive temperature dependence of strength (Fig. 32). At low temperatures only brittle strength is observed in the area *AB*. Beginning with the temperature T_{or} (the temperature of beginning orientation) at which a neck begins to form and the material acquires the capacity for orientation, the specimen possesses two values of strength depending where the break takes place — in the thick part or in the neck. At temperatures above T_{br} (temperature of brittleness) the sample stretches considerably, passes through the neck and then breaks. The size of the amount

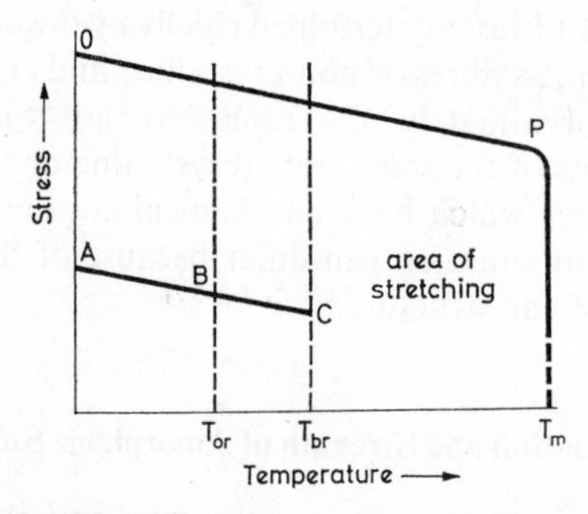

FIG. 32. Temperature dependence of strength of crystalline polymer under tension according to Kargin and Sogolova.[5]

of deformation of polymers at fracture changes sharply with temperature; at low temperatures it is small (brittle fracture), it increases with rising temperature and above T_{br} it remains approximately constant right up to the melting point T_m.

Orientated polymer (fibrous material) behaves differently. At any temperature below the melting point it is very strong (area *OP*) and its strength grows with decreasing temperature.

The structural changes during the stressing of crystalline polymers have been studied in great detail in recent times. It appears that the structural transformation under tension is a complicated process of recrystallization connected with the breakdown of secondary supramolecular structures (for instance, spherolites) and the formation of new fibrous crystalline structures. Defects of the structure in the shape of microvoids and microcracks are then formed. Besides, it was shown that the melting of the crystalline polymer has to be considered as a multistep phase transition of the first kind due to the presence of different forms of supra-molecular structures.

Crystalline polymeric materials are mostly produced in the shape of fibres which as a result of stretching acquire an anisotropic structure. The useful temperature range of crystalline fibres lies below the melting temperature and the technological range of stretching lies in the interval between the brittle and the melting temperature (see Fig. 32).

The strength of fibres changes in wide limits from 10 to 100 kg/cm^2. Crystallized, but not orientated, fibres have low strength. The strength of fibres is determined chiefly by the degree of molecular orientation, as fibres of non-crystalline and crystalline structure have approximately the same strength when orientated. Within the limits of the same state (crystalline or amorphous) the strength of fibres which have an identical chemical composition and degree of attenuation can differ because of the difference in the supramolecular structure.[6, 7, 16, 17]

2.2. Deformation and Strength of Amorphous Solid Polymers

In a polymer in the glassy state (see Fig. 29) there arise under low stress only elastic deformations with a Young's modulus 200–600 kg/mm^2 (for steel, Young's modulus is 20,000 kg/mm^2).

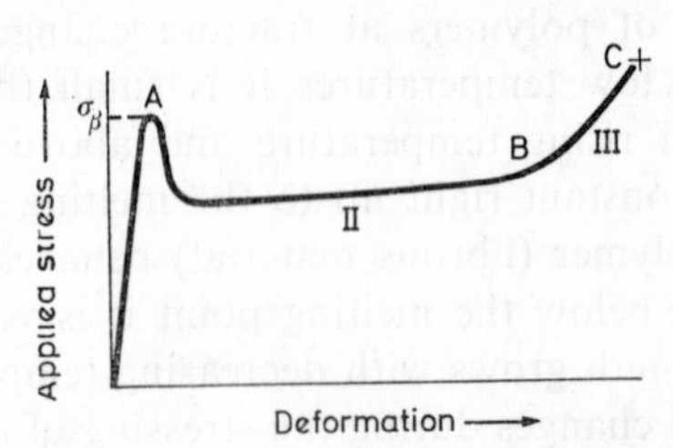

FIG. 33. Curve of tensioning an amorphous polymer in the glassy state (applied stress).

Under great stresses the deformation processes of amorphous polymers are more complicated.[8, 9, 16] In the glassy state in which plastics are at ambient, and rubber and elastomers at low temperatures, the elongation of the amorphous polymer (Fig. 33) appears outwardly like that of a crystalline one. When the stress† reaches the so-called limit of low elasticity σ_B (point A), a "neck" is formed in the weakest place into which the whole sample gradually passes (area AB). The thinned specimen stretches somewhat more up to fracture (area BC).

The limit of low elasticity σ_B is the higher the lower the temperature and the greater the speed of deformation. When starting at a rather low temperature (T_{br}) the value becomes greater than the strength of the polymer and the break is brittle, without reaching σ_B (Fig. 34). The state when the material shows brittle

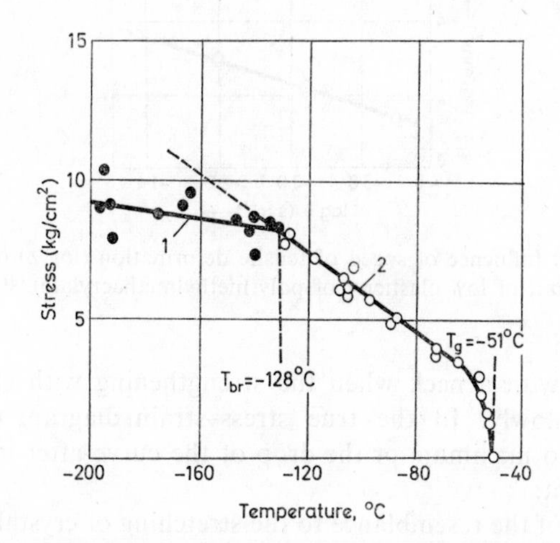

FIG. 34. Temperature of dependence of brittle strength (1) and limits of low elasticity (2) of Butadiene-styrene 30 rubber (SKS-30).

fracture is also attained with decreasing molecular weight owing to the decrease of strength (on section OA on Fig. 33). The limit of the low elasticity depends on the speed of deformation (Fig. 35), and is higher than the brittle strength of the polymer; therefore the temperature of brittleness depends on the speed of deformation.

After the whole sample has turned into a neck which leads to an increase of strength the orientated material breaks. At small attenuations the fracture of the specimen is similar to the ordinary

† The acting stress is numerically equal to the stretching force per unit of the initial cross-section of the specimen; the true stress is equal to the force per unit of the cross-section of the specimen after deformation.

brittle fracture; at large ones the specimen shows a brittle-fibrous break. For some polymers, in particular in previously stretched ones, the curve of deformation (see Fig. 33) has no maximum and the tensioning of the sample takes place without the formation of a neck. This is explained by the different character of the strengthening; this is based either on a stretching without a neck when the strengthening with the elongation proceeds quickly or on a

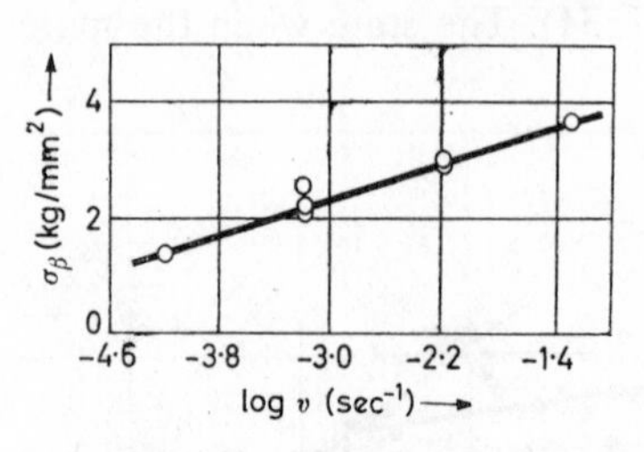

FIG. 35. Influence of speed of tensile deformation (log v) on the limit of low elasticity of poly(methylmethacrylate).[9]

stretching with a neck when the strengthening with elongation proceeds slowly. In the true stress–strain diagram there will either be no maximum or the drop of the curve after it becomes insignificant.

In spite of the resemblance to the stretching of crystalline polymers the causes for the elongation of amorphous solid polymers are somewhat different as the formation of "necks" is not connected with recrystallization.

In contrast to the visco-elastic state when visco-elastic deformation can be observed under any small stress, the polymer shows in the glassy state, at loads lower than σ_B, basically elastic deformation. The forced visco-elastic deformation ("cold flow") begins to develop only above σ_B, under the effect of considerable outside forces. This phenomenon of low elastic deformation[8, 9, 16] is explained by the relaxation theory. According to this theory the time of relaxation which characterizes the speed of regrouping the segments of the polymer chain depends on the absolute temperature T and the stress σ as follows:

$$\tau = Ce^{\frac{U - a\sigma}{kT}} \tag{2.1}$$

where C is a constant; U is the activation energy, defined as the barrier potentials which the segments overcome during the transition from one position of equilibrium to the other; α is the molecular weight of the segments; k is Boltzmann's constant.

From eqn. (2.1) follows that the critical time of relaxation τ_{cr} which is being compared with the given time of observation (or speed of deformation), during which the visco-elastic deformation develops, can be attained by raising either of the temperature to the temperature of vitrification T_v or of the stress to σ_B. From this follows that the values T_{cr} and σ_B must depend on the time of observation or speed of deformation. Below T_v the visco-elastic deformation remains "frozen" for an unlimited time after unloading but above T_v it disappears completely and the sample regains its former shape.

It also follows from eqn. (2.1) that a visco-elastic deformation develops also under stresses lower than σ_B but only slowly. One of the causes of creep of plastics is the slow accumulation of visco-elastic deformation; the other cause is the development of "silver" cracks (see Chapter 3).

It follows from the above that the failure of solid polymers at a temperature above T_{cr} is a complicated process consisting in the destruction of the shape of the specimen during the passing the limit of low elasticity and of the failure of the material on parting of the specimen. Deformation by shear is the first stage of failure without disturbing the integrity of the material. The molecular mechanism of the deformation by shear consists in the deplacement and orientation of segments of the polymer chains under the effect of outside forces. The second stage consists in the propagation of cracks in the orientated material and the appearance of shear deformations before the formation of a "neck" can be seen by examination of the stretched specimen in polarized light (Fig. 36, facing p. 66).

2.3. Visco-elastic Deformation of Polymers

Visco(or high) elasticity is observed in linear polymers of the most varied chemical structure: in typical hydrocarbons, for instance poly(isoprene), natural rubber, poly(isobutylene) in silicon-organic rubbers, for instance poly(methylsiloxane), in non-organic rubbers, for instance poly(phosphonitrilechloride). Linear polymers are in a visco-elastic state above T_v right up to T_m, and spatially arranged polymers are visco-elastic materials right up to the temperature of the chemical disintegration of the spatial network as their melting point is very high. Rubbers and elastomers are typical visco-elastic materials in the temperature range from -70 to $+200$°C and in some cases even outside these limits.

Visco-elastic deformation is a type of elastic deformation which is specific only to polymers. It is characterized by a small modulus of elasticity (1–10 kg/cm^2) and by great mechanical reversible deformations. In spatially arranged polymers (elastomers) obtained by cross-linking of linear macromolecules the visco-elastic properties are displayed in the purest form, as the knots of the net hinder the flow of the material. Therefore an elastomer recovers its shape after the removal of the load like elastic solids.

In other physical properties elastomers are similar to liquids. Elastomers, like liquids, are amorphous substances; they obey Pascal's law, the value of coefficients of thermal expansion, compressibility and a number of other characteristics of elastomers are more similar to values which are characteristic of liquids than to those of solids.

At the same time the nature of visco-elastic deformation of polymers differs from the nature of deformation of solids and liquids. Visco(or high) elastic deformation differs from the ordinary elastic one by the fact that the latter is connected with the change of the mean distances between the particles, and the visco-elastic one with the regrouping of the segments of flexible chains without change of the mean distance between the chains. Stress in the deformed elastomer, like the presence of compressed gas is proportional to the absolute temperature. The deformation of elastomers, like the compression of gas, is connected with the decrease of entropy. Thus the specially arranged polymer possesses a particular combination of the properties of solids, liquids and gases.

The flexibility of the polymeric molecules which is the cause of the visco-elastic properties is stipulated by the fact that in the chains of the main valencies there are bonds which are capable of turning or rotating relative to each other. The number of possible configurations of polymeric chains which arise through the rotating of bonds is limited by the interaction between the hydrogen atoms or the lateral groups. The thermal movement causes the transformation of one type of configuration into others; the frequency of these transformations depends on the size of the barrier potential of the rotation and the intensity of the thermal movement. The higher the stress, the easier the displacement of the segments into the direction of the force is achieved and vice versa.

The development of a deformation of the chain molecules takes place by the successive displacements of segments from one position into another, i.e. proceeds with time. Therefore a visco-elastic deformation lags behind a change of the outer stress. The process of regrouping of the segments is accompanied by the overcoming of the inner friction in the polymers and, consequently, by a dispersion of mechanical energy.

Amongst the various configurations of the chain molecules the most probable are the tightly coiled ones in which the distance between the ends of the macromolecules (or between the points of cross-linkage in the chains of the spatial network) is much smaller than its fully extended length. Under the effect of an outer force the chains will change their shape but after the cessation of the effects of the force the chain molecules will, as a result of the thermal movement, return again into the most probable state which means the tightly coiled one. In the language of thermodynamics the transition into a more probable state is connected with the growth of the entropy. Therefore, the return of the deformed elastomer specimen into the original non-deformed state is accompanied by an increase of entropy. On the other hand, the deformation of an elastomer is connected with a decrease of entropy. In this the deformations of elastomers differ from the deformations of solids, in which the entropy does practically not change.

In the first approximation it is assumed that the inner energy does not change during deformation of spatially arranged polymers. It follows therefore from thermodynamic laws that the outer force f depends only on the change of entropy S of the specimen. For the isothermic deformation in equilibrium: tension–compression the following equation applies:

$$= -T\frac{\partial S}{\partial \lambda} \qquad (2.2)$$

where f is the force per unit of the initial cross-section of the specimen; T is the absolute temperature; S is the entropy of the unit of the weight of the polymer (it decreases in proportion to the tension, therefore there is a minus in the right part of the equation).

From the conditions of incompressibility during deformation of visco-elastic materials follows a simple correlation between the impressed force f and the true stress σ, which refer to the actual section in the deformed state:

$$\sigma = \lambda f.$$

The statistical theory of the visco-elastic deformation in equilibrium,[18] not taking into account the interaction between the chains of the network, leads to the following expression for unidirectional tension or compression:

$$\sigma = G\left(\lambda^2 - \frac{1}{\lambda}\right) \tag{2.3}$$

where G is the modulus in shear.

As the statistical theory agrees only qualitatively with experimental results, various equations have been suggested, inferred from theoretical investigations which, whilst not strict, give a better description of the deformation of three-dimensional polymers.[19, 20] Common to all these theories is the following equation of the static deformation under tension:

$$\sigma = T\varphi(\lambda) \tag{2.4}$$

where $\varphi(\lambda)$ is the function of the tension which has a different actual meaning in different theories.

2.4. Strength and Deformation of Linear and Three-dimensional Polymers in the Visco-elastic State

The strength of linear and three-dimensional rubber-like polymers was investigated in the work of Bartenev and co-workers.[1, 10]

The curves of tension in linear polymers at low and high temperature in the range of visco-elasticity differ in principle. Above any temperature T_p (the temperature of plasticity) the stressed polymer is a ductile material as shown by Fig. 37. Up to point *A* the polymer shows in practice only visco-elastic deformation. Stress σ_p corresponds to point *A*; this is the yield point. Under stresses exceeding σ_p, there develops simultaneously with the visco(high)elastic a plastic (ductile) deformation. A uniform development of the remaining deformation along the specimen and its cross-section takes place until a contraction appears (point *B*). After this the stress grows, chiefly in the contraction. Then the failure begins (point *C*). Failure sets in in the described

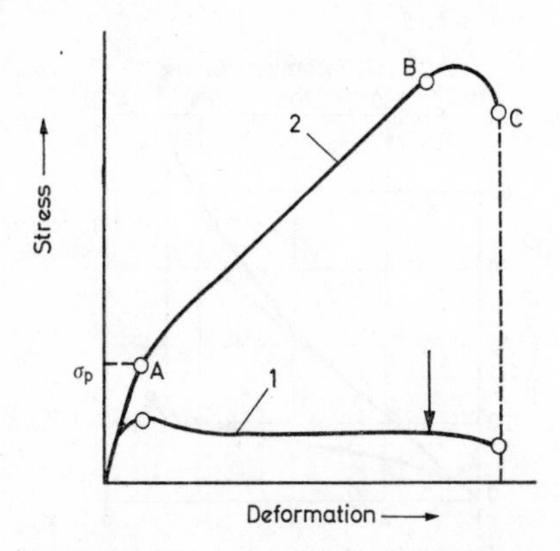

FIG. 37. Stress–strain diagram of amorphous linear polymer in plastic state. 1. Applied stress. 2. True stress.

stage of development of contraction because the flow in it cannot develop unrestrictedly, as the orientation of the molecules leads to a sharp growth of the viscosity of the material in the contraction.

The initial sections of the curves of tension of the polymer in the range of low elasticity (see Fig. 33) and in the range of plasticity (see Fig. 37, curve 1) are outwardly similar (in the referring stress-deformation section of the curves a maximum is observed). The difference consists in the fact that the growing branch (before the maximum) refers in the one case to the usual elastic and in the other to the visco-elastic deformation. The limits σ_B and σ_p signify that under these stresses a visco-elastic and plastic deformation respectively begins to develop in the material. The difference consists also in the fact that in the first case the decrease of stress after the maximum is connected with the formation of a "neck", in the second with the break-down of intermolecular and supra-molecular structures of the polymer, and the neck is formed only in the moment which precedes the break (on Fig. 37 the beginning of the formation of a neck is marked by an arrow). Below the temperature T_p when the linear polymer is a visco-elastic material it stretches up to the break without flow, as the limit of ductility lies in this case above the limit of strength. The diagram of tension in a linear polymer is in this field similar to the diagram of tension in a three-dimensional polymer (Fig. 38).

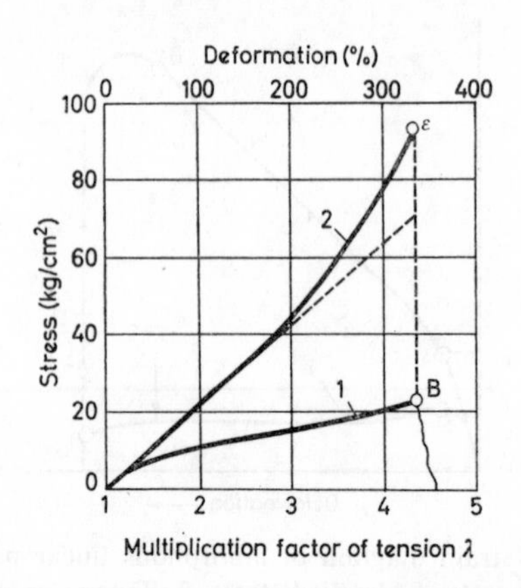

FIG. 38. Curve of tension of spatial polymer in visco-elastic state (pure rubber SKS-30). 1. Applied stress. 2. True stress.

The break of a visco-elastic material differs from that of a brittle one by the circumstance that a great deformation precedes it which is connected with the orientation and straightening of the polymer chains. Together with this, like in a brittle fracture, the nature of the section of the sample does not change from before the application of the load to after the break nor the relaxation of the specimen, and the surface of the fracture is, as a rule, normal to the stress vector. At the transition from a brittle to a visco-elastic failure the strength of the elastomer reaches its maximum value in the glass transition area (Fig. 39) and then decreases rather sharply with a rise of temperature.

Thus, the polymer shows above the temperature of the glass transition point two types of failure: visco-elastic, if the limit of plasticity (ductility) is higher than the strength, and plastic (ductile) if the limit of plasticity is lower than the strength (Fig. 40). In that case the transition from one kind of failure to the other is accompanied in an elastomer by a sharp change in the tensile strength (transition from curve 1 to curve 2 and Fig. 40).

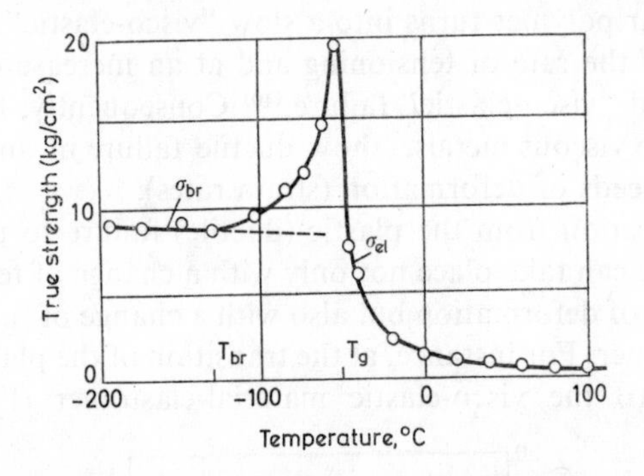

FIG. 39. Temperature of dependence of true strength of SKS-30 during tests under small rates of tensile strain (according to data by Bartenev and Voevodska).

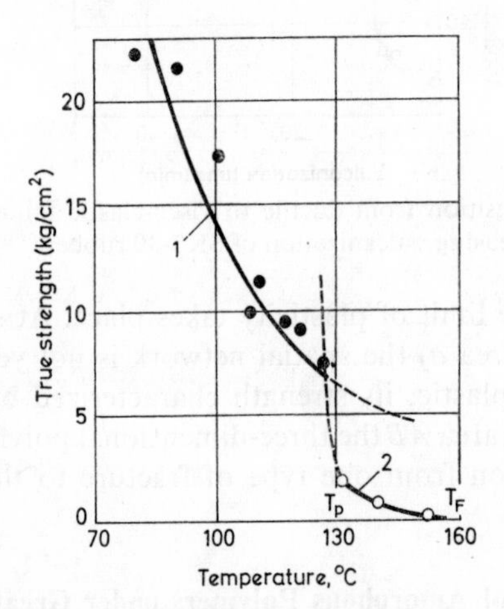

FIG. 40. Transition from the high elastic to the plastic failure of low-modulus elastomer from natural rubber. 1. True strength. 2. Yield point (according to data by Bartenev and Vishnitzkaya[10]).

Typical three-dimensional polymers have at all temperatures below the chemical break-up of the spatial network a high limit of plasticity which exceeds their strength. In this case the apparent plasticity is connected with the "chemical" flow of the resin.

The transition from one kind of failure to the other takes place with a change of speed of the deformation. The plastic failure of the linear polymer turns into a slow "visco-elastic" one at the decrease of the rate of tensioning and at an increase of the rate into a rapid "visco-elastic" failure.(10) Consequently, linear elastomers, like viscous metals, show ductile failure in an unlimited range of speeds of deformation (strain rates).

The transition from the plastic (ductile) failure to the "visco-elastic" one can take place not only with a change of temperature or of speed of deformation but also with a change of the structure of the polymer. For instance, at the transition of the plastic rubber mixture into the visco-elastic material-elastomer (Fig. 41), a

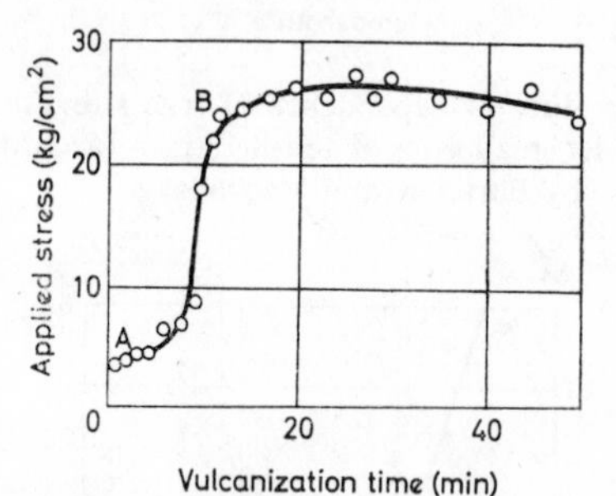

FIG. 41. Transition from ductile to visco-elastic failure with increasing vulcanization of SKS-30 rubber.(10)

sharp rise of the limit of plasticity takes place. At short times of vulcanization (area *a*) the spatial network is not yet formed and the material is plastic, its strength characterized by the limit of plasticity. In the area *AB* the three-dimensional polymer is formed, and the transition from one type of fracture to the other takes place.

2.5. State of Amorphous Polymers under Great Stresses

The above expounded experimental data on the strength of amorphous linear polymers make it possible to give a general picture of the properties in a great range of temperatures. The general diagram of the type of strength of amorphous polymers (Fig. 42) is more complicated than Joffé's diagram for simple solids (see p. 2). In amorphous polymers the brittle and the plastic areas are divided into two new areas: the low-elastic one in the interval from T_{cr} to T_g and the area where visco-elastic break is observed which lies in the interval from T_g to T_p. These fields are divided by the glass–transition temperature T_g which depends on the time rate of the experiment. Below the brittle temperature T_{br} the polymer shows a brittle break, and above it the break is

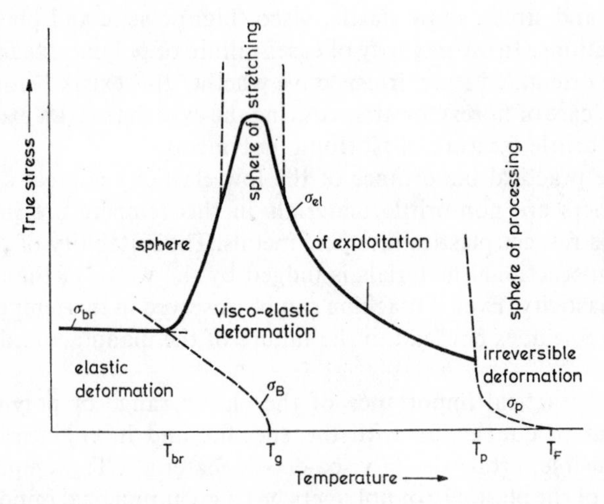

FIG. 42. Diagram of strength states of amorphous polymers (under tension).

preceded by a visco-elastic deformation which develops, beginning with the stress σ_B, depending on the temperature and speed of loading. Above T_g in the field of high elasticity the break is also preceded by a visco-elastic deformation but its development begins with the moment of the application of the load. Above the temperature of plasticity T_p, during the transition across the limit of plasticity σ_p, a plastic deformation develops until a contraction appears and failure begins. With a rise of temperature a state when $\sigma_p = 0$ is reached. This state can be observed at temperatures above the flow temperature T_F. Consequently, T_F is the lowest temperature when $\sigma_p = 0$. Such a definition agrees with the generally accepted one, according to which T_F is the temperature above which a viscous flow is observed under any stresses however small.†

It is an extremely important circumstance that the temperature limit of the strong ranges depend on the time rate of the deformation of the polymer. As a rule with increasing speed of deformation and shortening influence of the load, the limits of the range move towards higher temperatures. An especially great displacement can be observed at shock loading.

Polymers can, depending on the speed of deformation, temperature and stress, show elastic, visco (high) elastic and plastic deformations. In the majority of cases failure of polymers takes place in the orientated state, irrespective whether this exists before (like in the case of fibres), or arises during the experiment (an exception is the brittle fracture of isotropic polymers).

The practical importance of the low elasticity is great as solid polymers are non-brittle materials in this temperature interval. In this respect plastics resemble metals. The suitability of plastics as constructional materials is judged by the width of the field of low elasticity. Even if cracking can be observed in this temperature interval it does not lead to the failure of the manufactured products.

The practical importance of the elastic range of polymers is evident in connection with the specific, and in engineering indispensible, properties of visco-elastic materials. The temperature range of the plasticity of polymers has a great practical importance for the processing of polymers and production of articles from them. Finally, at the point of contact of the two temperature ranges in the region of the glass temperature T_g there is an interval where maximum deformations under small stresses can be observed. This temperature interval is often utilized in the technology of extrusion of polymeric materials.

The above experimental data (see Figs. 33–41) refer to the strength of polymers under tension with a constant rate of strain.

† Defining T_F from the beginning of the curve (Fig. 29) at high temperature without strict definition of the remaining deformation leads often to mistakes. In many cases the steep high-temperature branch is connected not with the true flow but with a partial softening of the polymer in connection with the breaking-up of unstable nodes of the spatial network, supramolecular structures, etc.[4, 21]

Between the time dependence of strength and the strength in experiments with a constant rate of stressing (tension) there is a well defined connection.†

Under loading at a constant rate $d\sigma/dt = w$ the strength of solids under tension is[22]

$$\sigma = \frac{1}{\alpha} \ln \alpha A + \frac{1}{\alpha} \ln w \qquad (2.5)$$

where α and A are the coefficients in the equation of durability (1.12) $\tau = Ae^{-\alpha\sigma}$. Taking into account that the coefficient $A = \tau_0 e^{U_0/kT}$ and the coefficient $\alpha = \gamma/kT$ (see Chapter 1) one can from eqn. (2.5) find the following temperature dependence of strength at constant rate of loading:

$$\sigma = a - bT \qquad (2.6)$$

where $a = U_0/\gamma$; $b = (k/\gamma) \ln kT/\tau_0\gamma w$.

The coefficient b is not a constant but, as the influence of temperature is small, b can be regarded as such. The formula is also applicable for a loading with the constant rate $d\varepsilon/dt = \text{const.}$ in those cases where the polymer obeys, in practice right up to fracture, the law of proportionality between stress and strain.

The strength of the solid polymer depends not only on the strain rate but also on the type of stressed state. For instance, during the transition from tension to compression the brittle strength and the limit of low elasticity increase but the former increases more rapidly.[9, 23] Therefore, other conditions of the experiment being equal, the polymer can show low-elastic deformation under compression but only elastic deformation and brittle fracture under tension.

2.6. Application of the Method of General Coordinates to Deformation and Strength Characteristics of Polymers

All mechanical and electrical processes in polymers are determined by the corresponding times of relaxation. The temperature dependence of the mechanical and electrical properties of poly-

† The time dependence of the strength of visco-elastic materials is investigated in Chapter 6, and the connection between the static strength and dynamic strength in Chapter 7.

mers are a consequence of a considerable influence of temperature on the relaxation times. Williams[24] showed that the temperature dependence of the mechanical and electrical properties and, consequently, of the relaxation times of amorphous polymers above the glass temperature can be described by some universal empirical function. Already earlier it was shown by Ferry[25] that a_T is the relation of the time of mechanical relaxation† at temperature T to the corresponding time of relaxation at some arbitrarily selected temperature T_s. Further b_T, the corresponding relation of the time of electrical relaxation remains, at a given temperature, constant for the whole spectrum of relaxation times of the same polymer. If one takes for the initial temperature T_s for different polymers any temperature outside the dependence on the glass temperature of those polymers the curves of dependence a_T on T intersect at some point, where $a_T = 1$ for all systems.

It was further shown[24] that one can select for each polymer a characteristic temperature T_s so that in a graph the curves of dependence a_T on $(T-T_s)$ for the various polymers coalesce; this confirms the universality of this dependence. And what is more, as shown in the work of Williams, Landel and Ferry,[26] the same universal function characterizes the temperature dependence of relaxation times of low-molecular fractions of polymers, inorganic glasses and other substances. The experimental viscosity values of various polymers lie on one common curve (Fig. 43).

For all amorphous substances the universal function is correct in the interval from the glass temperature T_g to a temperature approximately 100°C higher. Further on the curves diverge for the various polymers. Consequently, above this temperature interval the function a_T ceases to be universal as, probably, the difference in the temperature dependences of the electrical and mechanical properties which are caused by the actual structure of the polymer, makes itself felt.

The existence of a universal dependence of relaxation times on temperatures in substances with different molecular structure indicates that there exists in the first approximation a common mechanism of influence of temperature on the mechanical and electrical properties of materials.

† The relaxation times were calculated from experimental measurements of various mechanical and electrical properties.

A. Tobolsky[27] formulated, on the basis of extensive experimental material, the principle of mechanical equivalence of polymers at "respective" temperatures which are chosen in such a way that the difference $T-T_s$ is the same for the different polymers. If one equates the glass temperature with the temperature T_s, the elasto-viscous properties (viscosity, modulus of elasticity, time of relaxation, and others) of the different polymers coincide

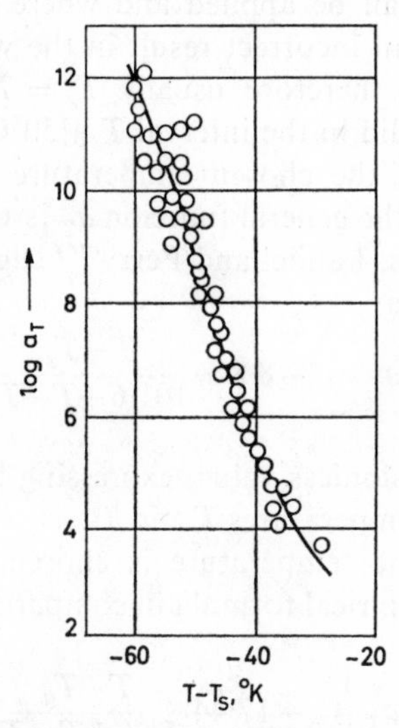

FIG. 43. Relative relaxation times, found from experimental values of viscosity of different polymers, lying on the universal curve.[26]

at corresponding temperatures and other comparable conditions for amorphous linear polymers with an identical distribution of macromolecules along their length, and for steric polymers with an identical degree of cross-linking and identical density of free chain ends. Crystalline polymers demand the fulfilment of much stiffer conditions in order to be mechanically equivalent: one must compare polymers with the identical T_g and T_p, identical molecular masses, equal degrees of crystallinity and identical supramolecular structures. The principle of mechanical equivalency of amorphous polymers is the theoretical basis for the investigation of the results of the work of Williams, Ferry and others.

Usually, for one polymer T_s is chosen arbitrarily and for the others T_s is found by comparing the corresponding graphs of a_T

and $(T-T_s)$ with the graph for the first polymer. When choosing the temperature T_s one could take the glass temperature. But this is not recommended as, firstly, a number of additional factors (the thermal history of the specimen, remnants of solvent, etc.) have a considerable influence on T_g and would lead to different values for it; secondly, it is rather difficult to measure mechanical properties near T_g; thirdly, T_g is the lower limit where the general relation (function) can be applied and where a small deviation from it can lead to an incorrect result in the whole temperature interval. One takes therefore usually $T_s = T_g+50°C$ and the general relation is valid in the interval $T_s \pm 50°C$.

The connection of the chosen temperature with T_g indicates that the existence of the general function a_T is connected with the glassy state. Williams, Landel and Ferry[26] suggested the following empirical formula:

$$\log a_T = -8{\cdot}86 \frac{T-T_s}{101{\cdot}6+T-T_s} \tag{2.7}$$

where a_T is a dimensionless value, expressing the relation of the relaxation times at temperatures T and T_s.

If the characteristic temperature is chosen equal to T_g, the coefficients of the empirical formula in comparison with eqn. (2.7) change:

$$\log a_T = -17{\cdot}44 \frac{T-T_g}{51{\cdot}6+T-T_g} \tag{2.8}$$

Equation (2.7) can also be obtained, starting from Fox's and Flory's[28] expression for the temperature dependence of viscosity and relaxation processes in polymers. Williams, Landel and Ferry showed that the use of Tobolsky's[29] expression for visco-elastic properties and Dienes' formula[30] of the temperature dependence of viscosity leads to the same formula.

According to Rouse's[31] and F. Bueche's[32] theories a_T characterizes the temperature dependence of the mobility of the monomers in the polymer chain on which the speed of the change of its shape depends. Bueche suggested that if concrete empirical constants for polystyrene and polyisobutylene are substituted for a_T the same result as from eqn. (2.7) is obtained.

Williams, Landel and Ferry[25, 26] connect the temperature dependence of mechanical and electrical properties of polymers with the temperature dependence of the free volume. As Fox and Flory indicated the relative free volume decreases sharply as the temperature drops and approaches the glass temperature; this, obviously leads largely to an increase of viscosity.†

Modifying Doolittle's[33] formula for viscosity which is applicable to supercooled systems and using the dependence of the free volume on temperature, Williams, Landel and Ferry deduced the following equation:

$$\log a_T = -\frac{1}{2 \cdot 3 f_g} \cdot \frac{T - T_g}{f_g / \Delta\alpha + T - T_g} \tag{2.9}$$

where f_g is the proportion of the free volume in 1 g of the material at the second order transition point; $\Delta\alpha$ is the difference of the coefficients of the volumetrical thermal expansion above and below the glass temperature.

Comparing eqn. (2.9) with the empirical equation (2.8) leads to the values f_g and $\Delta\alpha$ which agree with the data which have been obtained by other methods. Thus, according to eqn. (2.9) three values T_g, f_g and $\Delta\alpha$ define the glassy state and determine the temperature dependence of the mechanical and electrical properties of polymers above T_g.

One does not usually successfully obtain the dependence of the various properties of polymers on the observation time or the frequency of deformation in a wide range of these variables. In order to find such a dependence one moves on the graph the curves obtained at different temperatures and obtains a general curve at a selected temperature. Such a method, formulated in a general way by Tobolsky,[27] is widely employed and is based on the principle of a temperature–time connection, in particular on the temperature-cycle dependence of the deformation of polymers, first discovered by Alexandrov and Lasurkin in 1939.

According to this principle which is used by Williams, Landel and others, the data of mechanical tests, obtained at different temperatures, can be combined by a simple parallel displacement along the ordinate of the logarithm of time or of the frequency of

† According to Bachinsky's well-known furmula the viscosity of fluids is

$$\eta = \frac{C}{V - \omega}$$

where V is the specific volume; C and ω are characteristic constants, and $V - \omega$ is the free volume.

deformation. In this way one can determine the coefficient a_T which is one at T_s, less than one at higher, and more than one at temperatures lower than the selected one. The application of Williams', Landel's and Ferry's method (the WLF method) to different visco-elastic properties of polymers is investigated in detail in Ferry's book.[34]

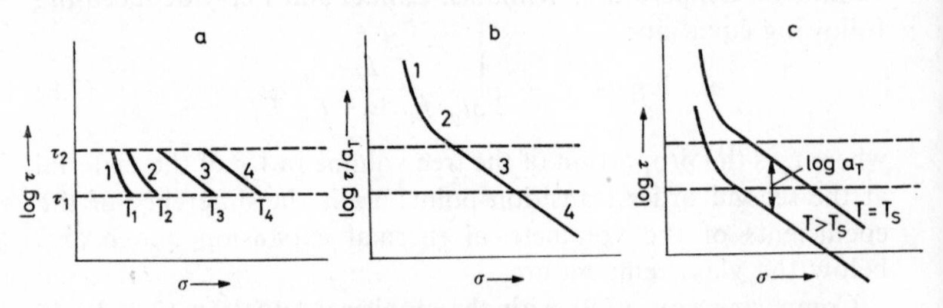

FIG. 44. Diagrams showing construction of the universal curve of durability of polymers with the method of the coordinated relations (temperature T_3 is taken as T_s).

The principle of connection is also applied to the study of the strength properties with the aim to obtain information about their behaviour under conditions which make a straight experiment difficult. In the interval of the measured times to failure from τ_1 to τ_2 curves of durability at different temperatures are constructed (Fig. 44a); then the temperature $T_s > T_g$ is chosen, and all points of each curve are transferred along the ordinates by the amounts $\log a_T = C_1(T-T_s)/(C_2+T-T_s)$ (the WLF equation in a general form, where the value of the constants C_1 and C_2 depends on the chosen temperature T_s). As a result a curve (Fig. 44b), composed of short initial curves, is obtained for the temperature T_s. Curves of durability at any temperature can be obtained from it (see Fig. 44c) through a displacement along the ordinate by the respective value of $\log a_T$. F. Bueche[32] obtained for poly(butylmethacrylate) above the second transition point a universal curve which defines the durability at different temperatures. The value a_T within the limits of experimental error agreed with the values which were obtained for characteristic visco-elastic properties of polymers by the WLF method.†

† A close connection between viscous flow and durability of elastomeric polymers was also confirmed by other experiments (see Chapter 6).

The method of coordinated relations was also employed in the preparing of data on the strength of elastomers at different speeds of deformation under tension,[32, 35] The coordinated relations

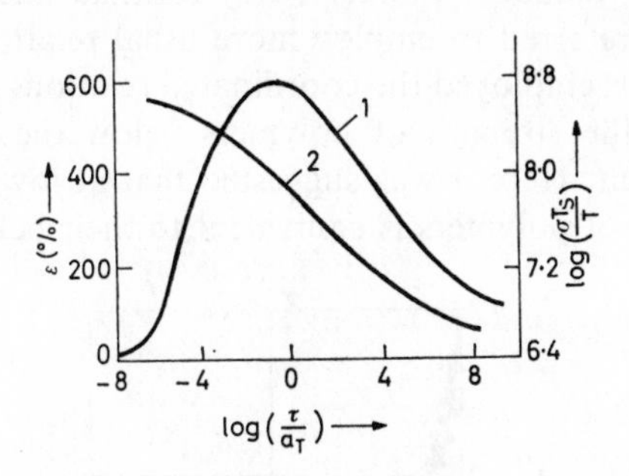

FIG. 45. Universal curves: strain in % – durability (in sec) (curve 1), and actual breaking stress (in dyn/cm²); durability in sec (curve 2), constructed from data which were obtained at different constant strain rates on SKS-30.[35]

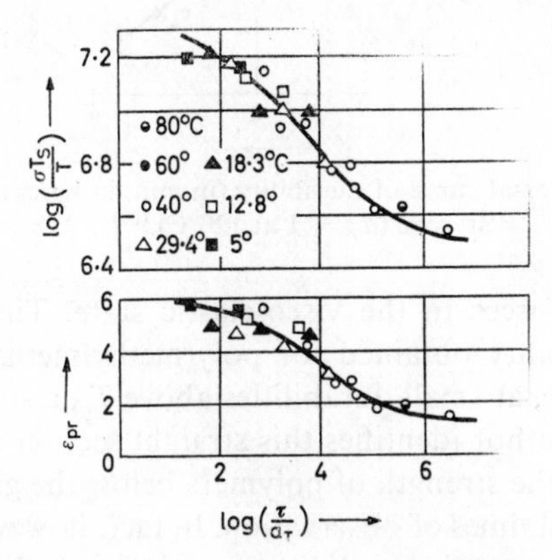

FIG. 46. Universal curve: maximum deformation in % – durability (in sec) and actual breaking stress (in dyn/cm²); durability (in sec) for butadiene-styrene rubbers.[35]

(Fig. 45 and 46) consist in the maximum deformation ε and the logarithm of the expression breaking stress $\sigma T_s/T$ as the ordinate and in log (τ/a_T) as the abscissa, where τ is the durability under the given speed of deformation. Within the limits of experimental error the values a_T tally with the values which were calculated from eqn. (2.7). However, in Smith's and Stedry's work where the coordinated relations were used in the study of the same elastomers which were examined at constant deformations, the deviation from the Williams–Landel–Ferry formula was so great that the authors preferred to employ more usual relations.

F. Bueche[36] employed the coordinated relations also to obtain the data on the strength of polymers below the second order transition point. Here it was suggested that at low temperatures the behaviour of polymers is equivalent to their behaviour under

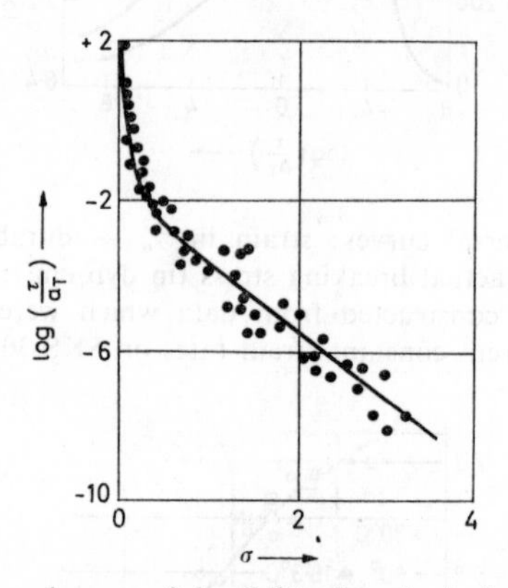

FIG. 47. Universal curve of durability (in min, in kg/cm²) of polystyrene ($a_T = 1$ at 120°C).[36]

rapidly acting forces in the visco-elastic state. The generalized curves of durability obtained for poly(methylmethacrylate) and polystyrene have, at small durabilities above T_g, a straight section (Fig. 47). The author identifies this straight section with the time dependence of the strength of polymers below the glass temperature during usual times of observation. In fact, however, the WLF method cannot be applied to the study of the strength properties of solid polymers as the process of their failure is only indirectly connected with the relaxation phenomena in the mass of the polymer. Kubshinsky and Bessonov showed (see Chapters 3 and 4) that the processes of deformation and failure of solid polymers are interrelated but their molecular mechanisms are different.

The application of the WLF method to the strength of elastomeric polymers can to some degree be justified as basically the mechanisms of their slow failure and viscous flow are extremely similar processes. The mechanism of brittle fracture of solid polymers is, however, different from the mechanism of the relaxation processes in polymers. The apparent success of the application of the WLF method to solid polymers can be explained by the circumstance that the relaxation time and durability [eqns. (2.1) and (1.13)] depend in an analogous manner on temperature and stress.

In connection with what has been said the application of the method of coordinated relations appears more correct for solid polymers as it derives from the time dependence of strength. Ivanov[37] applied to polymers the method of Nikitin's[38] which consits in the use of $T \log \tau - T$ as the ordinate (where τ is durability, T the absolute temperature). From the equation of durability follows:

$$T \log \tau = T \log \tau_0 + \frac{1}{2 \cdot 3k}(U_0 - \gamma\sigma),$$

According to this equation in Nikitin's coordinates straight lines parallel to each other are obtained for metals under various constant stresses; this is correct at $\gamma = \text{const}$. In reality[37] the structural coefficient γ for polymers (for instance, for polystyrene and celluloid) decreases with rising temperature, and the straight lines will not be parallel. In connection with this Ivanov suggested a more general method of obtaining the universal curve of durability. As the experimental straight lines in the ordinates $\sigma - \log \tau$ for different temperatures form a group converging in a point or pole (Fig. 48), the equation for the group looks like this:

$$\sigma - \sigma_k = \frac{2 \cdot 3kT}{\gamma}(\log \tau_0 - \log \tau)$$

where σ_k and $\log \tau$ are the coordinates of the pole.

Writing the right part of the equation as $(-\theta)$ one can combine the straight lines of the bundle in one line (Fig. 49), the equation of which is $\theta = \sigma_k - \sigma$.

For solid polymers one can employ for a_T the expression $e^{\gamma\sigma_k/kT}$ in which γ generally depends on the temperature. If one now enters $\gamma\sigma/T$ on the ordinate $\log(\tau/a_T)$, one obtains the values of the generalized curve of durability on the abscissa. The straight line thus obtained will be universal for all temperatures. As

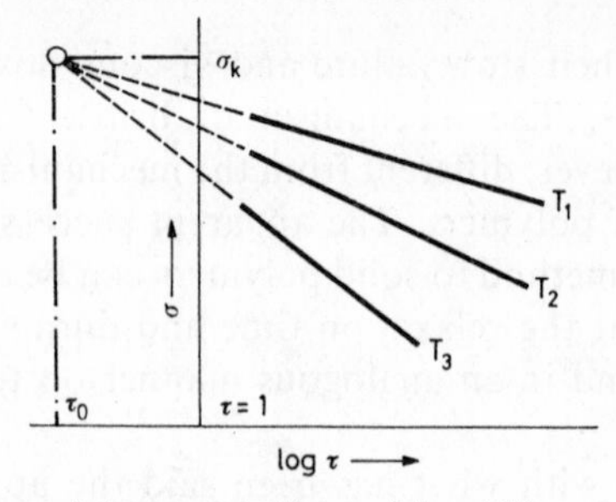

FIG. 48. Group of the straight lines of durability at different temperatures ($T_3 > T_2 > T_1$).

$U_0 = \gamma\sigma_k$ (see Chapter 1), the equation of this straight line will be

$$\log \frac{\tau}{a_T} = \log \tau - \frac{\gamma\sigma_k}{2{\cdot}3kT} = \log \tau_0 - \frac{\gamma\sigma}{2{\cdot}3kT},$$

where it is assumed that τ_0 does not depend on temperature.†

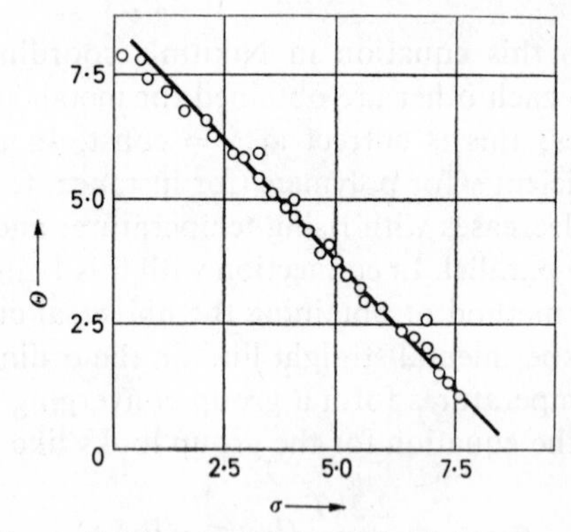

FIG. 49. Generalized curve of durability for polystyrene (θ and σ in kg/mm²).[37]

Literature

1. G. M. BARTENEV, Prochnost' i mekhanizm razriyva polimerov, *Uspekhi khimii*, **24**, No. 7, 815 (1955).
2. P. P. KOBEKO, YE. V. KUVSHINSKII, G. I. GUREVICH, *Izv. AN SSSR*, Ser. fiz., **6**, 329 (1937).

† According to the oscillation theory τ_0 is proportional to T but, as it appears only as its logarithm, this dependence is weak and can be disregarded.

3. V. A. Kargin, G. L. Slonimsky, *ZhTF*, **11**, 341 (1941); *DAN SSSR*, **62**, 239 (1948).
4. V. A. Kargin, G. L. Slonimsky, *Kratkiye ocherki po fiziko-khimii polimerov*, Izd. MGU, 1960.
5. V. A. Kargin, T. I. Sogolova, *DAN SSSR*, **88**, 867 (1953); *ZhFKh*, **27**, 1039, 1208, 1213, 1325 (1953).
6. V. A. Kargin, T. I. Sogolova, L. I. Nadareishvili, *Viysokomol. soyed.* **6**, 1272, 1407 (1964); T. I. Sogolova, *Doktorskaya dissertatsiya*, NIFKhI im. Karpova, 1963.
7. V. A. Kargin, *Sovremenniye prolemiy nauki o polimerakh*, Izd. MGU, 1962.
8. A. P. Alexandrov, *Trudiy 1-i i Z-i konferentsii po viysoko-molekuliarniym*, Izd. AN SSSR, 1945, p. 49.
9. Yu. S. Lazurkin, R. L. Fogel'son, *ZhTF*, **21**, 267 (1951); Yu. S. Lazurkin, *J. Polymer Sci.*, **30**, 595 (1958).
10. G. M. Bartenev, G. I. Belostotskaya, *ZhTF*, **24**, 1773 (1954); G. M. Bartenev, L. A. Vishnitskaya, *Viysokomol. soyed.* **5**, 1837 (1963).
11. V. A. Kargin, G. L. Slonimsky, O kristallicheskom sostoyanii polimerov, *Uspekhi khimii*, **24**, No. 7, 785, (1955).
12. A. I. Kitaigorodskii, *Uspekhi khimii i tekhnologii polimerov*, No. 2, Goskhimizdat, 1957, p. 191; *Vestnik AN SSSR*, No. 6, 35 (1958).
13. J. Miklowitz, *J. Coll. Sci.* **2**, 193 (1947).
14. P. V. Kozlov, V. A. Kabanov, A. A. Frolova, *DAN SSSR*, **125**, 118 (1959).
15. A. P. Maklakov, G. G. Pimenov, R. Ya. Sagitov, *Viysokomol. soyed.* **3**, 1410 (1961).
16. R. E. Robertson, *J. Appl. Polymer Sci.* **7**, 443 (1963).
17. M. D. Nosov, V. A. Berestnev, *Viysokomol. soyed.* **5**, 1080 (1963).
18. L. Treloar, *Fizika uprugosti kauchuka*, Izdatinlit, 1953.
19. G. M. Bartenev, T. N. Khazanovich, *Viysokomol. soyed.* **2**, 20 (1960).
20. G. M. Bartenev, L. A. Vishnitskaya, *Izv. AN SSSR, OTN (mekhanika i mashinostroyeniye)*, No. 4, 175 (1961); *Viysokomol. soyed.* **4**, 1324 (1962).
21. A. V. Sidorovich, Ye. V. Kubshinskii, *Viysokomol. soyed.* **3**, 1698 (1961).
22. S. N. Zhurkov, E. Ye. Tomashevskii, *v sb.* "*Nekotoriye problemiy prochnosti tverdogo tela*", Izd. AN SSSR, 1959, p. 68.
23. V. R. Regel', G. V. Berezhkova, *v sb.* "*Nekotoriye problemiy prochnosti tverdogo tela*", Izd. AN SSSR, 1959, p. 375.
24. M. L. Williams, *J. Phys. Chem.* **59**, No. 1, 95 (1955); *Problemiy sovremennoi fiziki*, **8**, No. 12, 17 (1956).
25. G. D. Ferry, E. R. Fitzgerald, *J. Coll. Sci.* **8**, 224 (1953); G. D. Ferry, M. L. Williams, E. R. Fitzgerald, *J. Phys. Chem.* **59**, No. 5, 403, 1955.
26. M. L. Williams, R. A. Landel, G. D. Ferry, *J. Am. Chem. Soc.* **77**, 3701 (1955); *Problemiy sovremennoi fiziki*, **8**, No. 12, 20 (1956).
27. A. V. Tobolsky, *Properties and Structure of Polymers*, John Wiley, New York (1960); *Svoistva i struktura polimerov*, "Khimia", 1964.

28. T. G. Fox, P. G. Flory, *J. Appl. Phys.* **21**, 581 (1950); *J. Polymer Sci.* **14**, 315 (1954).
29. A. V. Tobolsky, E. Catsiff, *J. Am. Chem. Soc.* **76**, 4204 (1954).
30. L. J. Dienes, *J. Appl. Phys.* **24**, 779 (1953).
31. P. E. Rouse, *J. Chem. Phys.* **21**, 1272 (1953).
32. F. Bueche, *J. Appl. Phys.* **26**, 1133 (1955); *J. Chem. Phys.* **21**, 1850 (1953).
33. A. K. Doolittle, *J. Appl. Phys.* **23**, 236 (1952).
34. G. D. Ferry, *Viscoelastic Properties of Polymers*, New York (1961).
35. T. Smith, P. Stedry, *J. Appl. Phys.* **31**, 1892 (1960); *J. Polymer Sci.* **99** (1958).
36. F. Bueche, *J. Appl. Phys.* **28**, 784 (1957).
37. Yu. M. Ivanov, *Zav. lab.* **27**, 455 (1961).
38. V. P. Nikitin, *Zav. lab.* **25**, 492 (1959); **27**, 71 (1961).

CHAPTER 3

MECHANISM OF FAILURE OF POLYMERS

THE MECHANISM of failure is the least studied side of the strength problems of polymers. Only in recent years have data been obtained which allow an approach to the understanding of the complicated failure mechanism.

At first the mechanism of failure of visco-elastic materials was investigated by one of the authors and co-workers in the Institute of Rubber Industries (Moscow), where at slow failures of elastomers the peculiar "fibrous" type of break was discovered, connected with the formation and breaking away of local highly orientated parts (strands). Next Kubshinsky and co-workers discovered in the Institute of Highly-Molecular Compounds, Academy of Sciences USSR (Leningrad), the special features of the structure of the "silver" cracks in plastics and showed that in contrast to ordinary cracks their folds are strengthened by strands—parts of highly orientated polymer material. Consequently, the formation of strands is observed both in the amorphous solid and in the high elastic state of the polymers. Only at low temperatures and great tensile strain rates has the indicated specific mechanism no time to appear, and the polymers fail through development of the usual cracks. In contrast to other polymers the polymeric fibres contain already in the initial state a highly orientated structure in the form of fibrils and have therefore an exceedingly high strength.

3.1. Mechanism of Failure of Polymer Fibres

In fibres the polymer molecules are in a highly orientated state. Therefore the fibres are strong along their axis and weak in the transversal direction. Fibres under tension fail not always precisely in the plane of their cross-section but often by a combination of breaks through parts of the fibres and splitting parallel to the axis (IV type of failure, see Fig. 2).

The failure of highly orientated polymers[1-6] assumes therefore often a peculiar "fibrous" character (Fig. 50). Sometimes the break of the fibres is accompanied by the formation of a small contraction.

FIG. 50. Fibrous break.

In natural and synthetic filaments and fibres the stress is, on account of the heterogeneity of their structure, distributed unevenly over the section and the fracture takes place along the weakest parts (not simultaneously over the whole section). The strength of the fibres is therefore considerably lower than the theoretical one.

The mechanism of failure in polymeric fibres was investigated both in static and in cyclic tests.

Under static tension a number of peculiarities of the failure depending on the type of fibre was discovered. For instance, there is on viscose fibres an ordered surface layer (orientated sleeve) in which the degree of orientation is higher than in the remaining mass of the fibre. The surface of the break on these fibres is rough-fibrous—mirror-like zones are rarely found. The fibres of natural silk are homogeneous throughout their cross-sections; their break is therefore less "fibrous". On nylon fibres swellings appear in parts of the fracture, obviously caused by a partial disorientation of the macromolecules due to the rising temperature during the deformation.

The mechanism of failure of polymeric fibres under repeated deformations which was investigated in detail by Kukin[3] has its own peculiar character: during the experiments a splitting of filaments into elementary fibres and their separation from each other takes place. The dynamic fatigue of textile materials appears as the result of the gradual "loosening" of the structure. During the repeating of the stress application, especially in the initial stages, the structure of the fibre gets often temporarily strengthened due to the processes of orientation and other phenomena. However, the following "loosening" of the structure of the fibre leads to a loss of strength.

As Kargin, Berestnev and others[4–6] have shown, under repeated tensioning the failure of the fibre is caused by the development of defects in it, the dimensions of which exceed by far the molecular ones. These defects (cracks) appearing anywhere on the fibre, develop in the direction normal to its axis. The propagation of such a crack can be stopped (obviously due to the orientation-strengthening of the material at the apex of the crack), and it turns into a so-called "split" which develops as a result of the slip of the elementary fibres parallel to the direction of the orientation—in the direction of the easiest slipping. From the slip new defects can develop which produce more slips, and so on, right up to the final break of the fibre. As a result the ends of the broken fibres have the most fantastic shapes. However, mostly the edges of the fibres have the shape of steps composed from transversal cracks and longitudinal slips (IV type of failure).

Under single loading to break the mechanism of failure of fibres is different. This is confirmed particularly by microphotographs[4–6] on which can be seen that the ends of the broken fibre are flat (there are no longitudinal slips). Under single tensioning to failure the break takes place in one, the weakest, place of the whole length of the fibre without any change in the structure of the fibre. Under repeated loadings, however, a great number of micro defects are gradually formed which spread through the whole volume of the fibre.

The failure of fibres under repeated loadings proceeds often in two stages. Elementary fibres which are yarned can after failure show both stepped and smooth ends. This implies that at first some elementary fibres fail due to the gradual development of macro defects with the formation of slips. When the load on the reduced number of fibres becomes equal to the ultimate strength of the remaining bundle, the break takes place and corresponds to the failure under single tensioning.

3.2. Mechanism of Failure of Solid Polymers

Solid polymers, subjected to tensile stresses, show under certain conditions (relatively high temperatures, small strain rates, etc.) a "cold flow" and thereupon fail in the orientated state like fibres. Brittle failure of polymers appears both in the classical form of brittle fracture and in the form of cracking of the material with the appearance of a great number of surface cracks without disrupting the specimen as a whole.

Under small constant tensile stresses leading subsequently slowly to failure a great number of silver cracks[7-11] are first formed in the solid amorphous polymer. Ordinary failure cracks can be formed only later; they break the specimen into pieces; under very small stresses they do not appear at all, and the specimen does not break for an infinite time. Under the influence of stresses which are relatively great for any given temperature, the failure of polymers is similar to a classical brittle one and leads in the first stage mainly to the growth of one or several of very critical cracks. There, like in the case of brittle failure, usually two surface areas of break develop: a mirror one and a rough one (Fig. 51).

The speed of crack propagation increases with rising temperature and failure takes place under lower stresses. In this sense the loads which are non-critical at low temperatures become critical at high ones. Therefore the character of the break under the same stress or speed of deformation, in the form of a surface break, approaches the classical brittle one with a rise of temperature—the two zones on the surface of the break are clearly expressed (Fig. 51b). Here the rule can be observed that the mirror zone is gradually displaced by the rough one with a rise of temperature, like in ordinary brittle material.

The microscopical investigation of the surfaces of brittle fractures of some plastics[1, 8, 12-15] showed that the break of this type takes place at relatively great loads and low temperatures (below T_{br}). It proceeds in several stages. The first one is characterized by a slow growth of the primary crack and by the formation of a mirror zone on the surface of the break. Subsequently, there appear in front of the primary crack secondary cracks and grow in various directions and in various adjacent planes; at the meeting of their fronts with the primary cracks they form "slip" lines, the geometrical form of which permits the assessment of the kinetic

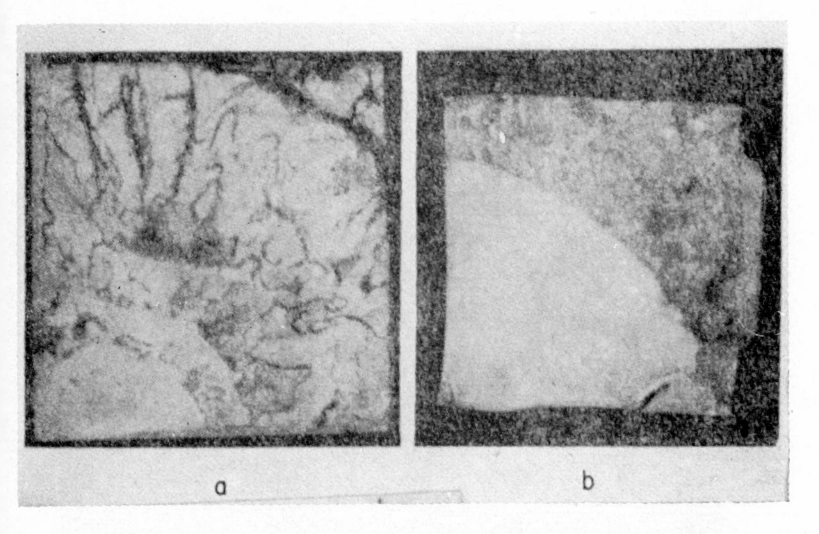

FIG. 51. Surface of failure in poly(methylmethacrylate) during breaking tests: (a) at 20°C; (b) at 70°C (before the break a small deformation was observed on the sample).

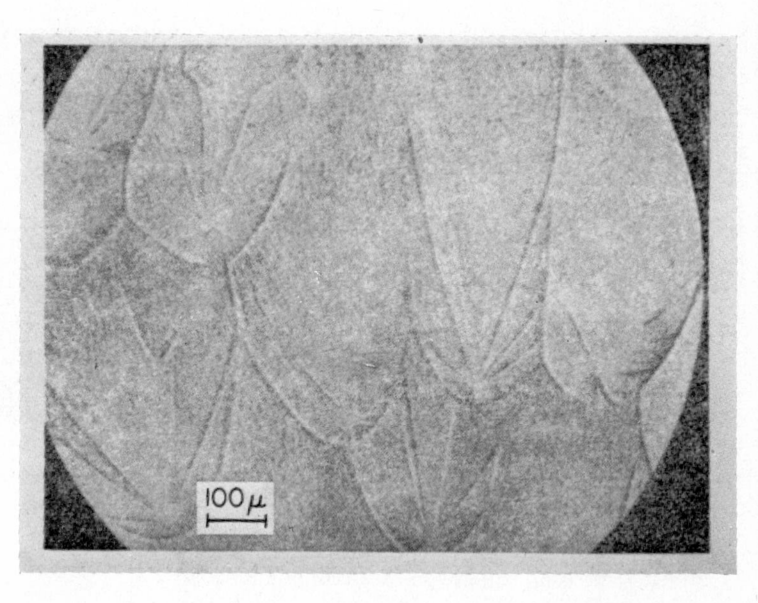

FIG. 52. Slip lines of brittle failure[13, 14] on the break surface of poly(methylmethacrylate).

growth of the crack. As a result more or less correctly defined "hyperbolae" are formed on the surface of the rough zone with their apices towards the centre of the mirror zone (Fig. 52). Each "hyperbola" formed by the slip line is a small step at the angle of 45° to the surface of the break. With increasing distance from the centre of the mirror zone the number of secondary cracks and the number of hyperbolae increases and they lie on top of each other in a disorderly manner and form a rough surface.

The shape of the slip line can be explained as follows. It is known that in a homogeneous material under uniaxial tension a crack has the form of a circle, or part of a circle; it spreads in a plane, perpendicular to the axis of stress. If two cracks would start growing simultaneously towards each other with equal speed from two centres *A* and *B* (Fig. 53, left sketch), the line of their junction would be straight. If, however, one of the cracks (*B* on Fig. 53, right sketch) begins to grow later, but with the same speed as the first one, a hyperbola forms. The later the second crack begins to grow, i.e. the less critical it is, the smaller is the angle φ between the asymptotes of each of the forming hyperbolae. One can therefore judge by the form of the hyperbolae in which sequence the various defects entered into the process of failure and with which speed cracks propagated.

If one puts the origin of the coordinates (see Fig. 53, right sketch) in the centre of the first crack, and places the x-axis on the straight line which joins the centres of the first and second

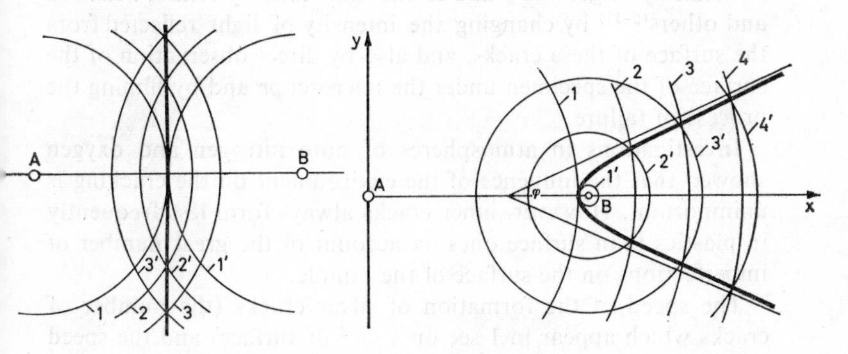

FIG. 53. Schematic drawing of forming slip lines at the meeting of cracks. 1, 2, 3, 4. Fronts of the growing crack *A*, grouped in sequence. 1′, 2′, 3′, 4′. Fronts of the growing crack *B*.

crack, the equation of the hyperbola [1, 13] will be

$$4(x_0^2-v^2\theta^2)x^2+4v^2\theta^2y^2-4x_0(x_0^2-v^2\theta^2)x+(x_0^2-v^2\theta^2)^2 = 0,$$

where x_0 is the distance between the centres of the cracks; v is the speed of growth of both cracks which is considered equal and constant. θ is the time interval between the beginning of the growth of the first and second cracks.

If the cracks begin to propagate simultaneously, $\theta = 0$ and the equation of a straight line $x = x_0/2$ ensues (see Fig. 53, left sketch). If the second crack begins to grow in the moment when the front of the first one reaches its centre, $\theta = x_0/v$, which leads to $y = 0$, which means the hyperbola degenerates into a straight line which coincides with the x-axis. With a still greater interval the θ front of the first crack moves forward, and no intersection of fronts takes place.

It was shown[12] that the slip lines in poly(methylmethacrylate) meet this equation sufficiently but in polystyrene the slip lines are often not hyperbolae and can even be closed curves.[13, 14] This indicates that in poly(methylmethacrylate) the speeds of propagation of secondary and primary cracks are really constant and equal, but in polystyrene they are different or even change in the process of failure.†

At small loads only a cracking of the material takes place and the very smallest silver cracks are formed. This phenomenon, discovered by Alexandrov in 1932, was subsequently investigated in detail by Regel[1, 7, 8] and at the same time by Hsiao, Maxwell and others[9-11] by changing the intensity of light reflected from the surface of these cracks, and also by direct observation of the surface of the specimen under the microscope and by filming the process of failure.

Investigations in atmospheres of pure nitrogen and oxygen showed that the influence of the environment on the cracking is unimportant. However, inner cracks always form less frequently in plastics than surface ones on account of the great number of imperfections on the surface of the sample.

The speed of the formation of silver cracks (the number of cracks which appear in 1 sec on 1 cm² of surface) and the speed

† This is evidently connected with the fact that at 20°C mainly silver cracks develop in poly(methylmethacrylate) and failure cracks in polystyrene.

of their propagation increases considerably with a rise of stress and temperature. The general principle of the course of the cracking process remains the same throughout: the number of cracks reaches after a time some limit, and does not further increase; the average size of the cracks increases continually, at first rapidly, then slower. This can be explained by the fact that as the silver cracks appear on the surface of the sample, the surface layers of the specimen become partially unloaded. Besides, all defects on which at the given stress cracks can form are gradually used up and, as a consequence, the number of cracks cannot increase unrestrictedly. With increasing stress the number of cracks per 1 cm^2 of surface grows and their average size decreases.

The number of cracks and their average size depends substantially on the mechanical preparation of the surface. If the sample is annealed or compressed after cracking, the silver cracks disappear, the specimen "heals" but the defects remain, and at subsequent tension the original picture of cracking is reproduced.

The phenomenon of cracking is closely connected with the low-elastic deformation of polymers which proceeds simultaneously with the formation of cracks even at apparently brittle failure. On account of the low-elastic deformation the cracks open rather wide (about 0·5 μ and more). The cracks open the wider the greater the influence of the relaxation processes of the deformation. As the speed of relaxation grows exponentially with the stress, the relaxation processes influence also considerably the magnitude of overstress at the root of the crack. The relaxation processes lead to a decrease of the overstress in the apices of the most critical microcracks. This prevents the primary growth of some isolated crack and partially explains the constancy of the speed of propagation of the silver cracks. The usual failure cracks which develop under great loads and low temperatures (below T_{br}), show in contrast to silver cracks accelerated growth.

Thus under great loads and low temperatures the process of failure of solid amorphous polymers takes place like in ordinary brittle materials. At small stresses and high temperatures the process of failure is specific and consists of two basic stages: slow development of silver cracks in the first stage, and formation of ordinary open failure cracks in the second. The first stage lasts the main part of the time to failure and the principle of the time dependence of strength is determined predominantly by this stage.

The nature of the silver cracks was studied in detail by Kuvshinsky, Bessonov and Lebedev[16, 17] and afterwards by Berry[18] and others.[19] In these papers it is shown that the silver cracks have a structure different from that of ordinary cracks. They are wedge-shaped regions which in places separate into layers and strongly deform the polymer (Fig. 54) which undergoes a considerable "cold" elongation and consolidation. The toughened parts (strands) strengthen the folds of the silver cracks which, in contrast

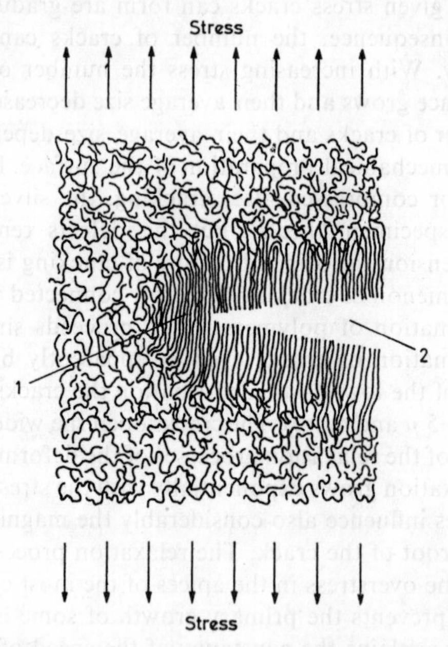

FIG. 54. Schematic drawing of a silver crack in a polymer; behind the silver crack (1) follows the failure crack (2).

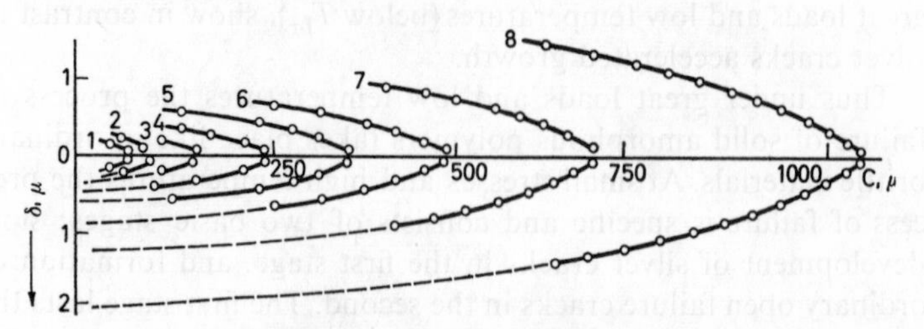

FIG. 55. Change of configuration of a silver crack (in section) with its growth (δ half width of crack).[16]

to the ordinary failure cracks, can be called pseudocracks. The process of the opening of the folds of the pseudocracks has been investigated for poly(methylmethacrylate) by the interference method by studying the pits and their shape in the material (Fig. 55). The formation of strands in the silver cracks is probably connected with the fibrous structure of the polymers.[10-20]

The strands of the silver cracks carry a part of the load and the overstress of the material near the roots of the silver cracks is therefore not as great as at the apices of failure cracks in brittle material and this is one more reason for the constancy of speed of propagation of the pseudocracks. Although the silver cracks do not unload the specimen as a whole, their growth leads to a

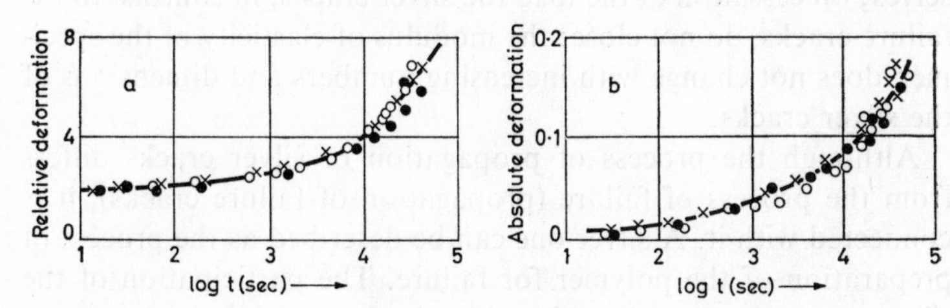

FIG. 56. Values of creep (a) and growth of the absolute dimensions of silver cracks (b) in specimens of poly(methylmethacrylate) of various diameters (from 1·4 to 3·0 mm) lie on one curve.[16]

redistribution of local stresses. Each pseudocrack discharges the area of the material which lies immediately above and below it while the regions to the right and left in the same cross-section are found to be overloaded. This leads to a deceleration of the growth of any given crack and an acceleration of the growth of adjoining ones, i.e. to a general equalization. The pseudocracks grow not only in each level of the cross-section of the sample, but also simultaneously in its height.

It follows from the constancy of the speed of propagation of the silver cracks that it is determined basically by the mean stress in the specimen and cannot depend on scale factors. In contrast to ordinary cracks the change of the absolute dimensions of the pseudocracks with time at a given stress does indeed not depend on the dimensions of the sample; this is also characteristic for creep of polymers (Fig. 56).

The speed of the propagation of silver cracks grows considerably with increasing load but the depth of their maximum penetration into the specimen then decreases, whilst failure cracks develop. With decrease of stress the depth of the pseudocracks in the specimen increases at all temperatures up to the moment of transition to the stage when failure cracks form.

Silver cracks therefore must not be identified with ordinary failure cracks. These pseudocracks grow with constant speed under constant tensile load and, although they reach dimensions which are comparable with the dimensions of the specimen, they do not lead to rapid failure:† the speed of growth of the silver cracks is determined by the stress on the whole section of the specimen, independent from the dimensions of the cracks themselves; on cessation of the load the silver cracks, in contrast to the failure cracks, do not close; the modulus of elasticity of the specimen does not change with increasing numbers and dimensions of the silver cracks.

Although the process of propagation of silver cracks differs from the process of failure (propagation of failure cracks), it is connected with it. A silver one can be described as the process of preparation of the polymer for failure. The participation of the silver cracks in the failure is evidently determined by the extent to which the weakening of the material as a result of emerging of microstratification is compensated by strands which reinforce the folds of the silver cracks. Failure cracks propagate behind the silver cracks through the breaking of the strands (see Fig. 54). Consequently, in front of the failure cracks are the silver cracks, i.e. the region of the material which is partially divided into layers and strongly deformed (has undergone "cold" elongation). At the root of the failure cracks is a region of strands (see Fig. 54). The area of strands in front of the propagating failure crack renews itself continuously.

The failure of polymers is, therefore, preceded by a complicated local change of the material, including its partial lamination and deformation. Thence follows the connection between the durability τ and the speed of creep $\dot{\varepsilon}$, expressed by the formula

$$\tau\dot{\varepsilon}^m = \text{const.},$$

where the exponent $m \neq 1$.

† Besides, we must note that at the basis of the phenomenon of creep (slow increase of the length of the sample), the formation of a great number of silver cracks accompanies the process of low-elastic deformation.

This shows that although failure and deformation of solid polymers take place simultaneously they are not processes which proceed according to an identical mechanism. The same kind of relation connects the durability and relative speed of growth of pseudocracks where the rate indicating exponent is also different from 1. We must not conclude from the fact that the silver cracks grow actually during the whole time whilst the specimen is under stress that the "silver process" is the only one which leads to the failure of plastics. Failure and cracking—these are two different processes, although they are, obviously, closely and in a complicated way, interrelated.

The structure of the silver cracks permits it to formulate the conditions of their emergence in polymers. It is determined by the relation, under certain experimental conditions, between the speed of the low-elastic deformation in the overstressed places of the material and the speed of propagation of the failure cracks. With a lowering of temperature and increase of speed of deformation the limit of low elasticity grows to such an extent that the sample fails quicker than the strands can form and the process proceeds only by propagation of failure cracks. The transition from one failure mechanism to the other takes place gradually: the areas of the silver cracks in front of the failure cracks will contract more and more and at low temperatures and great rates of deformation they disappear altogether. At the usual ambient temperature (20°C) the formation of silver cracks precedes the failure in some plastics, poly(methylmethacrylate), poly(vinylacetate), in others (cellulose acetate) one observes only brittle failure.

Silver cracks usually form in the temperature range of low elasticity under prolonged stresses which are below the limit of the low elasticity of the polymer. With an increase of stress (or rising of temperature) in areas of overstress (lamination of the material into strands in the silver cracks) a transition from the local processes of low-elastic deformation takes place to the low-elastic deformation which proceeds simultaneously in the whole section of the sample (formation of "neck"). The discoloration of the "neck" which can be observed in the specimen indicate that the silver process proceeds also in low-elastic deformations of solid polymers.

Investigators who develop various theories of the strength of solid polymers must, naturally, start from the failure mechanisms and this includes examination of both failure cracks and silver cracks. However, a great number of theoretical papers on the strength of solid polymers[21, 22] have appeared in recent times which start from the old hypothesis of the failure of polymers which was first proposed by Kuhn as early as 1946 (see p. 105). In these papers the failure of polymers is explained as before as a process of break under the effect of stress and thermal oscillation of chains in the whole mass of the material, without accounting for the (above discussed) actual failure mechanism of solid polymers.

A criticism of Kuhn's mechanism of failure which excludes the development of cracks was repeatedly expressed by various investigators. Kuhn's hypothesis itself was not confirmed by anybody up to now experimentally. This does not, of course, mean that in a stressed polymer no individual chains in the whole volume of the specimen break at all. The process of thermal degradation of polymer chains can often be observed in stressed polymers but it is not the chief one. Although the mean stress in the sample is always considerably lower than stress concentrations near the microcracks and other structural defects, one can also observe there the basic process of breaking polymer chains which is also a process of thermo-mechanical degradation but is localized in places of overstress.[23] Into the expression of thermo-mechanical degradation of the polymer one can introduce a modification which takes into account that the speed of crack propagation depends on the number of breaking chemical bonds per unit of the length of the crack, and that the number of the non-broken chains decreases. However, the coefficient β of stress concentration at the apex of the crack is in solid polymers approximately 10 and one can therefore reckon that the probability of a break of chains in the mass of the material is negligible. The thermo-mechanical degradation of chains which takes place in the polymer mass leads therefore only to a negligible increase of speed of crack propagation and has little influence on the durability.

The process of failure of individual chains in the polymer mass can consequently only play a secondary role as it has, apparently, the more influence on the durability, the more homogeneous the material is and the fewer defects there are in the specimen.

The mechanism of failure which has been investigated in this chapter refers basically to amorphous solid polymers. The mechanism of failure of crystalline solid polymers has been, with the exception of fibres, studied but little. Recently, Preuss[24] remarked that the brittle break of crystalline polyethylene is accompanied by a "fusion" of the surface of the fracture. The author suggests that at brittle failure of polymers the local temperature may rise up to 300°C, which leads to a change of character of the supramolecular structure. However, it is more probable that under the influence of stress concentrations a chemical change, and not melting, takes place.

The questions of failure of crystalline polymers are touched upon in Chapter 4 in connection with the research into the influence of supramolecular structures on the strength of polymers.

3.3. Mechanism of Failure of Visco-elastic Polymers

From the theoretical point of view the failure of elastomeric polymers is an extremely difficult question. The complexity of the phenomenon prevents at present the creation of any satisfactory theory of the strength of these materials. W. Kuhn and H. Kuhn[25] proposed a mechanism for the failure of elastomers which was based on a statistical model of the network of non-Gaussian chains. It is assumed that under tension each chain undergoes an appropriate deformation. The chain breaks if its tension exceeds some critical magnitude. Owing to the presence of chains of different length in the spatial polymer the chains break one after the other in proportion to the increase of the tension. This process proceeds, at some tension it becomes catastrophic, and the specimen breaks. F. Bueche's papers on the theory of the strength of elastomers are based on a similar hypothesis of failure.

Such a concept of the mechanism of break is very widespread but it gives rise to objections. Treloar,[26] for instance, considers this mechanism as scarcely probable, in so far as the assumption of a related deformation of the chains in the region of great tension does not tally with the actual picture of the deformation process. The chain which suffers a specially great stress can, instead of breaking, displace the corresponding nodes into a new position of equilibrium. Indeed, already the first investigations of one of the authors of this book and his co-workers,[27, 28] led to the conclusion that the actual mechanism of failure of elastomers differs from W. Kuhn's concepts. Patrikeev,[29] as well as other investigators, have come to the same conclusions.

Fundamental results were obtained in careful experiments with pure rubber under tension under constant stresses.[27, 28, 30] Subsequently, the mechanism of break of visco-elastic materials was studied by a number of investigators[31-36] also at other rates of deformation (finite deformation, constant rate of strain, repeated deformation). It was established that the mechanism of failure of rubbers has common features at all methods of testing.

The failure of a visco-elastic material under constant stress takes place in due course, and is characterized by two stages of break-down: a slow and a rapid one. The slow stage gives a rough zone on the fracture surface (dark on the photograph, Fig. 57), and the rapid one a mirror zone (light), whereas in a brittle fracture, on the contrary, the slow stage gives the mirror—and the rapid one the rough zone. The smaller the static stress (deformation) the longer is the process of failure and the clearer expressed is the rough zone (Fig. 58). At sufficiently slow failure the mirror zone is practically absent (Fig. 58d). With a rapid break, on the other hand, the rough zone has no time to develop, and almost the whole surface is occupied by the mirror zone (rapid visco-elastic fracture).

The increase of strain rate is equivalent to a lowering of temperature. A rapid fracture is therefore equivalent to a low-temperature one, and a slow one to a high-temperature fracture of visco-elastic materials. With experiments on the testing equipment with a constant strain rate of 500 mm/min, two zones of fracture surface cannot usually be observed. This can be explained not by a different kind of failure mechanism, but by the fact that the loading time of the specimen at strain rates of 500 mm/min is not sufficient for the development of the first (rough) zone. Only the mirror-zone of the fracture surface can therefore be observed (Fig. 59a) and the rough zone is in an embryonic state. If the strain rate decreases by one order of magnitude, the rough zone of the fracture surface appears clearly (Fig. 59b). A further lowering of the strain rate leads to an almost complete displacement of the mirror zone (Fig. 59d). The usual experiment on the testing machine pertains therefore to the rapid failure of elastomers with the formation of practically only one mirror zone on the surface of the fracture.

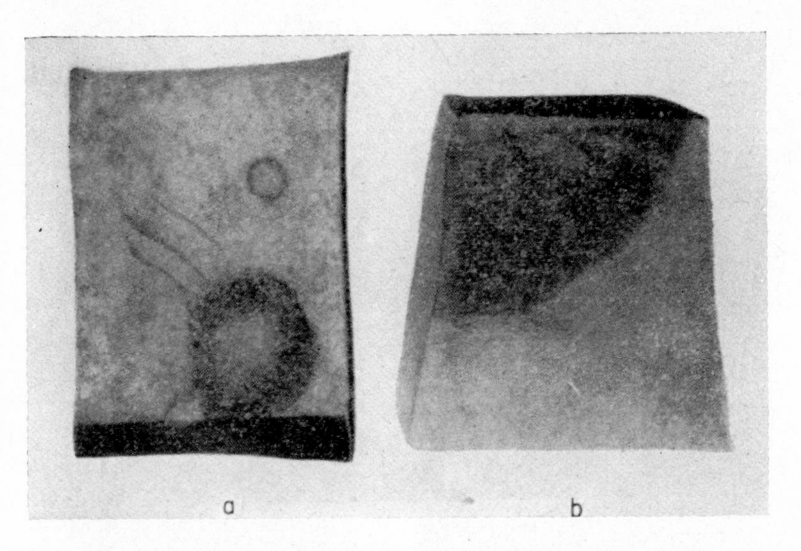

FIG. 57. Surface of slow break of unfilled elastomers at 20°C[28, 30] (a) SBR-30 rubber. (b) SKB. Specimens have a cross-section of 6×6 mm.

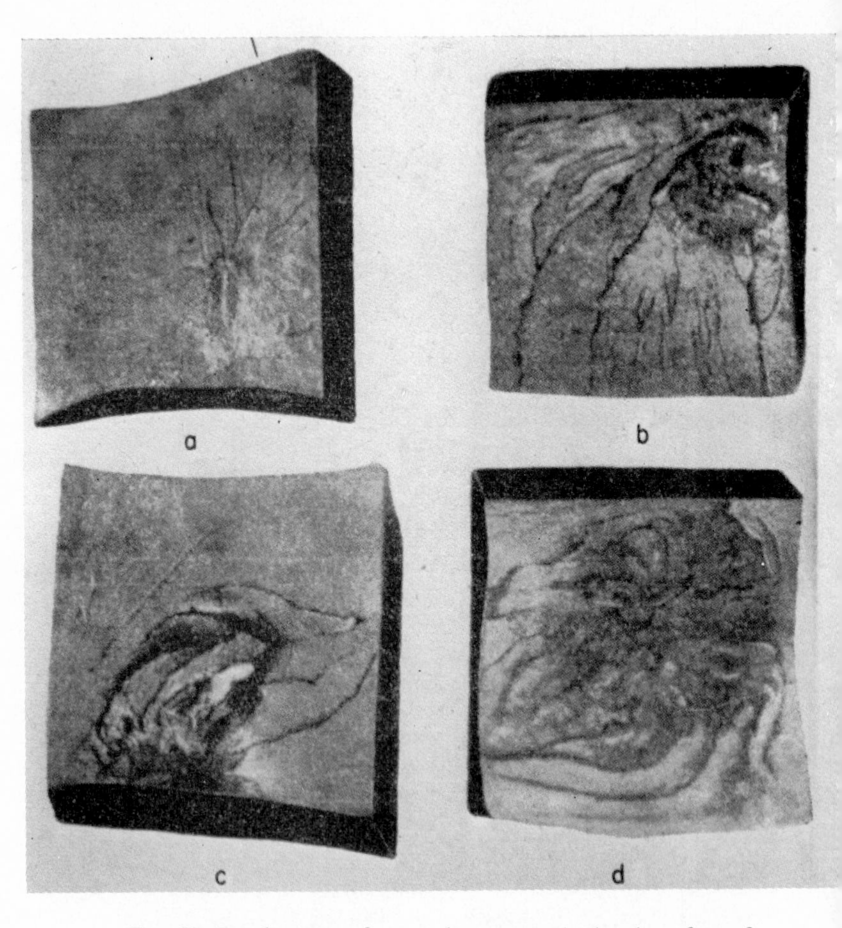

FIG. 58. Development of a rough zone on the break surface of SKS-30 rubber (20°C) during transition from great (a) to small (d) static tensile stresses. The time to failure increases from minutes (a) to weeks (d).

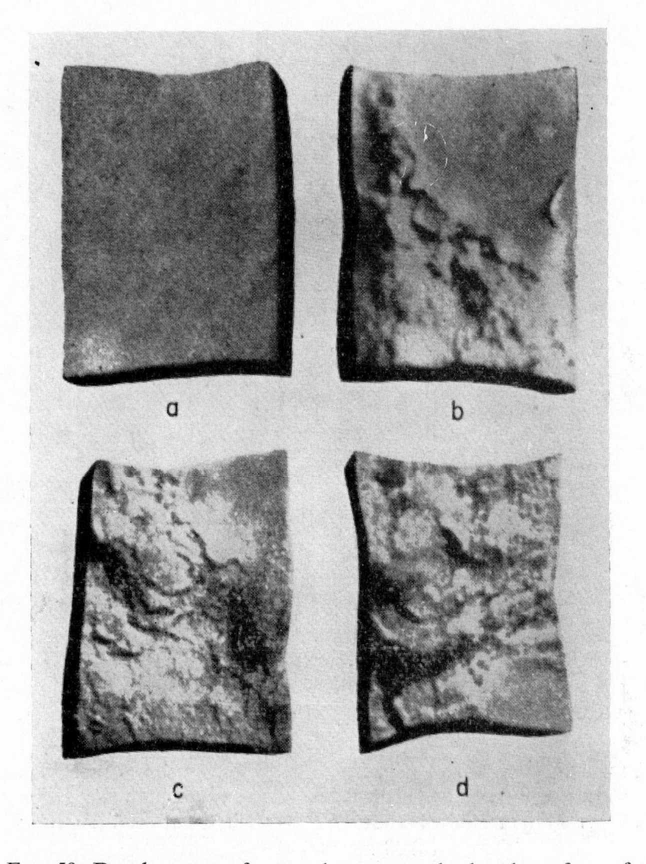

FIG. 59. Development of a rough zone on the break surface of lowmodular SKS-30 rubber at different strain rates.[28] (a) 500mm/min; (b) 50 mm/min; (c) 5 mm/min; (d) 0·5 mm/min.

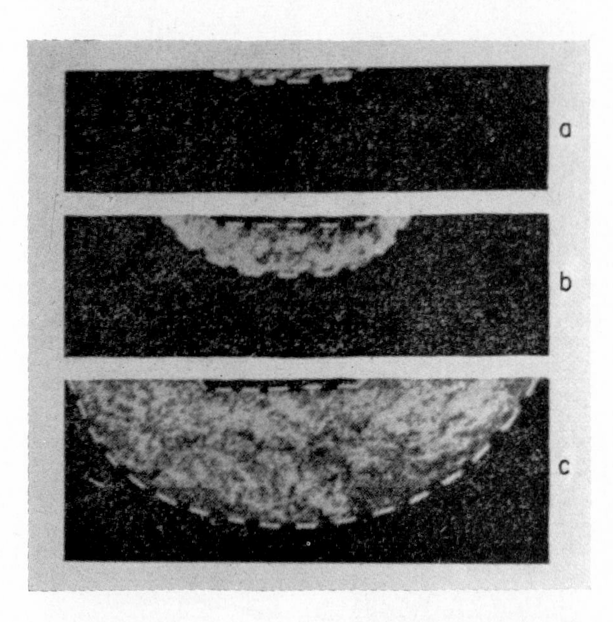

FIG. 60. Development of a rough zone during the growth of a notch during repeated tensile deformations (maximum elongation 125%) in a technical rubber.[33] (a) length of test 20 min; (b) 38 min; (c) 41 min.

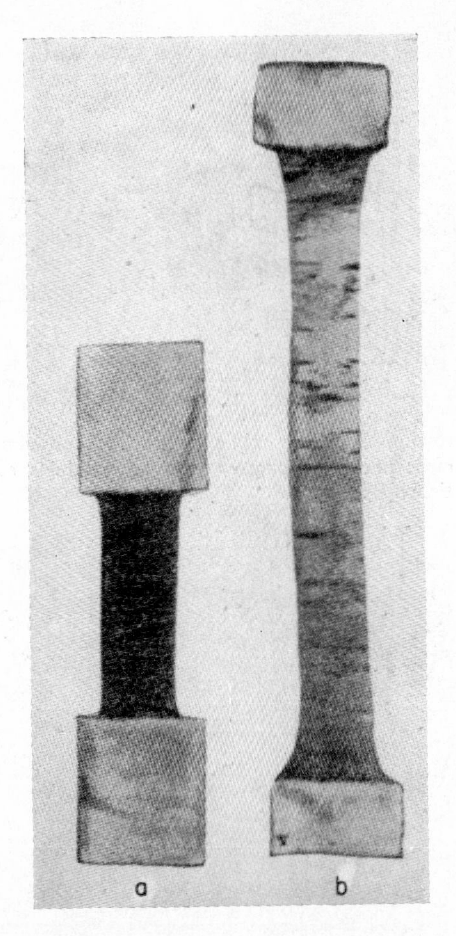

FIG. 61. Failure of rubber strips stretched for several months by 150% at 20°C. (a) Original sample. (b) Failed sample.

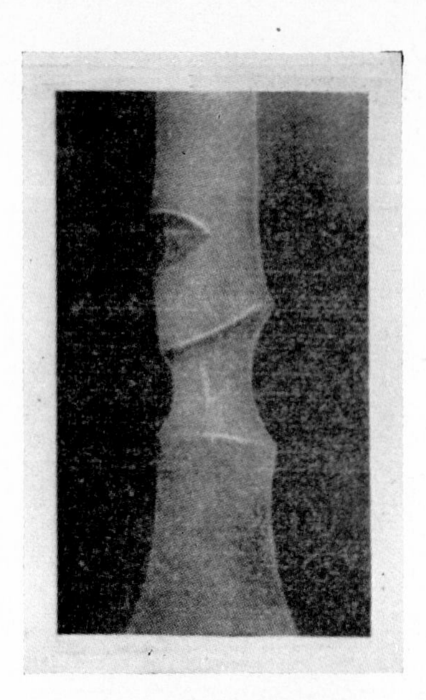

FIG. 62. Growing small tears in rubber specimens under prolonged small static tensile stress.[28]

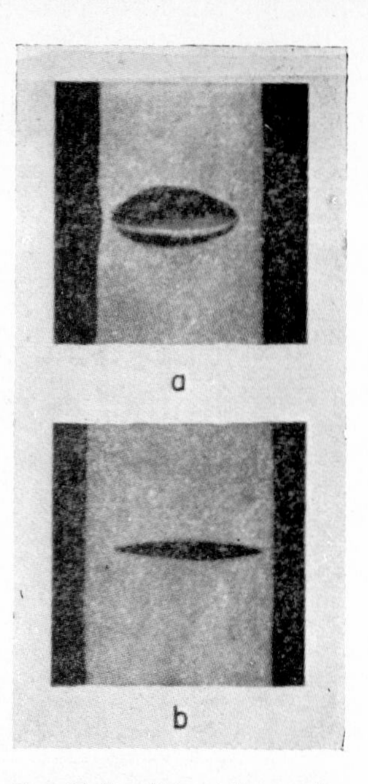

FIG. 63. Small tears in samples of low-modulus rubber after partial (a) and full (b) load.[28]

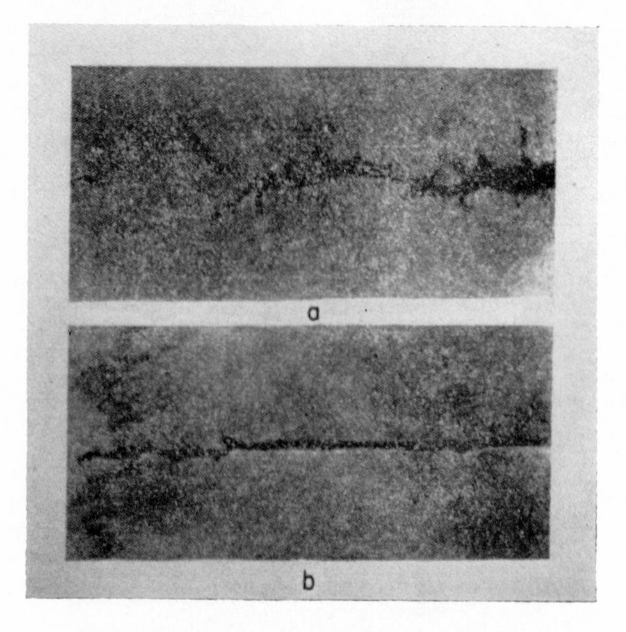

FIG. 64. Small tears in specimens of NK rubber (a) and carboxylate rubber (b) after frequent tensile deformation and subsequent unloading of specimen.[33]

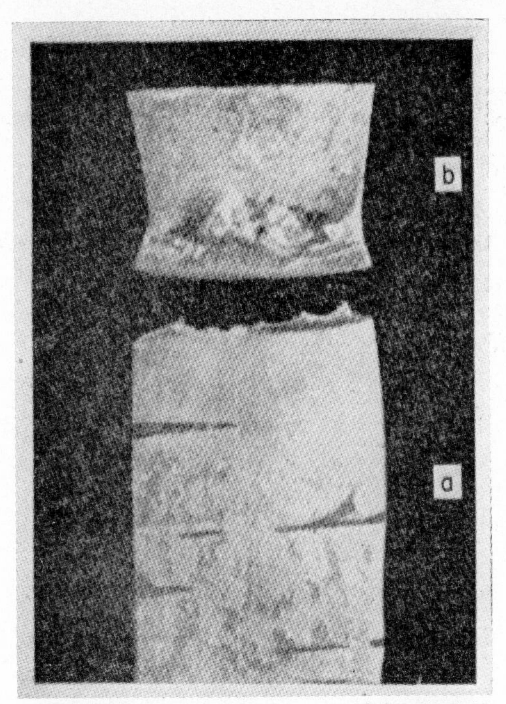

FIG. 65. Slow break of sample of low-modulus SKS-30 rubber. (a) One of the ends of the sample (side view). (b) View of the break surface.

FIG. 66. Growing small tear in SKS-30 rubber with sparse spatial network.[28]

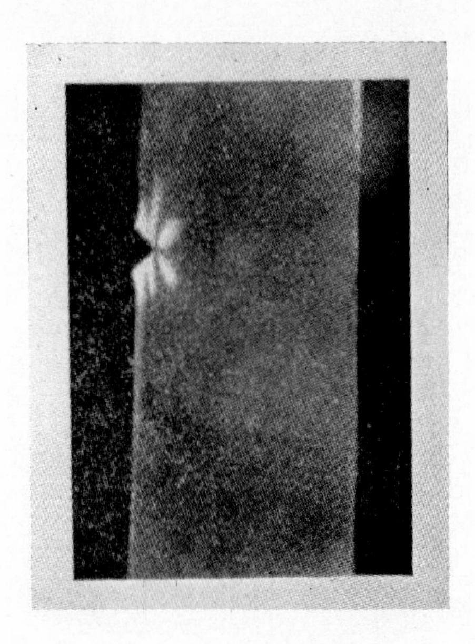

FIG. 67. Specimen of transparent elastomer with growing small tear, photographed in polarized light.[27]

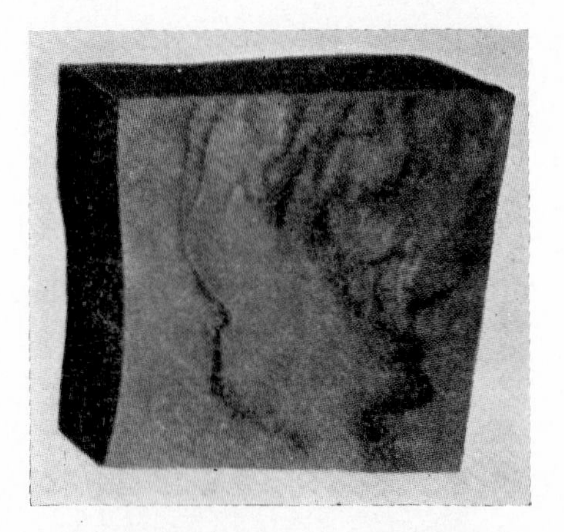

FIG. 68. Surface of slow break on low-modulus SKS-30 rubber at 25°C (large slip lines).

At repeated or cyclic deformations, on the other hand, a strongly developed rough zone of the fracture surface can be observed. The smaller the maximum tensile stress (or bending) in cyclic deformations, the slower is the process of failure and the clearer expressed the rough zone of the surface of the failure.[31, 33] On photographs of failure surfaces of technical elastomers it can be seen how the initial cut grows over a period at repeated tensioning, and forms a rough surface (Fig. 60). The mechanism of failure is more complicated with cyclic than with single deformations, as it is connected with the mechanical processes of fatigue of the elastomer (see Chapter 8).

The process of failure of elastomers at constant small tensile elongation differs from that of constant tensile strain rate. In this case the strain rate grows as the stress continuously increases in the remaining section; in the former the stress falls, as owing to the process of relaxation the material unloads itself and the failure process decelerates and even stops. The specimen does not part but it gets covered by a great number of "cracks" (Fig. 61). This process is analogous to the cracking of plastics under small stresses.

In all cases of slow failure of elastomers the first (slow) stage of fracture begins with the formation of centres of failure from which small tears grow, corresponding to the cracks in brittle materials.†

The small tears (Fig. 62) grow mainly in the transverse direction to the stretching forces, like the cracks in brittle bodies. On Figs. 63 and 64 the moment has been caught when the propagation of the small transversal tears has been partially or completely stopped by the unloading of the sample.

The small tears arise in the weakest places under the effect of the stress and the centres of failure appear both inside the material and on its surface. Their front is as a rule a circle, if the centre of failure is inside the material or a semicircle if that centre lies on the surface (see Fig. 57). The small tears appear and grow in different places of the specimen but amongst them there are always some more critical ones which lead to the parting of the

† The formation and propagation of small tears, like that of cracks, is determined by the true stress and not by the applied one. The physical characteristic of the strength of elastomers must therefore be the true breaking stress.

sample in two (Fig. 65). The strength of the elastomer is therefore determined by the probability of the formation of most critical small tears, analogous to the way in which the strength of brittle materials is determined by the most critical cracks.

The investigated mechanism of failure is specific for the spatial polymers† which are visco (high)-elastic. In low molecular rubbers with a sparse spatial network the fibrous structure can be most clearly seen at the roots of the small tears (Fig. 66). As the failure zone deepens one forms after the other and ultimately the strands break. The individual strands break in different places along their length. As the ends of the strands contract, small protuberances and depressions appear on the failure surfaces which in the aggregate form the rough surface.

The material is in an overstressed state near the apex of the small tear with the result that an additional deformation and orientation of the material occurs there. This is shown when samples of transparent rubbers are examined in polarized light (Fig. 67) and also by high-speed filming.[37-40]

The stress at the apex of the small tear grows as the remaining section of the specimen decreases; this leads to an acceleration of the propagation of the small tear. When the stress in the remaining section reaches a critical magnitude the break goes over into the rapid stage: ordinary cracks are formed (like in brittle bodies) which grow with a speed near that of sound in rubber.‡ This stage of failure lasts only a small fraction of a second.

The rapid stage of fracture in spatial polymers leads to the formation of mirror zones which are covered by slip lines which are reminiscent of the hyperbola on the break surface of solid polymers (see p. 97). At rapid fractures when there is no rough zone the fracture surface is mirror-smooth and covered with slip lines so fine as to be hardly noticeable with the naked eye. In slow breaks the mirror zone which forms in the second stage gets covered both with small and with big slip lines. In low molecular elastomers (Fig. 68) the slip lines are much less frequent but they are larger than in high molecular ones.

The presence of the slip lines confirms that in the second stage

† The failure of linear elastomeric polymers is investigated in § 3.5.

‡ The speed of sound in elastomers is approximately 50–100 m/sec, i.e. ten times less than in solid polymers.

of the fracture new centres form in the remaining section of the sample at different levels but near to each other from which cracks grow and, when joining, produce the slip lines. The slip lines sometimes resemble second order curves but the apices of the hyperbolae, parabolae and other curves are not always turned to the side of the primary centre of the break, as observed in solid polymers. To explain this one has to assume that the centres of new failures on the same level often form away from the growing small tear, in any place of the section and most probably on the surface of the sample.

Failure is possible under tension both by tear and by shear. On Fig. 69 two horizontal parts of the fracture-surface which are formed by tears and lie in adjoining planes are connected by a failure by shear (small step). The angle of the shear-surface is approximately 45° to the tear surface. This indicates that the shear takes place in the direction of the greatest tangential stress.

Thus the rapid break takes place without formation of small tears as a result of the propagation of failure cracks, whilst the slow one proceeds by formation and propagation of small tears.[28] In the former case the break-surface is smooth, in the latter it is rough. In the first stage of failure the defects grow in the shape of small tears, producing a rough zone on the failure-surface, in the second one defects in the shape of cracks producing a smooth zone. In accordance with this elastomers fail through the growth of two kinds of defects: "small tears" and "cracks". The mechanism of failure by crack propagation in elastomers corresponds to the failure of brittle bodies (direct break of bonds), by which the term "crack" for visco-elastic materials is justified. The formation of highly orientated strands in the first stage of failure is connected with the severing of intermolecular bonds. The molecular mechanism of the slow fracture of visco-elastic polymers consists therefore of elementary acts including the overcoming of intermolecular forces during the formation of strands and the break of chemical bonds.

A similar failure mechanism can be observed when a film of visco-elastic material is torn from the solid surface.[41] At small peeling speed a cohesive type of failure can be observed: at first strands form and then they break. On the surfaces of the film and the base remain traces of these strands (rough surfaces). At great speeds an adhesive type of failure is observed: no strands form and the film separates completely from the surface of the base (smooth surface).

The "fibrous" mechanism of the fracture of visco-elastic materials is probably connected with the directional structure of polymers. Electron microscopic investigations[42, 43] have recently shown that in linear rubber-like polymers the adjoining chains are not distributed completely at random, especially under great stress, but form more or less regular groups of molecules, orientated in a short-range order. The groups in their turn form secondary supramolecular structures in the shape of strips which are arranged disorderly in the rubbery mass. Strands get probably formed from them. The formation of strands can take place not only in connection with the directional structure "bundle" of the initial non-orientated polymer but also in the presence of other microheterogeneities of the structure of the material.

The formation of strands can be regarded as a process of stratification of the orientated polymer material in a heterogeneous field of stresses. It follows from Gul's and Chernin's[39, 40] observations that traces of strands begin to form in the section of the sample in front of the growing small tear. Like the zone of laminating material in the shape of a silver crack in front of a crack in plastics there is in elastomers a zone of material in front of a small tear which is ready for the stratification into strands.This has been confirmed by investigations,[37, 38] in which it was shown that the structure of the polymer material near the defect which grows during the breaking process differs considerably from the structure which is characteristic for the specimen as a whole. In essence it is not the original polymer which breaks but a material of a different structure which is orientated and possesses (in comparison with the original one) different relaxation properties. The changes which the material undergoes at the propagating small tear determine the character of the failure process of the specimen. With a significant change of the degree of subsidiary orientation change correspondingly all the characteristics of the strength of the material. High-speed filming of the fracture process[35–40] made possible measuring the subsidiary orientation in the place of the growing small tear, and defining the form and dimensions of the growing defect during rapid failure and the speed of growth of the small tear in the various stages of the process of failure. Towards the end of the failure process the speed of the propagation of the small tear increases rapidly and unevenly which is probably connected with the tearing of strands.

The transition from the failure mechanism of a polymer in the glassy state to the break of the same polymer in the visco-elastic state has been followed on poly(ethyleneterephthalate).[39, 40] In the glassy state the polymers fail along the planes of overstress which are located in front of the growing cracks. Visco-elastic polymers fail as a result of elongation and slipping along the axis of tension of those parts of the material from which strands are then formed.

The investigated mechanism of failure can be observed both in spatial, elastomeric polymers and in technical rubbers filled with carbon black.[33, 34] Knowing this mechanism has a great practical importance for the correct understanding of the processes of failure of technical rubbers which undergo in use a prolonged influence of constant or changing loads. Here the failure begins with the slow stage of break which determines completely the durability of the manufactured article. In this sense the standard experiments on the testing machine do not reflect the true picture of the failure of the product in use.

3.4. Influence of Structure and Temperature on the Failure Process in Visco-elastic Polymers

The process of failure in spatial, elastomeric polymers, the typical representatives of which are the uncompounded rubbers, is influenced to a very large degree by the number of cross-links (knots) or by the number of chains in the spatial network per unit of the volume of the elastomer, the polarity of the rubber and the temperature. The influence of these factors on the failure mechanism was studied in the papers.[28, 30]

The failure of elastomers with small or large static equilibrium modulus differs even in its outer appearance. In pure high-modulus elastomers from SKS-30 (equilibrium modulus 18 kg/cm^2) one can observe a rapid contraction of the specimen parts after the break- up to the original dimensions, and in low modulus ones (equilibrium modulus 0·2 kg/cm^2) the pieces will only contract completely after a long time. In specimens of elastomer with a small modulus the appearance and propagation of small tears can be beautifully observed (see Figs. 65 and 66).

The correlation between the rough and the smooth zones of the break surface depends greatly on the density of the spatial network of the elastomer (the numbers of cross-links per cubic centimetre), which can be judged by the size of the equilibrium modulus of the elastomer. The lower the modulus is (i.e. the sparser the spatial network), the quicker grows the rough zone and the sooner is the mirror zone of the break completely replaced by the rough one. With the increase of the cross-linking the speed of the formation and propagation of the small tears decreases, the strands which are forming become thinner and the rough surface of the first zone, characteristic for the low molecular polymer, changes into a dull one. The mirror zone which gradually replaces the rough one, gets more and more covered with numerous thin slip lines.

The two-zonality of the fracture surface is characteristic for all types of elastomers.[30] However, each polymer, depending on the number of its polar groups and the size of intermolecular forces, has its own peculiarities. In order to obtain traces of the rough zone in the non-polar rubbers the application of a load at usual temperatures for a few minutes is sufficient but for the polar ones a considerable time is required. For instance, for pure elastomers from SKN-26 with an equilibrium modulus of about 3 kg/cm^2 the rough zone begins to appear only after 3 hr under load at 20°C, and becomes noticeable only after 30 hr. This propensity exists also when the intermolecular forces are great. In elastomers on the basis of the strongly polar rubber SKN-40 even 500 hr are not sufficient for the rough zone on the break-surface to appear sufficiently clearly.

In spite of the fact that the fracture process of the three-dimensional rubber-like polymers with small and great intermolecular valency forces differ in kinetics and appearance of the break-surface, the mechanism of a slow break is the same in both cases. The difference is that vulcanized rubbers with great intermolecular valency forces have only at very prolonged loading a developed rough zone of the break-surface. Besides, elastomers from non-polar rubber have a very rough break-surface and the polar ones (SKN-26, SKN-40) a less rough one. On the other hand, the rough zone in SKN-30 with a high equilibrium modulus has the same dull surface as the rough zone of the low-modulus SKN-40. This can be explained by the fact that thinner strands form in the low-modulus SKN-40 due to its strong polarity, and in the high-modulus SKN-30 due to the great number of transversal bonds.

With an increase of polarity of the rubber and the density of the spatial network one can observe a tendency to a transition from the failure mechanism which is typical for elastomeric polymers to the mechanism of failure which is common to all solids. Here a mirror-surface of the break forms (Fig. 70) as the primary crack has time to propagate across the whole section before secondary ones arise.

For a more detailed investigation of the mechanism of slow failure the influence of temperature on the speed of failure of unfilled non-polar SKN-30 and polar SKN-40 was studied[30] under different constant tensile stresses. The character of the break was appraised by the relative size of the rough zone: $z = S_r/S$ (where S_r is the area of the rough zone on the break surface, S is the whole area of the cross-section of the sample) and the mean speed of the formation of the rough zone: $v \approx z/\tau$ (where τ is the durability or the time from the beginning of the application of the constant stress to the moment of the parting of the

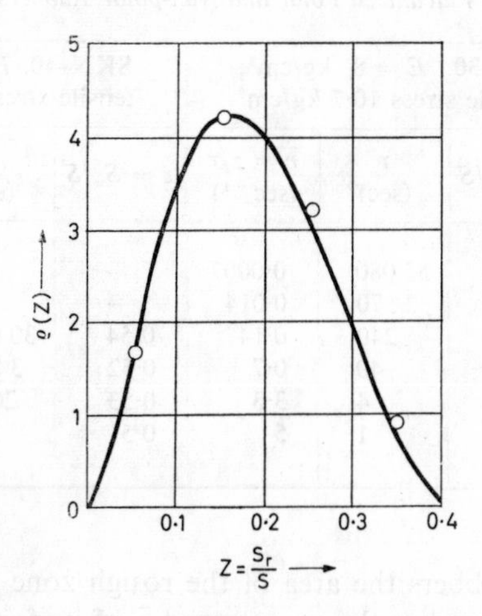

FIG. 71. Function of distribution $\varrho(z) = \frac{1}{N} \cdot \frac{\Delta N}{\Delta Z}$ (where N is the total number of specimens of SKS-30 rubber, tested at 100°C; ΔN is the number of specimens in the range Δz.[30]

specimen). A considerable scatter of z over a range of specimens under the same experimental conditions was found (Fig. 71). This indicates that identical specimens of elastomer have defects of different degrees of criticality on which the small tears form. Z was determined as the mean value of testing twenty samples.

The influence of temperature on the correlation of the rough and mirror zone of the break-surface is different for elastomers from SKN-30 and SKN-40. If one compares the break-surfaces of elastomers at different temperatures and equal durability, a common rule can be observed: with a rise of temperature the rough zone replaces gradually the mirror zone, independent from the type of rubber. If one compares the break-surfaces obtained under the same tensile stress one finds that the temperature influence on the relation of the zones is a different one for polar and non-polar rubbers. This can be seen on Table 4.

TABLE 4

Influence of Temperature on the Relation of the Rough and Mirror Zones of Vulcanized Polar and Non-polar Rubbers

Temperature) (°C)	SKS–30. $E = 8$ kg/cm²; tensile stress 10·7 kg/cm²			SKN–40. $E = 5$ kg/cm² tensile stress 5·1 kg/cm²		
	$z = S_n/S$	τ (sec)	$\bar{v} = z/\tau$ (sec^{-1})	$z = S_n/S$	τ (sec)	$\bar{v} = z/\tau$ (sec^{-1})
20	0·40	55 080	0·0007	—	—	—
40	0·37	2 570	0·014	—	—	—
60	0·34	240	0·14	0·54	30 000	0·002
90	0·28	40	0·7	0·52	3 840	0·014
120	0·13	4	3·3	0·53	2010	0·026
140	0·05	1	5	0·51	220	0·23

In SKS-30 rubbers the area of the rough zone decreases with rising temperature but the mean speed $\bar{v}$ of its formation grows. The explanation of this strange fact is this: whilst with a rise of temperature both the speed of formation and the propagation of the small tears increases, it is a peculiarity of this polymer that

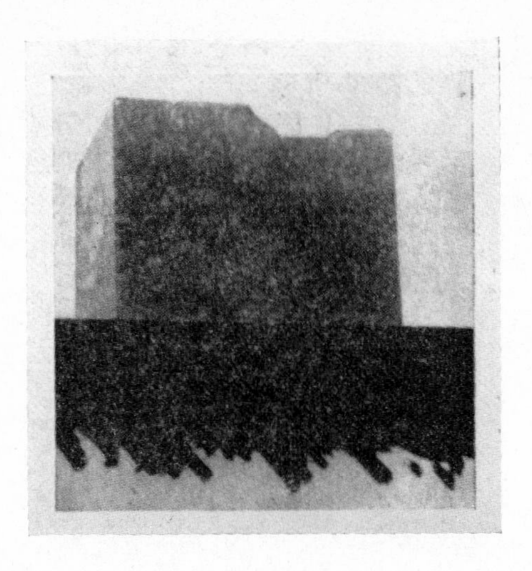

FIG. 69. Cross-section of break surface of rubber showing slip lines.

FIG. 70. Break surface, characteristic for high modulus elastomers and elastomers on a basis of polar rubber, of short term failures.

the increase in the speed of crack propagation outstrips the increase in the speed of propagation of the small tears. As a result the second stage of failure begins earlier at high temperatures than at low ones, and the mirror zone of the break-surface increases in size and displaces the rough one. Consequently, under the same stress at low temperature a slow, and at high temperature a rapid "visco-elastic" failure can be observed.

Generally in SKN-40 rubbers at 20°C no noticeable rough zone can be observed; the area of the rough zone which appears at 40°C does not increase with a further rise of temperature but the speed of its formation grows whilst the durability of the elastomer decreases. The constancy of the value z is also characteristic for brittle materials in which the first zone of the break-surface is a mirror one.

3.5. Ductile Failure of Polymers

The ductile failure of polymers is connected with their plasticity or fluidity, i.e. with their capacity to undergo viscous flow in certain conditions and under the effect of outside forces. This viscous flow can be divided into two basic types: the "physical" and the "chemical" flow of polymers. The first type of flow can be observed in linear, the second one both in linear and in spatial polymers. The "physical" flow is met with when processing linear polymers and the "chemical" flow chiefly in the use of polymeric products which have been under stress for a long time, especially at high temperatures.

The "physical" flow of linear polymers takes place through a continuous rearrangement of the intermolecular structure and a changing of position of segments of the macro molecules without any noticeable degradation and changes of the chemical structure of the polymer.

The liquid state of polymers is characterized by the following distinctive properties: (1) high viscosity in connection with a great molecular weight of the polymers; (2) the special role played by the stress which provides a decrease of viscosity during processing; (3) the temperature coefficient of viscosity is independent from many factors, for instance from the molecular weight and stress, and (4) the special role played by the visco-elastic deformation developing in the viscous flowing polymer.

The mechanism of viscous flow and the ductile failure of linear polymers must not be investigated without considering the intermolecular structure of the polymers and the conception of temporary nodes of the unstable spatial network of the linear polymer.[44]

Dogadkin[45] and Treloar[46] suggested the hypothesis of the presence of secondary transversal bonds in natural rubber which play the part of the temporary nodes of the spatial network of the polymer. Their existence in other polymers is proved by many experiments.[44, 47–51] Evidence of the formation of local intermolecular links in the polymers (hydrogen and Van der Waals bonds) is obtained by observing their molecular spectra. These secondary cross-links are temporary ones and break and get restored comparatively easily in the process of thermal movement, and the equilibrium between the broken and non-broken links shifts with the change of temperature. In polar rubbers the existence of temporary nodes in the network is evident as the polar groups of adjoining sections of chains form easily local cross-links. In F. Bueche's opinion, and in that of several other investigators, the role of the temporary nodes of the network can be played also by places of cross-overs and entanglements of the macromolecules.

According to Kargin's and co-workers' data[52] the molecular order is characteristic not only for solid amorphous polymers but also for polymers which are in a visco-elastic and viscous flowing state. The degree of molecular order (bundle structure) in the amorphous polymer depends on the temperature. The lower the temperature, the better develops the supra-molecular structure. It is obvious that at sufficiently high temperatures (in a viscous flowing state) the blocks, as fibrillar formations, become unstable due to the weak intermolecular forces between the polymer chains and to the intensive thermal movement, owing to which they break up and degradate. However, a lower degree of order is preserved at high temperatures. Thus, instead of fibril-blocks, which are much longer than the macromolecules, bundle clusters are formed or "microblocks".[53] They have the same properties of fluidity as liquids of similar viscosity but differ by a greater stability and greater order. Their life is only small compared with the time of observation but is considerably greater than the time of the transition of the free segments (which do not enter into microblocks) from one equilibrium position into another. After a sufficiently long time of observation the structure of the polymer at high temperatures can therefore be seen on the whole as a structure of randomly interlaced chains.

The molecular model of a linear polymer can thus be considered as a spatial network, the temporary nodes of which are microblocks. The sections of the chains which do not enter in any given moment into microblocks change their place with the speed of thermal movement (during the average life of the microblocks they can change their place many times). As the life of the microblocks is considerably longer than the time of the transition of free segments from one equilibrium position into another, the speed of the viscous flow depends chiefly on the speed of breaking and remaking of microblocks, and the viscosity depends on their average number in a unit of the volume of the polymer.

Thus the flow of a polymer presents itself as a rearrangement of the microblocks of the spatial network (their breaking in some places, and their formation in others) .Microblocks can be regarded as temporary nodes of the molecular network of linear polymer.

The breaking and restoring of the temporary nodes under the influence of thermal movement is probably one of the fundamental peculiarities of the viscous flow of linear and branched polymers[44] and of the process of slow physical relaxation in the spatial polymers.† On the other hand the mechanism of the transference of linear molecules of the polymer at viscous flow consists in frequent dislocations of small sections of macromolecules, called segments, under the influence of outer forces and thermal movement. To this points, firstly, the small activation energy of the process of flow compared with the activation energy of the flow of low molecular organic compounds and, secondly, the independence of the activation energy from the chain length. In essence the viscous flow of polymers is a diffusion mechanism of flow, complicated by the polymeric structure of the molecules and by the presence of the temporary nodes in the network of the linear polymer. The great viscosity of polymers is determined not by the segmental mechanism of the flow itself but by the great length of the macromolecules and by the structure of linear polymers—with their network formed by temporary nodes.

† The general theory of visco-elastic properties of linear polymers with a network formed by temporary nodes was suggested by Hayashi.[54]

The flow mechanism of actual linear polymers is as yet to a considerable degree only a hypothesis which has yet to be more accurately defined. One of the probable flow mechanisms which take into account the role played by the temporary nodes is investigated in the paper[55] and starts with the independence of the activation energy of the viscous flow from shear stress. It consists in the following.

The transfer of the segments from one equilibrium position to an adjoining one depends mainly on the frequency of the break of the temporary nodes under the influence of thermal movement. The stress promotes the moving of the segments chiefly in the direction of the outside force. At the same time the stress decreases the probability of a renewal of the broken nodes as it furthers the separation from each other of the active groups and parts which form temporary nodes. From this follows that with increasing stress the probability of a renewal of the temporary nodes decreases and the average number of broken nodes grows. As a result the viscosity which depends on the structure—in the given case on the number of temporary nodes—decreases. At the same time the activation energy remains constant as it is not determined by the number of temporary nodes but by their nature.

In the molecular mechanism of viscous flow of linear polymers the part played by the supramolecular structure has up to recent times not been taken into account. Former concepts of the flow have been preserved in the idea that the segment of the chain and not the microblock is the kinetic unit. The segment is the stable structural unit and the microblock the unstable one. The observed value of the activation energy relates therefore to the segment and not to the microblock for which, if it is considered a kinetic unit, the activation energy must be larger by 2 to 3 orders of magnitude. This means that in linear polymers we are dealing with a segmental flow mechanism. However, the question crops up now how to correlate the diffusion-segmental mechanism of the viscous flow with the role played by the temporary nodes which may be microblocks. For this one has to assume that the average length of the microblocks corresponds to the length of the segments (30–40 carbon atoms in the main chain), and that in its breaking and reforming the segments act as individual kinetic units.

The process of "physical" flow which we have now investigated is the basis for the understanding of ductile failure of linear polymers.

The "chemical" flow of polymers, and in particular of elastomers, proceeds according to two main mechanisms.

The elastomers flow either as a result of chemical reactions, including the break and renewal of chemical cross-links[44, 56] under the influence of relatively high temperatures or chemical reagents—or as a result of a break of the chains themselves under the action of sufficiently great stresses on macroradicals or their recombinations.[57, 58]

These mechanisms of "chemical" flow usually exist in elastomers either at high temperatures or under great stresses (see in more detail § 10.2 and § 10.3).

In order to understand the nature of ductile failure of elastomers it is imperative to study the processes of chemical flow.

It is known that the ductile failure of polymers is connected with the capacity of these materials to flow under the effect of stress under certain conditions which exceed the yield point or "plasticity" σ_p. Under the "yield point" we understand the stress above which the number of broken nodes is sufficiently great for the flow to become noticeable. The higher the temperature, the lower the yield point until at the temperature of flow T_F it reaches zero. Above this temperature the polymer flows under any small stress. The limits of flow of linear or spatial polymers are very different from each other because of the great difference in the strength of the chemical cross-links and the intermolecular bonds which are responsible for the formation of the temporary nodes of the network. The temperature of plasticity T_p is therefore higher in three dimensional polymers.

The existence of a yield point can be explained by the presence of temporary nodes of a different nature in linear polymers. The flow can also be observed below the yield point but in practice it proceeds with a negligible speed.

The ductile failure of polymers has been studied in detail[28] on rubber and elastomer compounds. Failure of this kind is accompanied by an irreversible flow of material and by the formation of a contraction, or "neck" before the break (Fig. 72a). Contractions form in several places of the specimen and the break takes place on the most critical one.

The ductile failure of polymers is outwardly similar to the break of viscous metals. Like on metals the ductile failure of polymers can be observed in a limited range of speed of deformation or time of action of the load. Under a small load or low strain rate a transition takes place to a visco-elastic break which is characteristic for elastomers. The explanation for this is that at stresses below yield point no contractions form, and the ductile break turns into a visco-elastic one; the presence of a spatial network in rubbers, formed by the temporary nodes, explains precisely this. The transition beyond the yield point is connected to the overstressing and breaking of these nodes.

Corresponding to the term "low elasticity" the transition from the visco-elastic state to a visco-flowing one with increasing stress could be called "low fluidity". It is understood that with a rise of temperature and a decrease of strain rate the yield point is lowered as the breaking of the temporary nodes is facilitated.

Under prolonged small tensile stresses many small tears appear in the specimen before the break (see Fig. 72b), and the parting of the specimen takes place on the most critical one. This is connected with the fact that under loads below the yield point the material behaves like a visco-elastic one, without displaying in practice any flow and breaks without contracting. The small tears of ductile failure on account of the relatively great strain rate have no time to propagate before contracting. The pronounced orientation of the molecules which arises at the same time strengthens the material and impedes the growth of small tears.

Such a phenomenon is similar to the failure of plastics in the range of low elasticity where under great tension "necks" form, and mainly silver cracks during the prolonged effect of small stresses.

The examination of the break surface permits it to define more accurately the characteristic features of the ductile failure and the transition, with decreasing stress, from one kind of break to another. With the tensile static stress (calculated on the initial section) decreasing from 5 to 0·2 kg/cm^2, the durability of SKS-30 rubber grows from 1 sec to 50 hr. The ductile failure turns here into a slow visco-elastic break which is characteristic for low-modulus elastomers.

This is evident from the fact that in a rapid break a considerable contraction forms in a specimen of SKS-30 rubber (Fig. 73a),

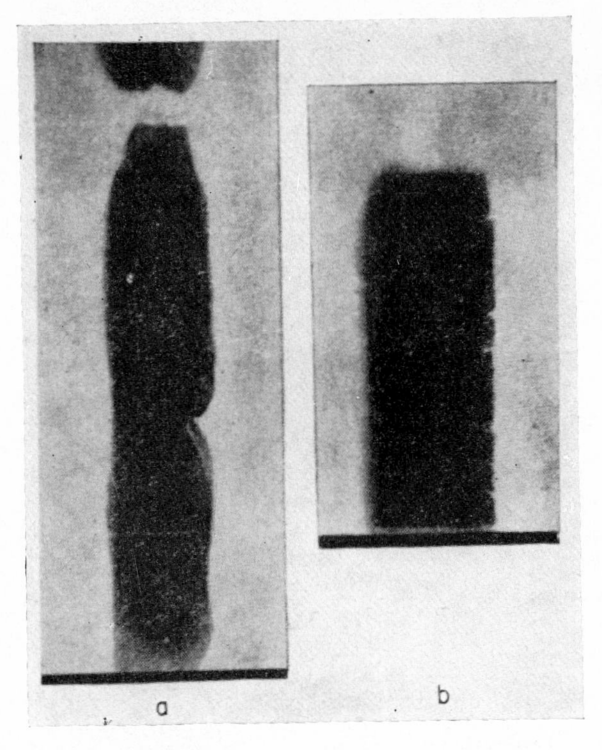

FIG 72. Specimens of plasticized SKS-30 rubber after rapid (a) and slow (b) break.[28]

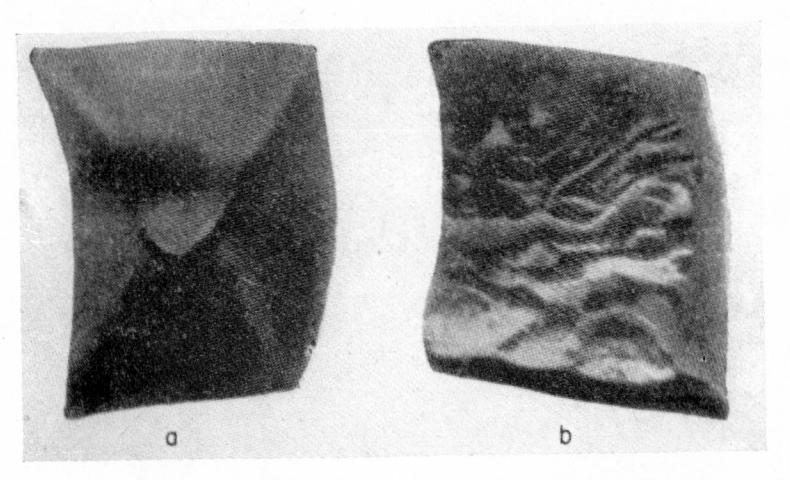

FIG. 73. Break surface of samples of unfilled mixtures of SKS-30 rubber.[28] (a) Rapid break (1 sec); (b) slow break (1·5 hr).

and that is where it will part; the surface of the break in the contraction consists only of a mirror zone. During the transition to a slow break under the influence of small loads a mixed type of failure can be observed—there will be only a tendency to contract and the break-surface consists of several mirror and rough zones (Fig. 73b). At a very slow break there is no contraction at all and the break-surface consists almost entirely of a rough zone. If, on the other hand, the rate of tensioning is vastly increased the material will at very great rates behave like a solid body, and the ductile failure turns into a brittle one.

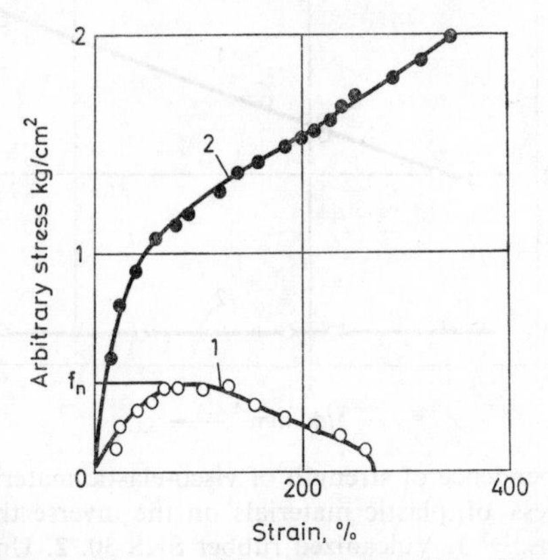

FIG. 74. Tension diagram of low-modulus NK rubber at a strain rate of 10% per minute. 1. At 140°C. 2. At 20°C.

During the processes of destruction which are characteristic for some rubbers the material becomes, during experiments at high temperatures, plastic on the outer layers but inside it remains visco-elastic. The break-surface acquires an unusual appearance as the outside layers undergo a plastic deformation before the break: both surfaces of the break appear as a "small cup".[30]

The limit of flow which is a strength characteristic of plastic materials can be determined from the strain–stress diagram (Fig. 74), and is shown by the maximum of the curve.[59, 60] The yield point can be expressed by the arbitrary stress f_p or by the true stress σ_p; here $\sigma_p = \lambda f_p$, where λ is a multiplication factor of the tension in the moment of reaching the stress maximum.

The breaking stress characterizes the strength of visco-elastic, and the yield point that of plastic materials. From the data of

Fig. 74 can be seen that an elastomer from natural rubber is a visco (or high) elastic material at ambient temperatures, and at high temperatures a plastic one.

As the deformation characteristics of a homogeneous material do not depend on the dimensions of the specimen, so does the yield point of plastic materials not depend on the scale factor which influences the strength of visco-elastic materials (see Chapter 5). This is confirmed by the data shown on Fig. 75.

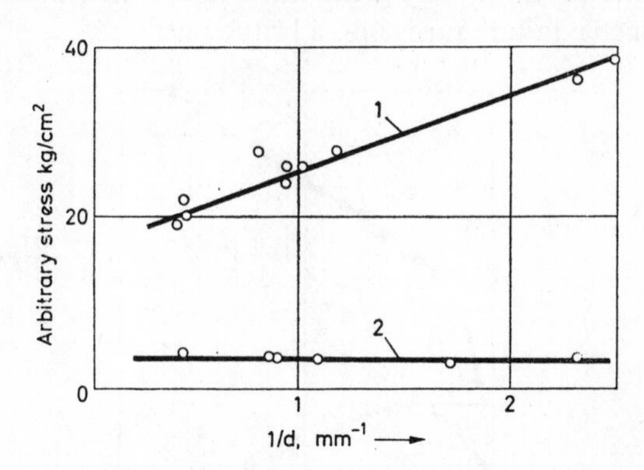

FIG. 75. Dependence of strength of visco-elastic materials and of the yield stress of plastic materials on the inverse thickness of the specimens.[28] 1. Vulcanized rubber SKS-30. 2. Unfilled mixture on the basis of the same rubber.

Literature

1. V. R. REGEL', *ZhTF*, **21,** 287 (1951).
2. V. V. LINDE, A. A. ROGOVINA, *ZhTF*, **23,** 1144 (1953).
3. G. N. KUKIN, *v sb.* "*Khimiya i fiziko-khimiya viysokomolekulianiykh soyedinenii*", Izd. AN SSSR, 1952, p. 289; A. N. SOLOV'YEV, G. N. KUKIN, *Tekstil'noye materialovedeniye*, Gizlegprom, 1955.
4. V. A. KARGIN, V. A. BERESTNEV, T. V. GATOVSKAYA, Ye. Ya. YAMINSKAYA, *DAN SSSR*, **122,** No. 4, 668 (1958).
5. V. A. BERESTNEV, T. V. GATOVSKAYA, V. A. KARGIN, Ye. Ya. YAMINSKAYA, *Viysokom. soyed.* **1,** 373 (1959).
6. A. V. ORLOVA, V. A. BERESTNEV, V. A. KARGIN, *Viysokomol. soyed.* **1,** 738 (1959); V. A. BERESTNEV, A. V. ORLOVA, Z. I. SULEIMANOVA, V. A. KARGIN, *Kauchuk i rezina*, No. 3, 10 (1963).
7. V. R. REGEL', *ZhTF*, **26,** 349 (1956).
8. V. R. REGEL', YU. N. NEDOSHIVIN, *ZhTF*, **23,** 1333 (1953).

9. I. SANER, H. HSIAO, *Trans ASME*, **75**, 865 (1953); *Plastics*, **19**, No. 198, 28 (1954).
10. C. HSIAO, I. SANER, *J. Appl. Phys.* **21**, 1071 (1950).
11. B. MAXWELL, L. RAHM, *Ind. Eng. Chem.* **41**, 1988 (1949).
12. A. P. ALEXANDROV, *Vestnik AN SSSR*, No. 7/8, 51 (1944).
13. A. SMEKAL, *Ergebn. exakt. Naturwiss*, **15**, 106 (1936); *Glastechn. Ber.* **23**, 57 (1950).
14. H. A. STUART, *Die Physik der Hochpolymeren*, Part IV. Springer-Verlag, 1956, p. 165.
15. H. SCHARDIN, *Glastechn. Ber.* **23**, 1 (1950).
16. M. I. BESSONOV, Ye. V. KUVSHINSKY, *FTT*, **1**, 1441 (1959); *Viysokomol. soyed.* **1**, 1561 (1959); *Viysokomol. soyed.* **2**, 397 (1960); *FTT*, **3**, 1314 (1961); *FTT*, **3**, 607 (1961).
17. G. A. LEBEDEV, Ye. V. KUVSHINSKY, *FTT*, **3**, 2672 (1961).
18. J. P. BERRY, *Nature*, **185**, 91 (1960); *J. Appl. Phys.* **33**, 1741 (1962).
19. R. P. KAMBOUR, *Nature*, **195**, 1299 (1962).
20. V. A. KARGIN, A. I. KITAIGORODSKII, G. L. SLONIMSKII, *Kolloid. Zh.* **19**, 131 (1957).
21. F. BUECHE, *stat'i v sb.* "*Fizika polimerov*", edited by M. V. Vol'kenshteina, Izdatinlit, 1960, 15.
22. A. I. GUBANOV, A. D. CHAVIYCHELOV, *FTT*, **4**, 928 (1962).
23. V. R. REGEL', T. M. MUNNOV, O. F. POZDNIAKOV, *FTT*, **4**, 2468 (1962).
24. H. H. W. PREUSS, *Plaste und Kautschuk*, **10**, 330 (1963).
25. W. KUHN, H. KUHN, *Helv. chim. acta*, **29**, 1095 (1946).
26. L. TRELOAR, *Fizikz uprugosti kauchuka*, Izdatinlit, 1953, str. 180.
27. G. M. BARTENEV, *DAN SSSR*, **84**, 487 (1952).
28. G. M. BARTENEV, G. I. BELOSTOTSKAYA, *ZhTF*, **24**, 1773 (1954).
29. G. A. PATRIKEEV, *DAN SSSR*, **146**, 402 (1962).
30. G. M. BARTENEV, L. S. BRYNKHANOVA, *ZhTF*, **28**, 288 (1958).
31. G. M. BARTENEV, F. A. GALIL-OGLY, *DAN SSSR*, **100**, 477 (1955).
32. G. M. BARTENEV, V. Ye. GUL', *Plast. massiy*, No. 1, 54 (1961).
33. R. KAISER, *Kautschuk und Gummi*, **8**, 133 (1955).
34. K. KRUSE, *Kautschuk und Gummi*, **6**, 202 (1953).
35. V. Ye. GUL', *Uspekhi khimii i tekhnologii polimerov*, Part 2, Goskhimizdat, 1957, p. 202.
36. V. Ye. GUL', Izvestiya viysshikh uchebniykh zavedenii, *Khimiya i khimicheskaya tekhnologiya*, **2**, 109 (1959).
37. V. Ye. GUL', G. P. KRUTETSKAYA, V. V. KOVRIGA, *Kauchuk i rezina*, No. 12, 1 (1957).
38. V. Ye. GUL', L. N. TSARSKII, S. A. VIL'NITS, *Kolloid. zh.* **20**, 318 (1958).
39. V. Ye. GUL', I. M. CHERNIN, *DAN SSSR*, **123**, 713 (1958).
40. V. Ye. GUL', I. M. CHERNIN, *Viysokomol. soyed.* **2**, 1613 (1960).
41. B. V. DERIAGIN, N. A. KROTOVA, *Adgeziya, ch.* **7**, *Izd.* AN SSSR, 1949.
42. V. A. KARGIN, G. S. MARKOVA, *ZhKhO im. D. I. Mendeleyeva*, **6**, 362 (1961).
43. V. G. ZHURAVLEVA, Z. Ya. BERESTNEVA, V. A. KARGIN, *DAN SSSR*, **146**, 366 (1962).

44. A. Tobol'skii, *Svoistva i struktura polimerov*, "Khimiya", 1964.
45. B. A Dogadkin, R. Uzina, *Kolloid. zh.* **9**, 97 (1947).
46. L. Treloar, *Rubb. Chem. Technol.* **17**, 813, (1944).
47. Z. N. Zhurkov, B. Ya. Levin, *v sb.*"*Khimiya i fiziko-khimiya viysokomolekuliyarniykh soyedinenii*", Izd. AN SSSR, 1952, p. 280.
48. G. M. Bartenev, *Kolloid, zh.* **12**, 408 (1950).
49. G. M. Bartenev, B. A. Dogadkin, N. M. Novikova, *ZhTF*, **18**, 1282 (1948).
50. N. I. Shishkin, M. F. Milagin, *FTT*, **4**, 2681 (1962).
51. L. A. Laius, Ye. V. Kuvshinskii, *FTT*, **5**, 3113 (1963).
52. V. A. Kargin, V. G. Zhuravleva, Z. Ya. Berestneva, *DAN SSSR*, **144**, 1089 (1962).
53. G. M. Bartenev, L. A. Vishnitskaya, *Viysokomol. soyed.* **6**, 751 (1964).
54. S. Hayashi, *Suppl. Progr. Theoret. Phys.* No, 10, 82 (1959).
55. G. M. Bartenev, *DAN SSSR*, **133**, 88 (1960).
56. B. A. Dogadkin, Z. N. Tarasova, *Kolloid. zh.* **15**, 347 (1953).
57. V. A. Kargin, T. I. Sogolova, G. L. Slonimskii, Ye. V. Reztsova, *ZhFKh*, **30**, 1903 (1956).
58. V. A. Kargin, T. I. Sogolova, *DAN SSSR*, **108**, 662 (1956); *ZhFKh*, **31**, 1328 (1957).
59. G. M. Bartenev, N. V. Zakharenko, *Kauchuk i rezina*, No. 1, 24 (1961).
60. G. M. Bartenev, L. A. Vishnitskaya, *Viysokomol. soyed.* **5**, 1837 (1963).

CHAPTER 4

INFLUENCE OF MOLECULAR WEIGHT, STRUCTURE AND MOLECULAR ORIENTATION ON THE STRENGTH OF POLYMERS

THE structure of polymeric materials has a great influence on their strength. For the spatial polymers (for instance elastomers) the chief structural factor is the degree of cross-linking, and also the structure which is formed by active fillers. The orientation of the chains which is permanent due to the restraining properties of the relaxation processes is one of the main structural factors which sharply raise the strength of solid polymers. The influence of the molecular orientation on the strength is specific only for polymeric materials. The processes of manufacturing synthetic fibres, films and orientated organic glass are based on this property.

In recent times great importance has been attached to the influence of the supramolecular structures on the mechanical properties of polymers. The polymers which after their synthesis possess certain structural properties can acquire a different complex of properties with the rearrangement of their supramolecular structures. The strength of orientated polymers depends not only on the completeness of their molecular orientation but also on the character of the supramolecular structure. The great diversity of supramolecular structures permits different properties within the limits of each physical state of the polymer: a crystalline, a glassy and a visco-elastic one.

In view of the general interest in these questions other polymeric materials are investigated in this chapter side by side with the visco-elastic ones.

4.1. Strength and Molecular Weight of Polymers

With an increase of the molecular weight polymers acquire above T_g not only good visco-elastic properties but they also become strong materials. For the understanding of the following it is important to have in mind that the temperature of vitrification does not depend on the molecular weight of the polymer, unless one counts the very lowest polymerhomologs with a

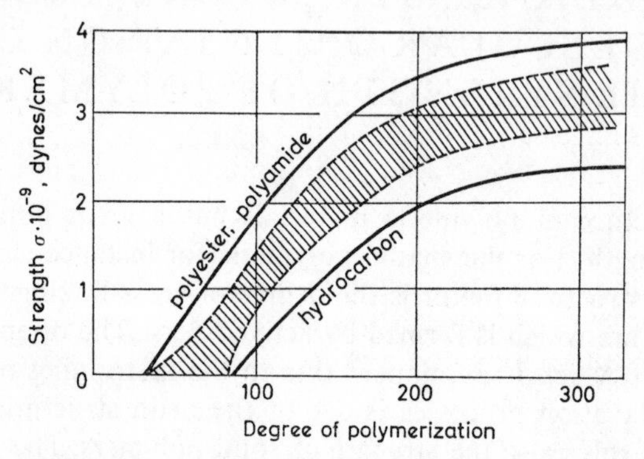

FIG. 76. Influence of degree of polymerization on the strength of the polymers (according to Mark[3]).

molecular weight below that of a segment of the macromolecule.[1, 2] One can reckon therefore that solid amorphous high polymers of practically all molecular weights are in a glassy state in the same temperature range. The temperature of flow and the width of the temperature range of high (visco) elasticity depends in linear polymers on the molecular weight; in spatial polymer, obtained as a result of the "cross-linking" of linear polymer chains, it does not in practice depend on the molecular weight.

If the molecular weight of the linear polymer is sufficiently high (in practice, above 50,000) its influence on the strength of a solid or visco-elastic high polymer is unimportant. The tensile strength of solid polymers depends on the molecular weight approximately as it is illustrated in the well-known diagram by Mark (Fig. 76).

According to the ideas of Mark and Meyer (see, for instance, the book by Alfrei[3]), the polymer with a low molecular weight breaks as a result of a slipping of chains and a surmounting of intermolecular forces. Above a certain degree of polymerization, which is the lower the stronger the interaction of the polymer chains, the material becomes strong: the polymer fails as a result of a break of the chemical bonds of the macromolecules. With a further increase of the molecular weight the strength does practically not depend on the length of the chain any more. According to Sookne and Harris[4] the strength of cellulose acetate grows with increasing molecular weight approximately to a degree of polymerization of 200 and then grows very slowly. Kargin and Mikhailov[5] showed that polymer fibres of great molecular weight do not, owing to their high viscosity, fail through a slipping of chains but through a straight break of chemical bonds.

The strength of a high polymer depends little on the molecular weight distribution but is basically determined by the mean of the molecular weight. In other words, the weight range has only little effect on the mechanical properties of polymers with a great molecular weight. For polymers of low and medium molecular weight the weight range has a considerable influence on the strength.[6]

In solid polymeric glasses the reduction of strength with a decrease of the molecular weight is connected not only with an increase of the part played during the breaking process by the intermolecular forces but also with a growth of brittleness on account of the loss of ductility of the short polymer chains. Low molecular amorphous linear polymers with a molecular weight of 10,000–20,000 crumble easily and give cracks, and their brittle temperature moves upwards. Solid polymers obtained by cross-linking linear ones, possess a greater strength and lower brittleness.

The strength of crystalline polymers is considerable already at low molecular weights (10,000–20,000); this is explained by their structure.

Linear polymers with decreasing molecular weight turn at certain temperatures above T_g gradually from the visco-elastic into the visco-flowing state in which the failure of the material proceeds through a slipping of adjoining chains up to their separation.

The influence of the molecular weight on the strength of orientated solid polymers has its specific features,[7] and is separately investigated in § 4.6.

The strength of spatial polymers does almost not change at temperatures below T_g but above the glass transition point it differs from the strength of the original linear polymer, and is the greater the lower the molecular weight of the latter. Decreasing molecular weight of the original polymer leads unavoidably to an even greater number of cross-links of the spatial network.

Experiments with butyl[8] and SKS-30 rubber[9] showed that the strength of vulcanized rubber decreases from some point with decreasing molecular weight of the original material. According to Flory, the fraction of the rubber with a molecular weight too small to form a continuous spatial network during vulcanization has a strength, when vulcanized, which is practically zero. If, however, the length of the original rubber molecule is considerably greater than the chain length in the network of the vulcanized rubber, the strength does not in practice depend on its molecular weight.

Flory assumes that the strength is determined only by the "active" chains, i.e. the parts of the chain which enter the spatial network. Only these parts of the chains participate in the orientation processes and crystallization under tension, whilst the terminal units of the chains and the chains which are not included in the network do not carry any load. According to Flory's theory[8] the proportion of the "active" chains is

$$N_a = \frac{M - M_s}{M + M_s} = 1 - \frac{2M_s}{M + M_s} \qquad (4.1)$$

where M is the molecular weight of the rubber fraction; M_s is the molecular weight of the chain in the spatial network of vulcanized rubber (a part of chain between adjoining cross-links).

If the molecular weight is lowered to $M = M_s$, the spatial network is not formed and the strength is low. The test shows that at $M_s = \text{const.}$ there exists a linear dependence between strength and the value $1/(M + M_s)$. It appeared, however, that for butyl rubber the strength is near zero not at $M = M_s$ but when the fraction of the "active" chains constitutes 40% of the mass of the whole polymer. Flory connects this discrepancy with the influence of the degree of crystallinity, assuming that the crystallization under tension of the vulcanized butyl rubber begins only at a certain minimal quantity of material which is capable of orientation.

The influence of the molecular weight on the strength of rubbers based on different elastomers was also studied by Johnson.[10]

In the study of the influence of the molecular weight on the true strength of non-filled vulcanized rubbers, fractions of non-crystallizing butadiene-styrene rubber SKS-30A (Fig. 77) with the same equilibrium modulus were compared in the above mentioned work[9] but with a different quantity of added sulphur, the more, the lower the molecular weight of the fraction† plus an equal

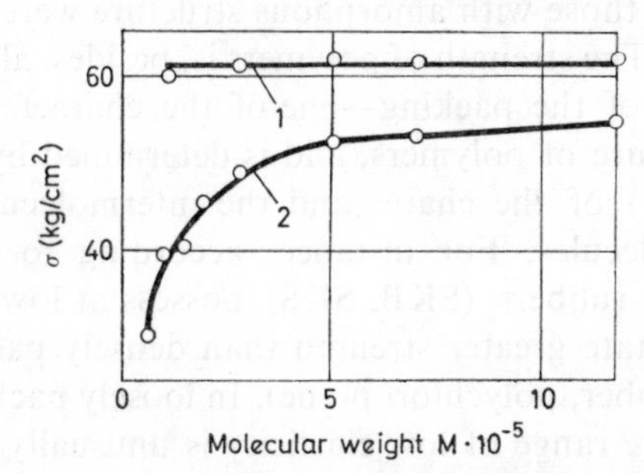

FIG. 77. Influence of the molecular weight of fractions of SKS-30-A rubber on the true tensile strength of non-filled vulcanized rubbers: (1) strength of vulcanized rubbers with an identical equilibrium modulus (13 kg/cm²); (2) strength of vulcanized rubbers with an identical quantity of combined sulphur. (2% on rubber.)

quantity of added sulphur (2% on rubber) for all fractions. The change of strength with the change of molecular weight of the fractions of vulcanized SKS-30A rubber with an identical quantity of added sulphur takes place, like for vulcanized butyl rubber, up to a molecular weight of 500,000 (see Fig. 77, curve 2); a further increase of the molecular weight has no noticeable influence on the strength. The molecular weight of the fraction does not influence the strength of vulcanized rubbers with the same equilibrium modulus (see Fig. 77, curve 1). The fact that the equilibrium modulus for vulcanized rubbers of all fractions has the same value indicates that the number of "active" chains of the spatial network is always the same. The equal quantity of combined sulphur, on the contrary, does not imply an equal number of

† The equilibrium modulus was 13 kg/cm² for all fractions from 56,000 to 1,170,000 which corresponds to 1·8% of fixed sulphur for high-molecular fractions, and to approximately 4% of fixed sulphur for low-molecular ones.

"active" chains in vulcanized rubbers from fractions with a different original molecular weight. Fractions with a low molecular weight have a sparser spatial network, and therefore also a lower strength.

4.2. Strength and Structure of Polymers

The chief differences between the strength of polymers with crystalline and those with amorphous structure were examined in § 2.1 and § 2.2. The strength of polymers is, besides, also influenced by the density of the packing—one of the characteristics of the primary structure of polymers, and is determined by the flexibility (or rigidity) of the chains and the intermolecular forces of the chain molecules. For instance, according to Lazurkin,[11] loosely packed rubbers (SKB, SKS) possess at low temperature in the glassy state greater strength than densely packed rubbers (NK, butyl rubber, polychloroprene). In loosely packed polymers the temperature range of low elasticity is unusually great (about 100°C), whereas in densely packed polymers brittle failure can be observed only at temperatures 20–25°C below the glass transition point. Dipole and hydrogen bonds increase the brittle strength of the polymer and therefore lower the temperature of brittleness. This can be seen especially clearly on samples of polyvinylchloride, SKN rubber, polyvinyl alcohol. Gul′, Sidneva and Dogadkin[12] showed that the strength of SKN rubber also grows above the glass transition temperature as the number of dipole bonds increases.

To a lesser degree explained is the influence of the chemical structure of the polymer molecules on the strength of the polymers. The influence of the type of chemical bonds in the polymer chains on the strength and durability of solid polymers is obvious. The basic difficulty in the examination of this question is the fact that the chemical structure of the chains is not the only feature which influences the strength of polymers. So do, for instance, the mechanical properties of the same polymer differ greatly in their dependence on the character of the supramolecular structure. This shows up especially clearly in crystalline polymers.

In recent years (see reviews 13–15 and references to Chapter 2) many forms of supramolecular structure have been discovered in polymers: blocks of molecules; blocks folded in ribbons and laminae; spherulites and monocrystals; ribbon-shaped and laminar formations from spherulites and others. All these forms can be found in crystalline polymers, but in amorphous polymers only primary forms (blocks and ribbons composed of blocks). The most characteristic supramolecular structures in amorphous polymers are the blocks, and in the crystalline the spherulites which have the dimensions of colloid particles.

In agreement with Sogolova's data the presence of chemically non-active fillers and of some plasticizers, as well as great deformations, do not change the crystalline structure of polymers—which is shown by X-ray diffraction—but considerably change the supramolecular structure and influence the mechanical properties. One of the important reasons of the change of the mechanical properties is the break-up of some of the supramolecular structures and the formation of others as a result of great deformations. The properties of crystalline polymers can be modified within wide limits by changing their supramolecular structure in various ways whilst preserving their chemical composition. The supramolecular structure is therefore extremely sensitive to thermal and mechanical treatment of the polymer.

Although at present the important part played by the supramolecular structure in the obtaining of the different mechanical properties of crystalline polymers is clear, the establishment of quantitative correlations between the characteristics of the supramolecular structure and the strength of the polymer is as yet a thing of the future.

The supramolecular structures have the greatest influence on the strength of solid polymers. In visco-elastic materials chemical cross-links are of the greatest importance and sometimes have an even greater influence on the strength than the type of supramolecular structure of the polymer.

The influence of the process of cross-linking on the strength of elastomeric spatial polymers was studied in a number of papers.[8, 16-20] Dogadkin[16] and Gee[20] showed that in blends with a large content of vulcanizing agent the strength of the crystallizing rubber changes with the increase of the degrees of vulcanization, and follows a curve with a maximum. Increasing cross-linking impedes in the beginning the plastic flow and leads to a rise of strength right up to a maximum. The decrease of strength after the maximum (apart from the influence of thermal and oxidizing processes) is connected with the circumstance that the quantity of material in the crystalline phase depends, in the moment of break, on the density of the spatial network. A sufficiently great number of cross linkages prevents the crystallization under tension and, consequently, the formation of highly orientated phases.[19] In blends with a small content of vulcanizing agent the chemical processes of degradation which decrease the

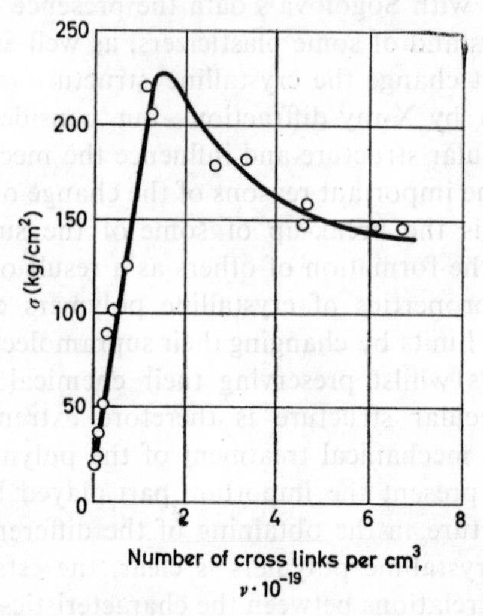

FIG. 78. Change of the true strength of butadiene-styrene rubber (Buna-S) with the increase of the number of cross-linkages.

density of the spatial network of the vulcanized rubber, can also lead to a lowering of strength beyond the maximum. This phenomenon was observed by Dogadkin during vulcanization of natural rubber with a small proportion of sulphur.

The curve of strength with a maximum can also be observed in non-crystallizing (amorphous) rubbers. In that case the strength falls after the maximum (Fig. 78) with increasing numbers of cross-links during chemical[16, 17] and radiation vulcanization. Such a principle of change of strength can be explained in the following way. Two factors influence the strength of the spatial amorphous polymer. Firstly, due to the irregular distribution of cross-links there are in the elastomer separate microsections without continuous network; they are those "weak places" where failure starts. The increase of strength to the maximum (see Fig. 78) is explained by the fact that with an increasing number of cross-linkages the number and size of the individual weakened parts in the material decreases. Secondly, the orientation of the chains, which decreases with increasing density of the spatial network influences and eventually lowers the strength. With a small number of cross-linkages the first factor has the greater influence; with a large number the second.

The molecular orientation changes the structure of the polymer during elongation or visco-elastic deformation. In view of the exceptionally great influence of the orientation on the strength of polymers this question is specially examined in the following sections.

4.3. Strength and Molecular Orientation of Polymers

The great dependence of strength on the molecular orientation is the fundamental feature which distinguishes the mechanical properties of polymers from the laws governing the strength of other solids. This dependence is particularly apparent in fibres and films. Stretching causes in synthetic and modified natural materials either a uniaxial orientation (in fibres) or a biaxial one (in film and sheet materials).

Natural fibres have an orientated structure before their processing. In articles made from rubber and plastics in which the material is practically isotropic, the orientation ,and usually a negligible one, appears only during deformation. In these articles small deformations can usually be observed or such kinds of stressed state (for instance, compression) in which no noticeable strengthening of the material occurs. Other methods are therefore employed for the strengthening of rubbers and plastics, for instance the introduction of various fillers.

The influence of orientation on strength† under uniaxial tension of amorphous and isotropic polymers in different physical states was examined in Chapter 2 (see Fig. 42).

In the visco-elastic state the strength of polymers grows with lowering temperature and reaches its maximum value near the

† Here, and further in all cases where it is not specially mentioned, we understand under strength the stress per unit of the area of the *broken* cross-section of the specimen, i.e. the so-called *true strength*.

glass transition point. Approximately here is reached the maximum elongation and molecular orientation. With a further lowering of temperature the strength decreases until that is reached at the brittle temperature which is characteristic for the polymer in the brittle state. Such a character of temperature dependence of strength is explained by the fact that under tension two processes take place simultaneously: the molecular orientation and the growth of cracks or small tears. The first process leads to a strengthening, and the second to a weakening of the material. With a lowering of temperature the speed of growth of cracks and small tears decreases and the process of failure decelerates. This leads to an increasing deformation of the break and consequently to an increase of the molecular orientation which in its turn decelerates the growth of cracks even more. All this leads to a sharp temperature dependence of strength of a material in the visco-elastic state.

With a lowering of temperature the development of visco-elastic deformation in the glassy state becomes difficult, and the growth of cracks takes place in a less and less orientated material. Therefore the speed of crack propagation grows in spite of the lowering temperature, and the strength decreases.

In the brittle state the speed of crack propagation and the strength of the polymer depends only on the temperature [according to eqn. (2.6), p. 81]. The temperature of brittleness T_{br} (see Fig. 42, Chapter 2) is the arbitrary boundary which divides two strength states of the solid polymer. Thus the process of brittle failure appears as the cracking of the specimen and its becoming white, the appearance of silver cracks, a.s.o., at temperatures a little above T_{br}. On the other hand, at temperatures a little below T_{br} a local low-elastic deformation can be observed in overstressed places of the sample (heterogeneities, defects, cracks) which leads to an added orientation of the material. Altogether the character of failure depends on the relation of the speeds of the processes of low-elastic deformation and failure.

In the glassy (or crystalline) state an originally orientated polymer preserves its orientation for an infinite time. The brittle strength and the limit of low elasticity of an originally orientated polymer depends on the degree of this orientation. As below the temperature of brittleness the original orientation does not change during the experiment, the influence of the degree of orientation on the strength of the polymer is shown best through the value of the brittle strength.

The influence of orientation on the strength of crystalline and amorphous solid polymers is studied in detail in the works of Lazurkin[11, 21] and in the works of Kargin and Sogolova which were examined in Chapter 2. According to the data of those authors it is difficult to establish different degrees of orientation in crystalline polymers under stress, due to the intermittent character of the process of recrystallization:† in the pivot of the rearrangement a crystalline phase forms with an almost complete orientation of chains in the direction of the tension. The strength of the orientated polymers depends on the angle between the tensile force and the direction of that previous elongation. For instance, orientated samples of crystalline polyamide possess at low temperatures (−80°C) a sharply anisotropic strength. Whilst strong in the direction of the orientation, they are brittle and weak under the influence of forces perpendicular to the direction of the orientation. At the temperature of −80°C all specimens, except those orientated in the direction of tension, disintegrate during failure into very small particles.

Similar to this, the molecular pre-orientation of amorphous solid polymers leads to a considerable strengthening of the material in the direction of orientation and to a weakening in the transversal direction (Fig. 79). The strength in the direction of orientation can in this way be increased manyfold.

The influence on the brittle strength and the yield stress in amorphous solid polymers is explained by the following factors: (1) by the magnitude of the pretensioning (degree of elongation) and (2) by the angle between the direction of the tensile force during testing and the direction of orientation (Fig. 80). It can be seen from the data on Figs. 79 and 80 that the brittle strength depends more on the degree of orientation and on the angle between the direction of orientation and direction of tensioning than does the yield stress. The modulus of elasticity depends on the orientation even less than the yield stress.

It has long been noticed[22] that the polymer changes with an increase of the extent of elongation from a brittle to a low-elastic state. Consequently, orientation influences the condition

† In amorphous polymers one can establish any degree of orientation by stretching them above the glass transition point, and subsequent cooling.

of a solid polymer like the rising of temperature. With an increase of the extent of elongation the brittle strength of the polymer grows faster than the yield stress (Fig. 81). At some value of the critical elongation the strength σ_{cr} becomes equal to and then exceeds the yield stress σ_B.

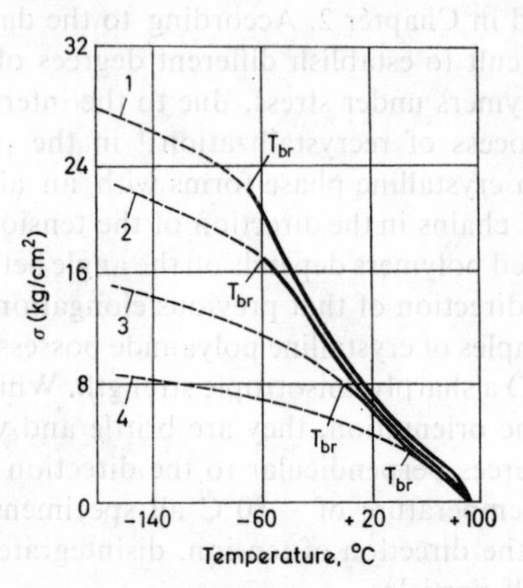

FIG. 79. Dependence of brittle strength (chain lines) and yield stress under tension (solid lines) on the direction of orientation for poly(methylmethacrylate) (according to Lazurkin). 1 and 2. Specimens orientated parallel to the tension (elongation 160 and 100%). 3. An isotropic sample. 4. sample orientated perpendicular to the axis of tension (elongation 160%).

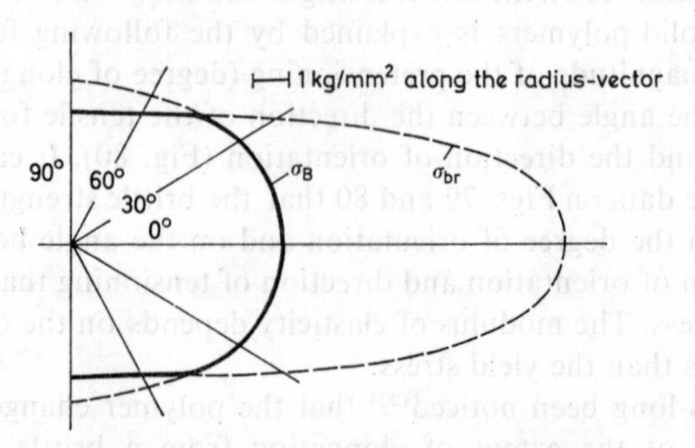

FIG. 80. Dependence of brittle strength σ_{br} and yield stress at a tension σ_B of poly(methylmethacrylate) prestretched by 200% on the angle between the direction of orientation and the direction of tension at 20°C (according to Lazurkin).

The transition from the brittle failure to the low-elastic deformation of the orientated polymer can also be observed during the change of the angle between the direction of tensioning and the direction of orientation (see Fig. 80). With the increase of the degree of orientation the strength grows markedly in the direction of orientation, and in the transversal direction to the orientation it decreases markedly. As a result a sharp decrease of T_{br} can be observed at longitudinal orientation, and at the transversal one an increase of T_{br} (see Fig. 79).

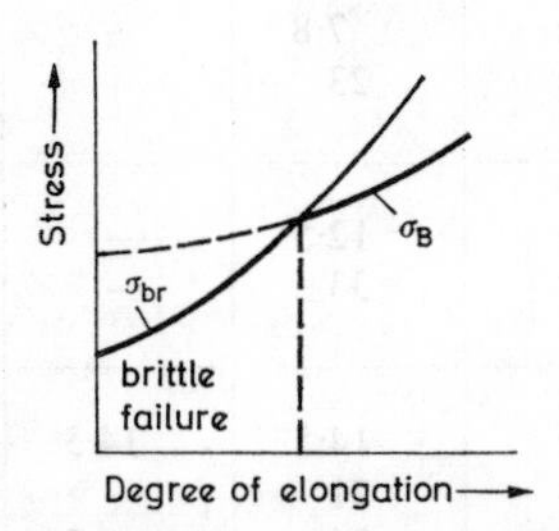

FIG. 81. Diagram of transition from brittle failure to low-elastic deformation with an increase of the degree of elongation during tests-to-break parallel to the stress.

In Table 5 Lazurkin's data on the brittle strength and the temperature of brittleness of unfilled rubbers and plastics are shown. For the rubbers the strength was determined at a temperature of $-253°C$, for poly(methylmethacrylate) at $-140°C$. The experiments were carried out at a strain rate of $6{\cdot}4 \times 10^{-4}$ sec^{-1}.

The influence of the angle between the direction of tension and the direction of the orientation of the chains on the strength is considerable also for fibres,[(23)] as in fibres the orientation of chains does not always coincide with the axis of the fibre. The strength under tension changes rather strongly in dependence on the angle between them (Fig. 82). The strength of crystalline and amorphous fibres is much lower in the transversal direction. At the first glance it seems that this is immaterial in practice as the fibres are normally used under tension and flexure. However, the strength of the fibre in the transversal direction plays an important part in the phenomena of fatigue under repeated deformations.

TABLE 5

Strength and Temperature of Brittleness of some Polymers in Dependen ce on the Direction of Orientation

Material	Elongation (%)	Brittle strength σ_{cr} (kg/mm^2)		Temperature of brittleness T_{br}, (°C)	
		In direction of orientation	Perpendicular to orientation	In direction of orientation	Perpendicular to orientation
NK rubbers	0	7·8	–	– 80	–
	300	23	–	–200	–
SKN–40 rubbers	0	12·5	–	–110	–
	250	33	–	–200	–
Poly (methylmethacrylate)	0	14·5	14·5	15	15
	100	21	–	– 50	–
	160	26	8·5	– 75	60
	200	28·5	–	– 90	–

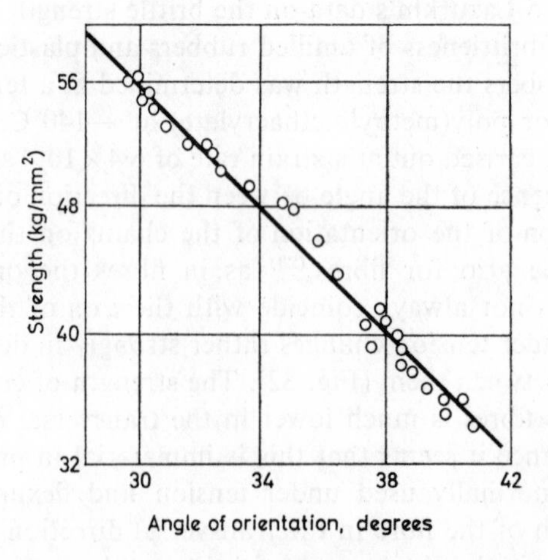

FIG. 82. Influence of the angle between the direction of tension and the direction of orientation on the strength of cotton fibres.[23]

The strength of cellulose fibres, with their obvious amorphous structure, has been examined in detail by Kargin and co-workers.[5, 24–26] In these papers it is shown that the orientation of chains during the stressing of the fibres under circumstances which allow a flow of the material is a condition of the increase of the strength of the fibre. High amounts of elongation of cellulose fibres swelled in water must unavoidably be accompanied by a straightening of the cellulose chains which in its turn is also reflected in the relaxation processes. In proportion to the elongation the chain molecules straighten out and become stiffer as the straightening is connected with a decrease of the number of possible configurations of the chain. Within this limit a completely straight chain can only exist in one configuration and will therefore be completely rigid. Consequently, as the polymer is stretched its chain will become stiffer and the viscosity will increase. In polymers with stiff chains, as for instance, celluloses, these effects are especially great right up to the complete loss of the visco-elastic properties. In this case a peculiar transition of the swelled polymer into the glassy state can be observed, due to the increase of the rigidity of the chains. After the removal of the stress the orientated fibre will hardly shrink as the viscosity, owing to the straightening of the chains, has grown enormously.

Kargin, Kozlov[27] and co-workers showed that crystalline fibre forming polymers are characterized by various supramolecular structures. The strength of crystalline fibres depends therefore not only on the molecular orientation but also on the character of the supramolecular structure.[28]

It is very important to appraise correctly the degrees of molecular orientation and the connection between the latter and the degree of elongation (or a proportional factor of tension) of the polymer. X-ray diffraction[29] and optical methods are the most effective means for the determination of the orientation of macromolecules in fibres. Polymeric fibres acquire by stretching a uniaxial optical anisotropy which is revealed by birefringence. This method of determining the degree of molecular orientation is very widely used.

As Shishkin and Milagin[30] showed, the extent of elongation of the polymer cannot correctly characterize the degree of orientation of the molecules of a solid polymer, even if there is no irreversible flow under stressing. It follows from their experiments with poly(methylmethacrylate) that the higher the temperature, the smaller the molecular orientation under the same amount of elongation. In other papers the same authors[31] established that the strength of fibres is clearly connected with the magnitude of the birefringence independent from the technology of the elongation. In itself the degree of elongation is not a uniaxial characteristic of strength, even if only a visco-elastic deformation occurred during the stretching.

4.4. The Causes of High Strength of Orientated Polymers

According to Kobeko the high strength of orientated polymers is caused by three facts: by the transition from failure of basically intermolecular bonds to failure of chemical bonds in the polymer chains in proportion to their orientation; by the slow crack propagation owing to the difference in the values of the moduli of elasticity in the longitudinal and transverse directions (according to Stepanov[32]); and by the equalization of the stress concentrations and the healing of critical cracks and defects during orientation. Kobeko considered the first of these causes the chief one. The increase of the role played by the chemical bonds in the failure of polymers in proportion to the increase of the degree of orientation was also recorded by Kargin, Mikhailov and co-workers.[5, 24–26]

If Stepanov's hypothesis (second cause) were applied to polymers then, as a result of their elongation, there would be observed simultaneously with the anisotropy of strength, an anisotropy of the modulus of elasticity. It has, however, been noticed that in practice the orientation does not influence the modulus of elasticity of solid polymers. This question was again discussed in the recent papers of Lains and Kuvshinsky,[7] Bessonov and Kuznetsov.[33] In a wide range of elongation (change of the proportional factor of tension from 1 to 10), when a sharp rise of strength was observed, the authors did not succeed in discovering a difference between the "longitudinal" and "transversal" modulus of elasticity, nor a change of modulus with increasing elongation. From this the conclusion was drawn that the character of the molecular rearrangement during orientation leads to a change of the effective number of bonds which determine the strength of the polymer in the given direction but does not change its nature.

As for the hypothesis of the transition from the surmounting of the intermolecular forces during the break of non-orientated polymers to the breaking-up of the chemical bonds during the failure of orientated polymers, it is correct for linear polymers under certain conditions like relatively high temperature, low molecular weight and others (see § 4.1).

At very high temperatures or extremely low molecular weight the orientation of the polymer chains is small. Under these conditions the polymers show a "viscous" break, in which the macromolecules, on account of the small internal friction, slip over the adjoining ones without any break of chemical bonds. With all other types of failure (brittle, visco-elastic and, to a considerable degree, ductile) of orientated and non-orientated polymers the chemical bonds fail as well. This is evident for the three-dimensional polymers and is discussed for the linear ones in a number of papers.[3, 5, 8, 24–26, 34] It is shown, for instance, in one of those papers[34] that the activation energy of the failure process of the polymer does not depend on the factors which change the intermolecular interaction (orientation, plasticizing, introduction of solvents, and others), and that it corresponds in magnitude to the energy of the failure of the chemical bonds. This is confirmed by the data in Table 6 where the values of the constants of the formula of durability [see eqn. (1.13)] for little and highly orientated fibres is shown.

The practically equal value of the constant τ_0 and the presence of a common pole on the graph of time dependence of the strength

TABLE 6

Constants of the Equation of Durability for Fibres in Dependence on Orientation

fibre	log τ_0 (sec)	U_0 (kcal/mol)	$\gamma \cdot 10^{19}$ (mm³)
Nylon			
little orientated	−12	45	3·1
highly orientated	−12	45	2·1
Hydro-cellulose			
little orientated	−13	40	2·8
highly orientated	−13	40	1·8

of non-orientated and orientated polymers (Fig. 83) confirm the equal values of the activation energy of failure, independently from the extent of orientation. Consequently, the barrier energy U_0 is determined by the chemical structure of the polymer chain, and all changes of the strength properties at different orientations are caused by the change of the structure sensitive coefficient γ.

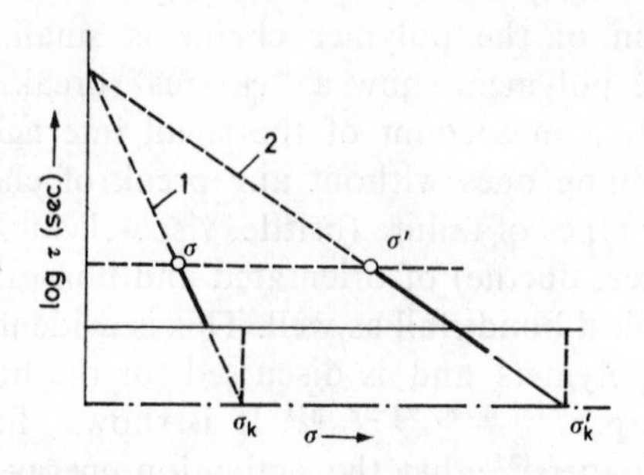

FIG. 83. Diagram of time dependence of strength of (1) non-orientated, and (a) orientated polymers.

The non-dependence of the activation energy on the orientation and the agreement of its value with the activation energy of the thermal destruction of the polymer in vacuum permits the following deduction. The main factor in the failure mechanism of solid polymers is the break of the chemical bonds independent whether the polymer is in a non-orientated or orientated state; in the elementary act of failure approximately one chemical bond breaks.

The probability of a break of the chemical bonds depends on the stress, and, in agreement with the oscillation theory of strength it is determined by the value of the activation energy of failure $U_0 - \gamma\sigma$. The greater the stress the more probable the process of failure. The decrease of the activation energy under the influence of the tensile forces is determined by the coefficient which is sensitive to the change of structure during orientation or plasticizing. The smaller this coefficient is (incline of the straight line of durability), the stronger is the polymer (see Fig. 83).

Zhurkov and Abasov[34] consider that it is precisely the intensification of the intermolecular forces during orientation which leads to the decrease of the structural coefficient γ and to a rise of strength. And on the contrary, the weakening of the intermolecular forces during plastification increases this coefficient and lowers the strength. In the opinion of those authors the participation of the intermolecular forces is shown in the change of the coefficient γ; when this decreases the probability of a break of the chemical bonds is lowered. However, as will be shown later, the change of γ is connected mainly with the change of the number of breaking chemical bonds in the cross-section of the specimen, and not with the change of the intermolecular forces. Neither can one agree with the definition of the structural coefficient in the papers[(34)] as the coefficient of overstress and of the coefficient τ_0 as period of oscillation of atoms in the polymer chain which was already indicated in Chapter 1. One has to pay attention to this circumstance as the incorrect interpretation of the physical sense of the coefficients in the equation of durability has become very widespread.

4.5. Theory of Strength of Orientated Solid Polymers

The influence of orientation on the strength of solid polymers can easily be understood from the oscillation theory of strength.

In a non-orientated material out of every three chain segments on a given cross-section of the sample, on the average one lies in the direction of tension, and two across it. Under tension the latter move only after surmounting the intermolecular forces and the former suffer a break of chemical bonds. In the fully orientated state all segments lie parallel to the direction of tension and they all break along chemical bonds.

The theoretical strength of polymers equals the number of breaking chemical bonds† per unit of cross section of the sample multiplied by the maximum value of the quasi-elastic force F_m of the chemical bond.‡ For the fully orientated state this definition of the theoretical strength is entirely exact, as the intermolec-

† In non-orientated polymers the same chain can intersect the cross-section of a specimen i times; therefore it can break on that section into $i+1$ pieces. In fully orientated polymers the chain can intersect the given section only once.

‡ The theoretical strength of solid polymers cannot be assessed quite correctly by eqn. (1.5) ($\sigma_m \approx 0{\cdot}1E$), as it depends mainly on the chemical bonds, and Young's modulus E on the intermolecular ones.

ular bonds in practice do not take part in the break. In non-orientated materials, however, not only does one-third of the chemical bonds break but also a part of the Van der Waals bonds, the contribution of which can be disregarded in a rough calculation. The theoretical strength of the fully orientated material σ'_m must exceed the theoretical strength of the non-orientated one σ_m by n times, where $n > 1$. Consequently,

$$\sigma'_m = n\sigma_m \tag{4.2}$$

If in the non-orientated and fully orientated state there are S and S' elementary planes in the cross-section of the specimen per segment which are orientated in the direction of tension, then

$$\sigma_m = F_m/S; \qquad \sigma'_m = F_m/S'$$

and therefore $n = S/S'$.

In non-orientated polymers two out of every three segments, lying in the plane of the cross-section, occupy an area of about $2r_0l$ (r_0 is the diameter of the segment, equal to the mean distance between the polymer chains, l is the length of the segment); one segment, orientated in the direction of tension, occupies an area of about r_0^2, and therefore

$$S \approx r_0^2\left(1+\frac{2l}{r_0}\right)$$

In fully orientated polymers $S' \approx r_0^2$. From this follows that $n = 1+2l/r_0$.

If $l = r_0$, then the maximum possible strengthening due to orientation is $n = 3$. Usually l exceeds r_0 several times. For instance for natural rubber $l = 12$ Å, $r_0 = 4$ Å. Consequently, its strength (below T_g) in the amorphous, non-orientated and the orientated state differ by a factor of 7. For rigid-chain, loosely packed solid polymers l, is greater and correspondingly greater is the strengthening under orientation (ten to fifteen times), but during elongation new critical defects emerge—microcracks which reduce the effect of the strengthening.

The theoretical strength is approximately equal to the critical overstrain at the root of the crack (see p. 19) In the calculations below O_h is therefore replaced by σ_m.

The durability under a tensile stress σ is, according to the oscillation theory, equal to

$$\tau = \tau \exp\left(\frac{U_0 - \omega\beta\sigma}{kT}\right).$$

As the constants τ_0 and U_0 do not depend on the molecular orientation, and as the durability has the same value $\tau = \text{const}$ for the fully orientated and non-orientated states, the products $\omega'\beta'\sigma'$ and $\omega\beta\sigma$ are equal. Consequently we have in the case of critical failure,

$$\frac{\sigma'_K}{\sigma_K} = \frac{\omega\beta}{\omega'\beta'} \tag{4.3}$$

where ω, ω' are the oscillating volumes, and β, β' are the coefficients of stress concentration in the root of the crack in non-orientated and in fully orientated material. On the other hand the critical stresses are equal to

$$\sigma_k = \frac{\sigma_m}{\beta} \quad \text{and} \quad \sigma'_k = \frac{\sigma'_m}{\beta'}$$

which gives, taking into consideration the relation (4.2):

$$\frac{\sigma'_k}{\sigma_k} = \frac{n\beta}{\beta'} \tag{4.4}$$

It follows from eqns. (4.3) and (4.4) that

$$\omega/\omega' = n.$$

One can also obtain this result by another method, based on the physical meaning of ω and ω'. In agreement with the theory $\omega = \lambda_1\lambda\lambda_0$, where λ is the path on which the microparticle with a length λ_0 in the front of the crack advances with each break of the chemical bonds. The values λ and λ_0 approximately equal the distance between successively breaking segments, taking into account that in each subsequent elementary act of break one chemical bond participates, and the number of breaking bonds depends on the number of segments which are orientated in the direction of the tension. The constant λ_1 equals the distance by which the chemical bond stretches during break. In the fully orientated state $\lambda\lambda_0$ is S/S' times smaller than in the non-orientated one. Consequently, in this case ω decreases n times which agrees with eqn. (4.4). From this the cause of the decrease of the constant $\gamma = \omega\beta$ with orientation (and its increase with introduction of plasticizer) is clear.

Thus the change of γ is connected mainly with the change of the number of broken chemical bonds per 1 cm of the crack, and only to a lesser degree with the change of the intermolecular forces.†

At the same durability the relation of the strength of fully orientated polymers σ' and non-orientated ones σ is (see Fig. 83):

$$\frac{\sigma'}{\sigma} = \frac{\sigma'_K}{\sigma_K} = \frac{n\beta}{\beta'} \tag{4.5}$$

In the orientated state the stress concentration coefficient will probably be greater than in the non-orientated one, as in the latter the stress concentrations in the apices of the cracks decrease owing to the low-elastic flow of the material. In general the coefficient β changes little with orientation, as the orientation has almost no influence on the modulus of elasticity. Considering this, one can draw the conclusion that the relation of strength of non-orientated and fully orientated polymers does not exceed the value n which depends on the flexibility of the chain molecules.

The strength of orientated crystalline polymers is approximately three times greater than that of non-orientated ones. The strength relation must be greater for amorphous polymers as the strength of isotropic amorphous polymers is lower than that of isotropic crystalline ones, and the strength of fully orientated polymers in the amorphous and crystalline states is in practice identical.[3]

As far as the activation energy of the breaking process is concerned, the calculation of the intermolecular bonds in the breaking mechanism must lead to a small difference of these values for orientated and non-orientated polymers; U_0 must be greater for

† This is confirmed by the fact that the intermolecular forces generally change little with the orientation, at least in solid amorphous polymers at a 200–300% elongation. According to Lazurkin's data the constants a and U in eqn. (2.1), and consequently, also the relaxation time do not, practically, change within these limits of elongation, whilst plasticizing leads to an increase of the effective volume of the segment a and to the decrease of the activation energy u of the "cold" flow of the polymer. A similar phenomenon[35] can be observed in elastomeric polymers in which right up to break (200–500% elongation) the change of the intermolecular forces between the chains is so small that no noticeable change whatsoever of the potential energy occurs.

the non-orientated ones owing to the additional influence of the intermolecular forces.

In fact, a careful analysis of Zhurkov's and Abasov's data[34] confirms the indicated difference for several polymers which is small and amounts to approximately 10%.

Methods of calculation of the strength of orientated solid polymers were suggested also by other investigators.[3, 36, 37] In those papers a parameter is introduced, which characterizes numerically the degree of orientation of the polymer. The orientation can be estimated, for instance, with the help of the included angle Ω of the half-space, within which 50% of all segments are orientated. If the specimen is completely disordered, then $\Omega = \pi$, at complete unidirectional orientation $\Omega = 0$. In particular the *parameter of orientation* χ is defined by the expression

$$\chi = \frac{1}{\pi}(\pi - \Omega)$$

Its value changes from 0 to 1 with the transition from the non-orientated specimen to a fully orientated one. While χ is small, the increase of strength with the orientation is also small, as the segments show a random orientation. With increasing orientation the change of χ causes a considerable increase of the strength and $d\sigma/d\chi$ grows.

In the first approximation

$$\frac{d\sigma}{d\chi} = c\chi$$

which leads to the following dependence of the strength of an originally orientated solid polymer on the orientation:

$$\sigma = \sigma(0) + \frac{c}{2}\chi^2 \tag{4.6}$$

where $\sigma(0)$ is the strength of the non-orientated polymer (i.e. $\chi = 0$) and c is a constant, called the ratio of orientation.

The dependence of the strength of viscose fibres on orientation has been calculated according to Schieber's data[38] (Fig. 84) which agrees well with eqn. (4.6).

Hsiao[36, 37] developed a number of variants of the statistical theory of the strength of orientated polymers based on the molecular model of the polymer which had been suggested by him.

According to his calculations the maximum strength (in uniaxial orientation) exceeds the strength of the non-orientated material by a factor of 6, whilst according to Alfrei[3] the strength increases only so much at simultaneous orientation and crystallization of the polymer (the strength due to orientation only increases three to four times). Hsiao calculated also the strength of the orientated material in the direction transversal to the

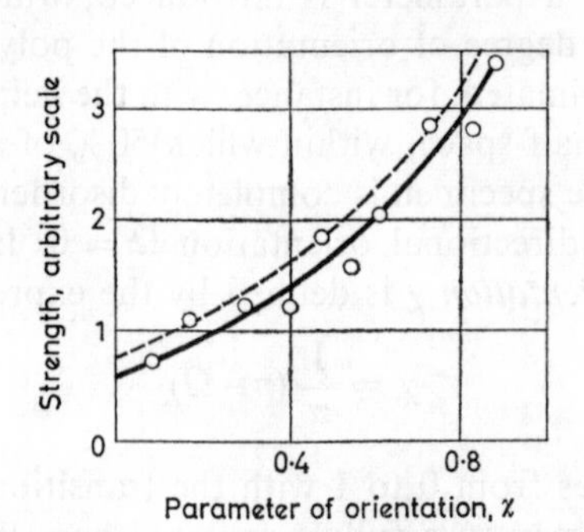

FIG. 84. Influence of orientation on the strength of viscose fibres (according to Alfrey[3]). Chain curve refers to a polymer with a somewhat higher degree of polymerization.

orientation, and showed that its decrease, found by calculation, is in agreement with the experimental data.

With uniaxial orientation the strength in the direction of the stress is, according to Hsiao's theory

$$\sigma = N_0 F_m l \varrho_0 f_0 \alpha^{3/2}(1+\alpha^{3/2}) \tag{4.7}$$

where N_0 is the number of segments in a unit of the volume of non-orientated polymer in a certain (included) angle; F_m is the quasi-elastic force which leads to the break of the polymer chains; l is the length of the chain segments; ϱ_0 is the probability of the distribution of the segments in the non-orientated state; f_0 is the fraction of non-broken segments; α is the amount of elongation of the polymer.

On Fig. 85 are shown the theoretical curves of the influence of uniaxial and biaxial† orientation on the strength under tension parallel and transversal to the direction of orientation according

† One has to keep in mind that biaxial tension corresponds to a uniaxial compression during which the strain assumes a negative value.

to Hsiao. On Figs. 86 and 87 the theoretical dependences are compared with the experimental data for polystyrene and polyethylene.

The calculation methods of Hsiao are of great practical interest and much more exact than others.

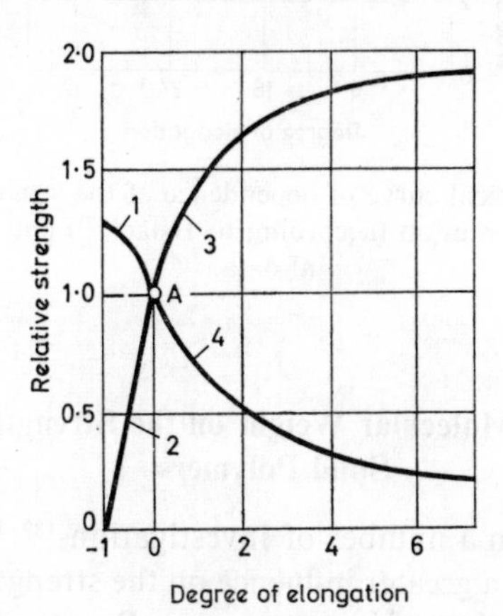

FIG. 85. Theoretical curve of dependence of relative strength of solid linear polymers on degree of orientation (according to Hsiao[37]). 1. Strength under tension of biaxially orientated sample along the plane of orientation. 2. The same, perpendicular to the plane of orientation. 3. Tensile strength of uniaxially orientated sample in the direction of orientation. 4. The same, perpendicular to the direction of orientation. Point *A* refers to the strength of non-orientated polymers, taken as basis.

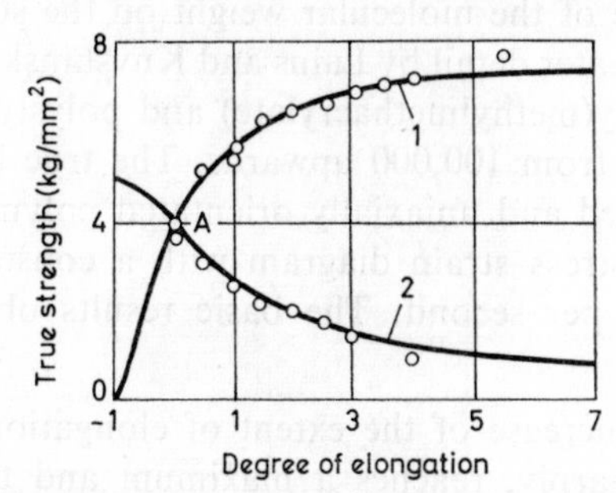

FIG. 86. Theoretical curve of dependence of the strength of polystyrene on the degree of orientation (according to Hsiao[37]) and experimental data. 1. Tensile strength in the direction of uniaxial orientation. 2. The same, transversal to the uniaxial orientation. The circles are experimental data.

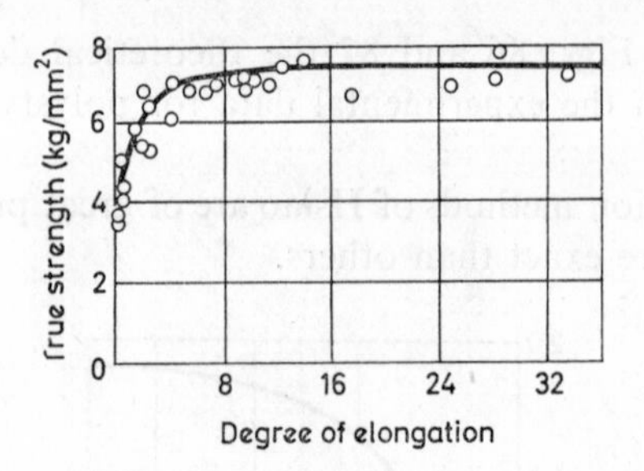

FIG. 87. Theoretical curve of dependence of the strength of polyethylene on orientation (according to Hsiao[37]) and experimental data.

4.6. Influence of Molecular Weight on the Strength of Orientated Solid Polymers

It was noticed in a number of investigations[39–41] that the molecular weight has a greater influence on the strength of orientated polymers than on that of isotropic ones. Barg and co-workers[39] found that the strength of highly orientated poly(vinylacetate) grows with the change of the molecular weight from 20,000 to 200,000, whereas the strength of the isotropic polymer is constant in the same range and is approximately 10 kg/mm². The greatest strength of orientated sample of polymer was measured with 120 kg/mm².

The influence of the molecular weight on the strength was investigated in greater detail by Lains and Kuvshinsky[7] on poly(vinylacetate), poly(methylmethacrylate) and polystyrene with molecular weights from 100,000 upwards. The true breaking stress of non-orientated and uniaxially orientated polymers was determined on the stress–strain diagram with a constant strain rate of 0·14–0·55% per second. The basic results obtained in that paper are:

1. With an increase of the extent of elongation the breaking stress grows sharply, reaches a maximum and then decreases, whilst the value of the maximum grows with increasing molecular weight (Fig. 88).

2. The influence of the molecular weight on the strength of non-orientated and orientated high polymer is different: the

strength of the non-orientated one does practically not depend on the molecular weight, and the strength of the orientated one grows (Fig. 89).

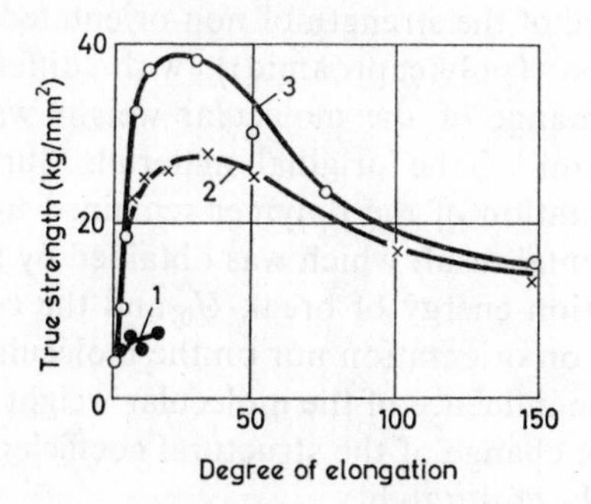

FIG. 88. Dependence of strength of poly(vinylacetate) of different molecular weights at room temperature on the degree of pre-stretching. 1. $M = 1{\cdot}5 \times 10^5$; 2. $M = 2{\cdot}5 \times 10^5$; 3. $M = 4{\cdot}5 \times 10^5$.

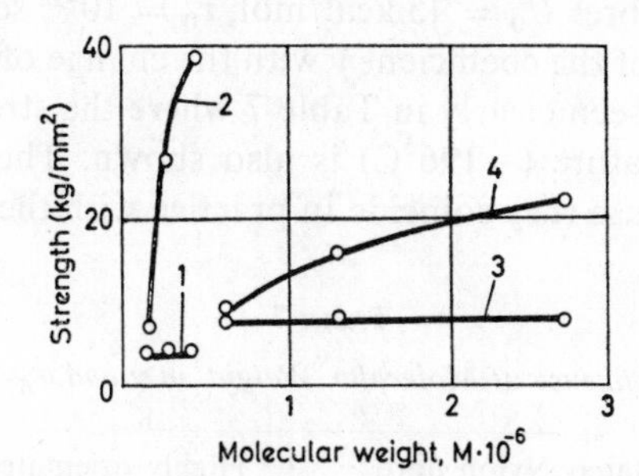

FIG. 89. Dependence of strength of poly(vinylacetate) (1, 2) and poly(methylmethacrylate) (3, 4) on the molecular weight.[7] 1 and 3. Isotropic. 2 and 4. Orientated specimens.

The authors assume that it is as yet not possible to give an explanation both of the effect of the strengthening of the polymers under stress, and of the influence of its molecular weight on it.

In Lains' and Kuvshinsky's paper the reason for the decrease of strength after reaching the maximum (see Fig. 88) remains unclear. One can, however, assume that under very great elongations a process of mechanical and thermal degradation takes place in the polymer which causes the formation of microcracks; these lower the strength so much that the increase due to the stretching cannot any more make up for the weakening.

The influence of the molecular weight on the behaviour of orientated polymers (see Fig. 89) is investigated below, starting from Zhurkov's and Abasov's data.[34] Those authors report data on the time dependence of the strength of non-orientated and orientated fibres of Nylon (poly(caproamid)) with different molecular weights. The change of the molecular weight was obtained by photodegradation of the original material, during which the degree of orientation of the polymer remained unaltered.

The fundamental result which was obtained by the authors was that the activation energy of break U_0 and the constant τ_0 does neither depend on orientation nor on the molecular weight of the polymer. All the influence of the molecular weight on the strength is caused by the change of the structural coefficient γ, introduced into the formula of durability

$$\tau = \tau_0 e^{\frac{U_0 - \gamma\sigma}{kT}} \tag{4.8}$$

For Nylon fibres $U_0 = 45$ kcal/mol, $\tau_0 = 10^{-12}$ sec.

The change of the coefficient γ with the change of the molecular weight can be seen clearly in Table 7 where the strength of fibres at low temperature (−196°C) is also shown. These values are called σ_K because they coincide in practice with the critical stress.

TABLE 7

Influence of Molecular Weight on γ and σ_K

Non-orientated Nylon fibre			Highly orientated Nylon fibre		
M	$\gamma \cdot 10^{19}$ (mm³)	σ_K (kg/mm²)	M_W	$\gamma \cdot 10^{19}$ (mm³)	σ_K (kg/mm²)
16950	12·9	150	14700	2·1	152
8800	24·6	78	5700	3·3	99
4700	29·0	42	4300	4·5	70
3700	32·7	33	3400	8·7	37

Table 7 shows that the value γ decreases with the increase of orientation and with the increase of the molecular weight of the polymer. Correspondingly, the time dependence of strength will change with an increase of the molecular weight as well as with an increase of orientation (see Fig. 83).

The structural coefficient γ is therefore the most important indicator of the strength of the polymer. The smaller it is, the greater will be, with other conditions equal, the strength and durability of the material.

The data shown in Table 7 lie on a straight line and agree with the equation

$$\frac{1}{\gamma} = \frac{1}{\gamma_\infty} - \frac{p}{\gamma_\infty} \cdot \frac{1}{P} \tag{4.9}$$

where γ_∞ is the structural coefficient for the same polymer with an infinitely large molecular weight; P is the degree of polymerization (a value proportional to the molecular weight of the polymer M); p is a constant.

At a constant temperature and a given time rate of the experiment (τ = const.) it follows from the time dependence of the strength that the breaking-stress σ is inversely proportional to γ. The strength of the polymer which depends on the molecular weight is therefore expressed by

$$\sigma = \sigma_\infty - \frac{c}{\gamma_\infty} \cdot \frac{1}{M} \tag{4.10}$$

This formula agrees with the experimental data (Fig. 90).

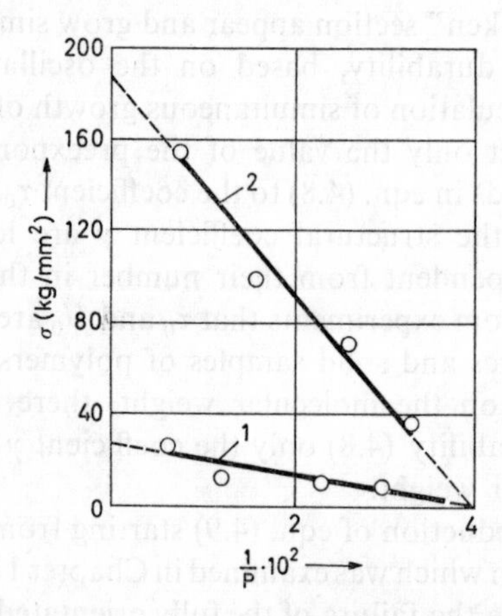

FIG. 90. Dependence of the strength of Nylon fibres at the temperature of liquid nitrogen on the degree of polymerization.[34] 1. Little orientated fibre. 2. Highly orientated fibre.

Zhurkov and Abasov[34] developed eqn. (4.9) for the fully orientated material and suggested that the failure of the polymer took place according to the mechanism of Kuhn and Bueche, and that the parting of the specimen takes place simultaneously all along a section of the specimen in some certain critical spot. The thickness of the spot is chosen so small that the ends of the broken macromolecules inside it can pull out of it spontaneously.

It is difficult to agree with these suggestions of the authors, as, firstly, the failure mechanism of the polymer is a different one (see Chapter 3) and secondly it is completely unclear in which way the critical spot of the sample separates. Besides, in solid orientated polymers the ends of the broken macromolecules will not pull out of the spot for the simple reason that for this it would be absolutely necessary that the whole macromolecule would slip as one whole, and that is impossible. In solid polymers the broken ends stay near each other after the elastic contraction. Consequently, the conclusions of eqn. (4.9) are physically unfounded.

In a number of experiments[34] the strength of fibres of various molecular weight was measured on samples of eighty parallel mono-filaments. The failure of such fibres is more complicated than the brittle failure of homogeneous materials with the development in the first stage of generally one (primary) crack. It is possible that in the fibres a considerable number of microcracks in the same "broken" section appear and grow simultaneously. In the formula of durability, based on the oscillation theory of strength, the calculation of simultaneous growth of several microcracks will affect only the value of the preexponential member which corresponds in eqn. (4.8) to the coefficient τ_0. The activation energy U_0 and the structural coefficient γ are identical for all cracks and independent from their number in the specimen. In fact, it follows from experiments that τ_0 and U_0 are approximately identical for fibres and solid samples of polymers. As τ_0 and U_0 do not depend on the molecular weight, there remains in the equation of durability (4.8) only the coefficient γ which depends on the molecular weight.

The correct deduction of eqn. (4.9) starting from the oscillation theory of strength which was examined in Chapter 1 is shown below.

Let us examine the failure of the fully orientated material in the form of strips with the given molecular weight M and assume that its spread is such a narrow one that the scatter can be disregarded.

In the breaking process the most critical crack cuts the specimen in two (Fig. 91).

Let us call H the length of the macromolecule, parallel to the direction of tension and h the thickness of the layer of polymer comparable with the dimensions of the micro-area of the overstress in the root of the crack. The ends of some macromolecules lie in the layer h and do not take part in the failure of the specimen. The probability of the end of the macromolecule lying in the spot h is h/H. The number of chains which pass through the spot is $S_0N(1-h/H)$, where N is the number of macromolecules per unit

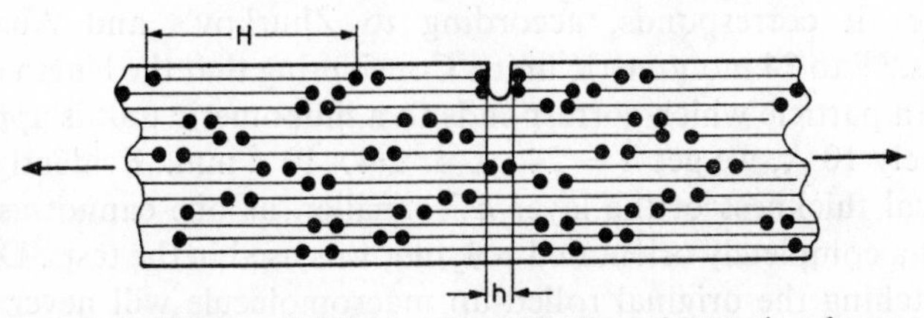

FIG. 91. Diagram of failure of an ideally orientated polymer. The dots represent the ends of the polymer chains.

of the area of the cross-section S_0. Further, we recall that $\gamma = \beta\omega = \beta\lambda_1\lambda\lambda_0$, where $\lambda\lambda_0 = S$ represents the cross-section of each chain which passes through h. The constants β and λ_1 do not depend on the length of the chains, and S depends on it in the following way (S_∞—value S with $M_{\to\infty}$):

$$S = \frac{1}{N(1-h/H)} = \frac{S_\infty}{1-m/M} = \frac{S_\infty}{1-p/P},$$

where $h/H = m/M = p/P$, where m is the molecular weight of the chain section with a length h, and p is the degree of polymerization which corresponds to it. The molecular weight and degree of polymerization are connected with the relation $M = m_1P$ where m_1 is the molecular weight of the monomer (for Nylon $m_1 = 113$).

The expression

$$\gamma = \frac{\beta\lambda_1 S_\infty}{1-m/M} = \frac{\gamma_\infty}{1-p/P} = \frac{\gamma_\infty}{1-m/M} \tag{4.11}$$

is obtained for the structural coefficient; from this, eqn. (4.9) can easily be obtained.

The structural coefficient γ with $M_{\to\infty}$ takes on the value $\beta\lambda_1 S_\infty$. It is remarkable that for Nylon fibres the value γ_∞, as found from experiments, coincides almost exactly with the coefficient which has been calculated from this formula. Therefore, according to the papers[34] $\gamma_\infty = 0{\cdot}24$ kcal mm^2/kg mol, or, after changing to the units of volume $\gamma_\infty = 1{\cdot}7 \times 10^{-19}$ mm^3. For solid polymers $\beta \approx 10$, $\lambda_1 \approx 1$ Å and $S_\infty \approx 16$ Å^2; as with a parallel packing of chains $\lambda \approx \lambda_0 \approx 4$ Å, λ is the mean distance between the polymer chains. The calculated value for $\gamma_\infty = 1{\cdot}6 \times 10^{-19}$ mm^3.

The thickness of the layer h can be calculated from the value of the empirical constant m, which enters eqn. (4.11). For Nylon fibres it corresponds, according to Zhurkov's and Abasov's data,[34] to 24 monomeric units. Considering that the length of the chain particle which corresponds to a monomeric unit is approximately 10 Å, we get $h = 240$ Å or $2{\cdot}4 \times 10^{-5}$ mm. Evidently, the actual thickness of the layer h is smaller, as one cannot assume that a completely orientated polymer was used in the tests. During stretching the original rolled-up macromolecule will never completely stretch into one straight line but will consist of several, i, parallel rolled-up parts, with a mean length of H/i (probably, $i \geq 3$). This means that the macromolecule in the fully orientated material can pass through the layer h several times. This is equivalent to dealing with parallel chains with a length i times smaller than H. When we calculate this, we obtain

$$\gamma = \frac{\gamma_\infty}{(1-i)m/M}.$$

Thence it follows that for Nylon fibres the probable thickness of the layer $h < 80$ Å. This value for h is an entirely reasonable one which corresponds to the linear dimensions of the area of overstress in the apex of the crack.

4.7. Influence of Molecular Orientation on the Strength of Visco-elastic Materials

In visco (high)-elastic polymers no orientation of chains can be found which will survive an indefinitely long time. Orientation and strengthening takes place in the process of stretching of the polymer. The orientation in the moment of break can therefore not be regarded as a parameter which does not depend on the conditions of the test.

The strength of rubbers, crystallizing and non-crystallizing ones, differs greatly. However, a number of data lead to the conclusion that it is not the crystalline state as such which is the basic reason of the high strength but the orientation of chains.

The influence of orientation and crystallization on the strength of natural rubber was studied by Mark and Valko,[42] Hauk and Neumann[43] and others. It was found that the strength of natural rubber, crystallized under tension (600% elongation) is, at a temperature of −80°C, six times greater than that of the non-crystallized amorphous one. At the same time the non-deformed rubber, crystallized during cooling, proved to be only twice as strong as the non-orientated amorphous one. From this follows that in the isotropic semi-polycrystalline state the strength of the solid natural rubber is only twice that of the amorphous non-orientated one. Already the first experiments with crystallizing elastomeric polymers showed, therefore, that only the orientated crystalline phase has a substantial influence on the strength of visco-elastic materials.

The study of the process of crystallization of various elastomers based on natural rubber under stress by X-ray diffraction showed that the strength depends on the proportion of the orientated crystalline phase which develops towards the moment of failure.

The capacity of elastomers from non-crystallizing rubbers for molecular orientation was then investigated by Lukin.[44] Under stretching of such rubbers the relative intensity of the amorphous halo changes on the X-ray photographs and textures emerge on it as a result. The emergence of textures on the amorphous halo is evidence of the orientation of parts of the molecular chains under the effect of an outside stress. After photometric evaluation of the X-ray photograph in two directions perpendicular to each other —along the equator and the meridian—the degree of orientation was determined according to the formula $\varphi = a/b-1$, where φ is the degree of orientation which changes from 0 for non-stretched specimens (a $= bh$) to ∞ (full orientation); a is the intensity of the amorphous halo along the equator, b along the meridian.

The capacity for orientation of elastomers was compared with their strength under tension. A non-filled SKS-30 rubber possesses little capacity for orientation and low strength. Under an elongation by 400% the degree of its orientation is defined by the value $\varphi = 0{\cdot}10$, and the tensile strength is 20 kg/cm². When filled with an active filler the degree of orientation grows sharply before failure, increasing eight to ten times. The tensile strength increases in the same degree (Fig. 92). The extent of orientation and the strength reach a maximum in rubbers which contain approximately 30 g carbon black on 100 g rubber. Further filling causes a decrease of the degree of orientation and a fall of strength.

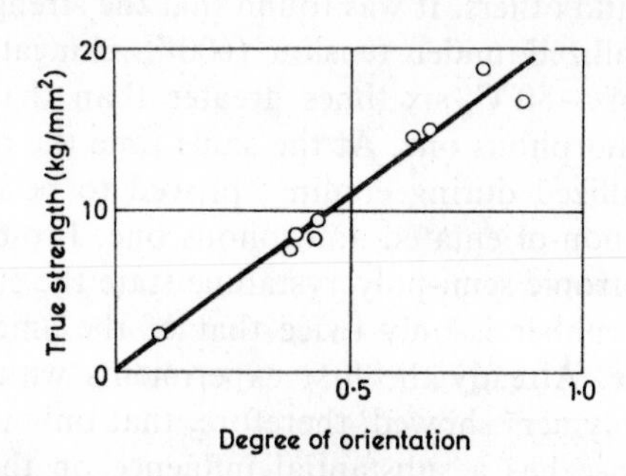

FIG. 92. Dependence between strength and degree of molecular orientation before failure in technical rubbers SKS-30 with different content of carbon black.[44]

Consequently, rubbers stretch under growing stress, as long as a further increase of the number of orientated chains is still possible, as well as an improvement of orientation. When a further change of position of the macromolecules has become impossible, and the molecular chains which carry the load cannot sustain the growing stress, the break of the specimen occurs.

The straight proportional dependence between the tensile strength and the degree of orientation confirms that the molecular orientation is the fundamental fact which determines the strength of the rubber.

Literature

1. V. A. KARGIN, T. I. SOGOLOVA, *ZhFKh*, **23**, 530 (1949).
2. V. A. KARGIN, G. L. SLONIMSKII, *ZhFKh*, **23**, 563 (1949).
3. T. ALFREI, *Mekhanicheskiye svoistva viysokopolimerov*, gl. 4, Izdatinlit, 1952.
4. A. M. SOOKNE, M. HARRIS, *Ind. Eng. Chem.* **37**, 478 (1945).
5. V. A. KARGIN, N. V. MIKHAILOV, *ZhFKh*, **14**, 195 (1940).

6. L. G. Gurevich, *Viysokomol. soyed.* **3,** 1062 (1961).
7. L. A. Lains, Ye. V. Kuvshinsky, *Viysokomol. soyed.* **3,** 215 (1961).
8. P. J. Flory, *Principles of Polymer Chemistry*, chap. 11, New York, 1953; *Ind. Eng. Chem.* **38,** 417 (1946); *J. Polymer Sci.* **4,** 435 (1949).
9. G. M. Bartenev, A. S. Novikov, F. A. Galil-Ogliy, *Kolloid. zh.* **18,** 7 (1956).
10. B. L. Johnson, *Ind. Eng. Chem.* **40,** 351 (1948).
11. Yu. S. Lazurkin, *Doktorskaya dissertatsiya, Institut fizicheskikh problem AN SSSR, M.*, 1954.
12. V. Ye. Gul', N. Ya. Sidneva, B. A. Dogadkin, *Kolloid. zh.* **13,** 422 (1951).
13. V. A. Kargin, G. S. Markova, *ZhKhO im D. I. Mendeleyeva*, **6,** 362 (1961).
14. P. H. Lindenmeyer, *J. Polymer Sci.*, Part. C, **1,** No. 1, 5 (1963).
15. L. B. Morgan, *Progress in High Polymers*, London, vol. 1, 1961, p .233; *Khimiya i tekhnologiya polimerov*, No. 11, Izdatinlit, 1962, p. 3.
16. B. A. Dogadkin, *v sb.* "*Issledovaniya v oblasti viysokomolekuliarniykh soyedinenii*", Izd. AN SSSR, 1949, p. 102; *Trudiy NIIShP*, 2, Goskhimizdat, 1950, p. 21; *sb.* "*Dokladiy 3-i Vsesoyuznoi Konferentsii po kolloidnoi khimii*", Minsk, 1953; *Kolloid zh.* **9,** 348 (1948).
17. P. Flory, N. Rabjohn, M. Shaffer, *J. Polymer Sci.* **4,** 435 (1949).
18. P. Mason, N. Wookey, *The Rheology of Elastomers*, London, 1958, p. 122.
19. V. I. Kasatochkin, B. V. Lukin, *ZhTF*, **19,** 76 (1949).
20. G. Gee, *J. Polymer Sci.* **2,** 451 (1947).
21. Yu. S. Lazurkin, *J. Polymer Sci.* **30,** 595 (1958).
22. H. Penkert, *Zs. VDI*, **95,** 119 (1953).
23. A. I. Belen'kii, R. A. Dulitskaya, Ye. G. Eiges, *v sb.* "*Khimiya i fizikokhimiya viysokomolekuliarniykh soyedinenii*", Izd. AN SSSR, 1952, p. 250.
24. N. V. Mikhailov, V. A. Kargin, V. N. Bukhman, *ZhFKh*, **14,** 205 (1940).
25. N. V. Mikhailov, V. A. Kargin, *Trudiy IV Konferentsii po viysokomolekuliarniym*, Izd. AN SSSR, 1948, p. 138.
26. V. A. Kargin, N. V. Mikhailov, V. I. Yelinek, *v sb* "*Issledovaniya v oblasti viysokomolekuliarniykh soyedinenii*", Izd. AN SSSR, 1949, p. 315.
27. V. A. Kargin, P. V. Kozlov *i dr. Viysokomol. soyed.* **1,** 182, 1848 (1959); **2,** 284, 931, 1109, (1960); **3,** 139, 1100 (1961); **5,** 1156 (1963).
28. P. V. Kozlov, G. L. Berestneva, *Viysokomol. soyed.* **2,** 590, 601 (1960).
29. L. G. Kazarian, D. Ya. Tsvankin, *Viysokomol. soyed.* **5,** 976 (1963).
30. N. I. Shishkin, M. F. Milagin, *FTT*, **4,** 2681 (1962).
31. M. F. Milagin, N. I. Shishkin, *FTT*, **4,** 2689, 3578 (1962).
32. A. V. Stepanov, *ZhETF*, **19,** 973 (1949).
33. M. I. Bessonov, M. P. Kuznetsov, *Viysokomol. soyed.* **1,** 761 (1959).
34. S. N. Zhurkov, S. A. Abasov, *Viysokomol. soyed.* **3,** 450 (1961); **4,** 1703 (1962); *FTT* **4,** 2184 (1962).
35. G. M. Bartenev, *Kolloid. zh.* **13,** 233 (1951).

36. C. C. Hsaio, J. E. Osborn, D. B. Rozendal, *J. Phys. Soc. Japan*, **16,** 459 (1961).
37. C. C. Hsaio, *J. Appl. Phys.* **30,** 1492 (1949); *J. Polymer Sci.* **47,** 251 (1960); *J. Polymer Sci.* **44,** 71 (1960).
38. W. Schieber, *Angew. Chem.* **52,** 487 (1939); **52,** 561 (1939).
39. E. I. Barg, D. M. Spitkovskii, N. N. Mel'T'yeva, *DAN SSSR*, **84,** 257 (1952).
40. V. A. Sokolova, Z. A. Rogovin, *Khim. volokna*, No. 5, 45 (1959).
41. S. Tachikawa, *Rayon and Synthetic Textiles*, **32,** No. 3, 31, 43 (1951); **32,** No. 7, 32, 42 (1951).
42. H. Mark, E. Valko, *Kautschuk*, **6,** 210 (1930).
43. V. Hauk, W. Neumann, *Z. Physik. Chem.* (A) **182,** 285 (1938).
44. B. V. Lukin, *v sb.* "*Rentgenovskiye metodiy issledovaniya i ikh primeneniye v khimicheskoi promiyshlennosti*", Goskhimizdat, 1953, p. 43.

CHAPTER 5

STATISTICAL THEORY OF STRENGTH AND SCALE EFFECT

THE results of tests concerning the strength of all materials are non-reproduceable when dealing with separate samples, and the determined strength depends on the dimensions of the stressed area of the material.

The test results concerning the first property show a considerable scatter which goes far beyond the limit of errors of measuring. Owing to this the strength of elastomers is usually characterized by an average strength value† and it is indispensible either to indicate the mean square value or variation, or an equivalent property which characterizes the scatter of the results of the individual experiments. What has been said refers also to the durability of the material. The data on strength and durability are therefore usually presented as mean values, obtained by experiments on from 3–5 to 20 and more specimens, depending on the demands of accuracy of the determination of these values.

The statistical theory of strength explains the scatter of experimental results and partly the dependence of the strength on the scale factor.

5.1. Statistical Theory of the Strength of Solids

In the homogeneously stressed state the material begins to fail in the weakest places which can be obvious defects of structure (microcracks) or various kinds of structural heterogeneity and mechanically weakened places. For instance, in solid polymers the blocks which lie parallel to the tension are strong places of the material and those lying perpendicular are weak ones.

† Or by a more exact indicator as the most probable value of strength which may be derived from the curve of strength distribution.

In the first stages of development of the statistical theory of strength the model of the solid body was described as a uniform homogeneous medium in which defects were distributed (micro-cracks, extraneous enclosures, etc.). This model can be regarded as the first approximation to the actual material as the defects have a considerable and often decisive influence on the strength. In any material the defects are heterogeneous and irregularly distributed in the volume of the material and on its surface. The strength of the specimen is determined mainly by the most critical defects, and in the different samples of the given material the most critical defects differ in their turn according to the degree of criticality. In strength measurements a definite scatter of data from specimen to specimen is therefore observed which is not connected with the errors of measuring but which appears due to the actual structure of the real material.

Starting from these general concepts, different statistical theories of the strength of brittle bodies were suggested,[1-4] according to which their failure takes place when the stress exceeds the strength of the weakest part. For the calculation of the brittle strength simplified models were examined; they presented the solid body as a combination of parallel lines of different strength or as a combination of a succession of connected links of different strength (problem of the strength of the chain).

The fundamental aspect of the statistical theory of strength which was developed in those papers can lead to the following.

1. In samples of the same material is a great number of defects of different origin and of different degrees of criticality, and outwardly identical samples which were obtained by identical methods can have defects of different criticality.

2. The strength of the sample is determined by the most critical defect.

3. The greater the volume (or surface) of the specimen, the more probably will a most critical defect be found. From this follows that the strength must decrease with an increase of the number of specimens.

The idea of the statistical theory of strength is confirmed visually by the well-known experiments of Powell and Preston[5] on the failure of glass by the method of impressing a little steel ball. When the diameter of the sphere decreases the destructive tensile stress which appears on the surface grows from the usual value (about 5 kg/mm^2) to a maximum one (approximately 200 kg/mm^2), which is near the theoretical strength. Even more striking results were recently obtained by Bonkin.[6] With a radius of the indentations of 0·136 and 0·010 mm the strength of glass proved to be 700 and 1040 kg/mm^2 respectively, i.e. with the transition to an area of loading with a radius of about 1 μ the strength of glass coincides in practice with the theoretical strength.

These results can only be explained by the fact that defects of different degrees of criticality are distributed on the surface of the glass. In a small volume of stressed glass the probability of finding critical microcracks or weak places is extremely small. The strength of the stressed material in dependence on the dimensions of the area can therefore have a value which changes from the macro-strength which is characteristic for large areas to microstrength which is characteristic for micro-areas.

As solid state physics developed, the concept of the defects of structure broadened. In monocrystals dislocations were discovered, in amorphous solids structural microheterogeneities (micro-volumes with somewhat different structure, density, chemical composition, etc.). On the other hand it became clear that even such perfect solids as monocrystals possess a "block mosaic" structure, not to speak of the polycrystal and amorphous bodies, in particular of the polymeric materials, where the microhetero-geneity of structure is especially clearly expressed. From this follows that a material which is ideally homogenous in its micro-scopical structure and mechanical properties does not exist. All known materials are microheterogenous.

In connection with this another very important side of the statistical theory of strength was developed in the works of Afanas'yev[7] and Volkov;[8] it consists of the following. A material may have no obvious defects, for instance, in the form of micro-cracks but because of the microheterogeneity of the structure the macroscopically homogenous stressed state of the specimen (from the point of view of the theory of elasticity) is in reality heterogenous when one examines the structural microvolumes (grains in polycrystalline materials, blocks and other elements of supra-molecular structure in polymers, micro regions of stratification in non-organic glass, etc.). The microstress which appears under pure tension in different microvolumes of the material is irregularly distributed. Parts both more and less stressed can be found and even (very rarely) also parts under compression. If the strength of the material is the same for all microvolumes the most over-stressed microvolumes begin to fail; if the mechanical strength of the microvolumes is not uniform then the structurally weakest and simultaneously most stressed ones will be the first to fail. When compressing such a "flawless" material there can appear even tensile stresses in individual microvolumes which lead to microbreaks and to the formation of microcracks.

According to Volkov overstresses appear in microvolumes even in "flawless" material due to irregularly distributed stresses caused by different kinds of imperfections of the structure (secondary type of stress). The failure of the material in the microvolumes originates mainly in tensile stresses, as they lead to the breaking of bonds.

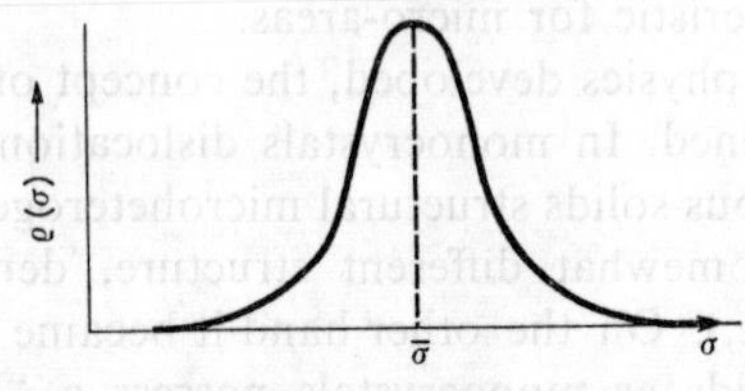

FIG. 93. Distribution of secondary stress in the cross-section of the specimen under tension ($\bar{\sigma}$ is the mean stress in the cross-section).

If the structure approaches that of the ideal homogenous material, the distribution curve of secondary stresses in the material (Fig. 93) narrows infinitely. This can in practice be observed in such homogenous objects as thin flawless glass fibres and thin metal fibres ("whiskers").

The statistical theory of strength occupies an intermediate position between the classical theory of elasticity and the molecular mechanics of solids. At the present time the statistical theory of strength has not yet reached that degree of development when it could establish an exact connection of the structure with the strength properties for each type of material. In different existing statistical theories this is considered as the introduction of some hypothetical function of the distribution of defects or of secondary stresses.

5.2. Statistical Nature of the Strength of Elastomers

The basic assumption of the statistical theory of strength that the strength is determined by the most critical defect or by the most overstressed part of the specimen is applicable also to elastomeric polymers and rubbers.

For the understanding of the nature of strength it is very important to know what the primary defects in the original non-stressed material are. They can be: microscopic cracks (especially on the surfaces—the most vulnerable part of the specimen) which appear as a result of thermal, mechanical or other influences during manufacturing and processing of the material; defects and imperfections of structure (microparts with different mechanical properties and different orientations, molecular weight, etc.); places of residual stress concentration which are always found in the material, etc. Such microdefects and microheterogeneities are origins of failure in the material under load.

From the basic assumption of the statistical theory of strength follows in particular the possibility of studying the distribution of defects in the material by the curves of the distribution of strength and durability. Although this is an indirect method, yet it is at present the only one which permits to judge the character of the distribution of defects or of the places of overstress. The most critical defect or the overstressed part of any specimen is numerically characterized either by the destructive stress (under certain conditions of experiment) or by the time to failure.

If a great number of specimens are tested, a spread of strength values is discovered. Consequently, the strength of the specimen lies with a certain determined probability within given limits and is characterized by the curve of distribution.

The curve of distribution of strength expresses graphically the function of distribution (or the degree of probability) of strength (Fig. 94). The function of distribution $\varrho(\sigma)$ or $\varrho(f)$ is determined from the equation

$$\Delta N = N\varrho(\sigma)\,\Delta\sigma; \qquad \Delta N = N\varrho(f)\,\Delta f \tag{5.1}$$

where ΔN is the number of specimens with a strength in the range from σ to $\sigma+\Delta\sigma$ (from f to $f+\Delta f$); N is the number of tested specimens of a given series (not less than 100); σ is the true strength; f is the strength on an arbitrary scale.

The interval $\Delta\sigma$ and Δf is so chosen that the curve, based on the individual values $\varrho(\sigma)$ is smoothed out. It can be seen from Fig. 94 that with an increasing thickness of the samples the curves of distribution move to the side of lower strength, and the absolute scatter of strength decreases.[9] If the data of all three curves are drawn in comparative scales, it is found that they lie on a common curve (Fig. 95). This proves that the relative scatter of the data does in practice not depend on the thickness of the elastomeric

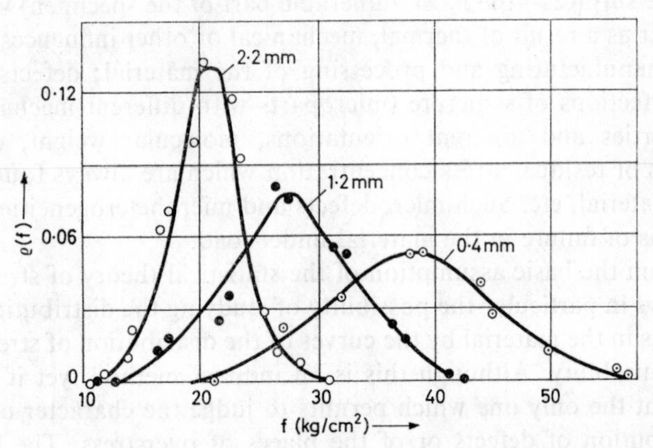

FIG. 94. Strength distribution curves, calculated on the initial section for samples of unfilled SKS-30 rubbers of different thicknesses.[9]

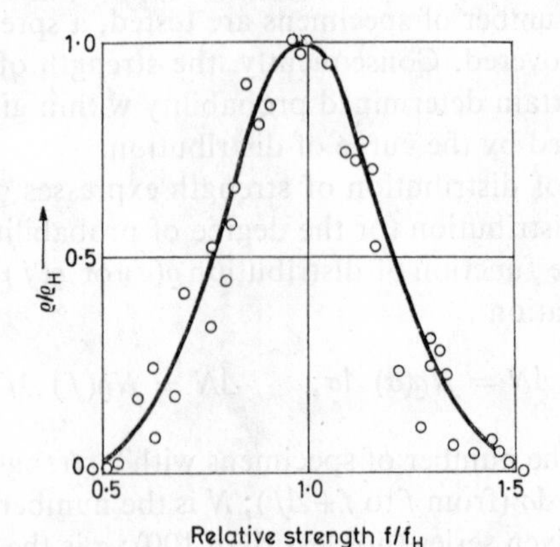

FIG. 95. The values of Fig. 94 in relative coordinates.

sheet from which the samples were cut with a punch. Here f_H is the most probable and f the expected strength of the specimen (calculated on the original section), ϱ_H is the maximum value of the function of distribution ϱ at $f = f_H$.

The symmetrical curve of distribution of strength is characteristic for unfilled elastomers from non-crystallizing rubbers; for SKS-30 rubbers (see Fig. 95) it is described by the following approximate equation of distribution:

$$\varrho(f) = \varrho_H e^{-15(1-f/f_H)^2} \tag{5.2}$$

Symmetrical curves of strength distribution can also be observed for unfilled elastomers from crystallizing rubbers.[10] Consequently, for all unfilled elastomers the greatest probable strength can be calculated as the arithmetical mean.

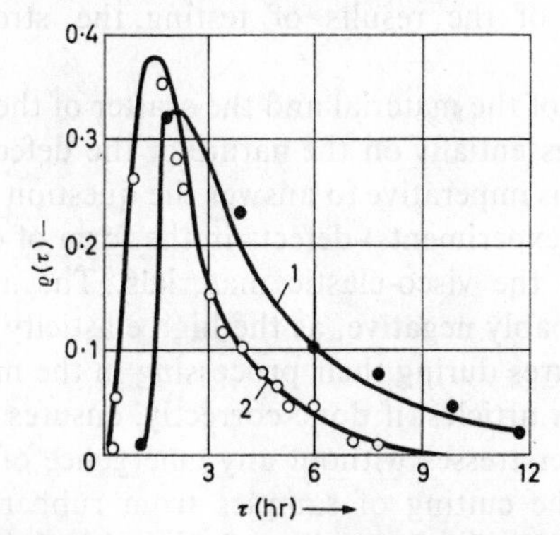

FIG. 96. Distribution curves of durability at constant tension for unfilled SKS-30 rubbers. 1. High-modulus rubber. 2. Low-modulus rubber.

On Figs. 94 and 95 data are shown which were obtained on the testing machine at a constant strain rate (500 mm/min). In these experiments the failure stress changes from specimen to specimen. On Fig. 96 are shown data which were obtained on rubber at slow failure under the effect of a constant static tensile stress. In these experiments the stress for all specimens of any series was predetermined, and the durability changed from specimen to specimen. These curves of distribution are unsymmetrical owing to the non-linearity of the connection between stress and durability which is expressed by the formula of durability.

If one compares rubber samples of the same durability one can observe a scatter of the static tensile stresses around some mean value.

The curves of distribution characterize the strength of rubber at certain experimental conditions. It is, therefore, absolutely necessary to test a great number of samples: this is only possible in special investigations. For standard or current experiments, and also for quick determinations, one usually applies methods based on the normal laws of distribution to evaluate the results of testing a few specimens (but not less than five). By these methods[11-13] one can calculate the deviation of the experimental results from the mean value, the accuracy of the determination of the mean value and other indispensible data which characterize the statistical picture of the phenomenon. They are widely applied during the evaluation of the results of testing the strength of rubbers.[14-16]

The strength of the material and the scatter of the experimental results depend essentially on the nature of the defects. In connection with this it is imperative to answer the question whether there are (before the experiments) defects in the form of cracks, microbreaks, etc., in the visco-elastic materials. The answer to this question is probably negative, as the high elasticity and plasticity of rubber mixtures during their processing in the manufacture of technical rubber articles, if done correctly, ensures a rapid relaxation of the overstresses without any emergence of microbreaks. Neither does the cutting of samples from rubber sheet with a sharp knife lead to the formation of any critical surface defects. For instance, it was found that the strength of specimens with identical dimensions, some of which were vulcanized in the mould and others cut out of a sheet which was produced under the same conditions of vulcanization, is identical.[9] There are, however, data[17] which indicate the influence of the configuration of the punch on the strength of sheet from crystallizing rubber.

The basic reason for the statistical nature of the strength of elastomers is probably the presence of structural heterogeneities which appear as a result of unequal vulcanization in the microvolumes of the rubber. In filled rubbers the structural heterogeneity and the secondary stresses appear already as a result of the presence of the filler particles. In the rubber which undergoes tension or other types of deformation there appears, therefore, an irregular distribution of secondary stresses which is probably the reason for the scatter of experimental results, in conformity with Volkov's theory. We shall from now on understand under-defects of the structure of rubber not only cracks, microbreaks and other defects, but also its most critical heterogeneities.

5.3. Statistical Theory of the Strength of Rubber

Almost all technical rubbers contain active fillers. Filled rubbers, as distinguished from the unfilled ones, possess an unsymmetrical curve of strength distribution (Fig. 97). For all rubbers

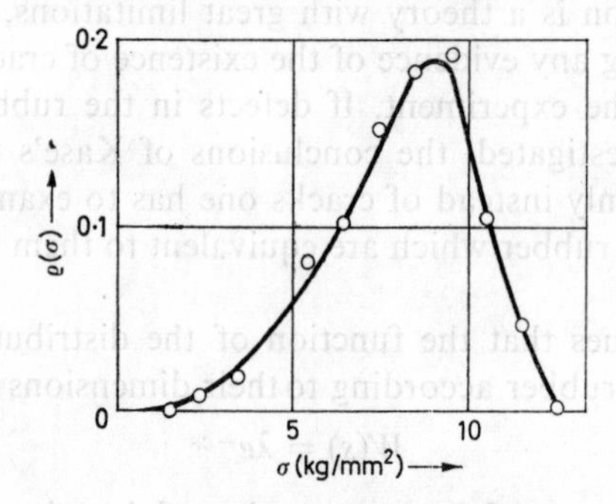

FIG. 97. Distribution curve of true strength of carbon black filled SKS-30 rubber.

filled with carbon black a more or less clearly defined "tail" can be observed on the curves of distribution in the region of low strength, like the one shown on Fig. 97. The presence of a relatively large number of low-strength values in technical rubber can be explained by the presence of various types of heterogeneities, inclusions and other defects. Heterogeneity in the structure can appear as the consequence of an insufficiently careful mixing of filler and rubber or because of the presence of coarse filler particles in it. Accidental damage which may occur during the cutting of the material, etc., is also possible. In so far as the presence of a "tail" on the curve of strength distribution can be entirely attributed to the presence of filler in the rubber, a method suggests itself to determine the quality of the mixing of the filler and the rubber by the size of the "tail" and also by the width of the curve of strength distribution. Well-processed rubber mixtures are characterized by a distribution curve which approaches symmetry and is extremely narrow.

The application of the statistical theory of strength to filled rubbers was developed by Kase.[18] He investigated the strength of elastomers at rapid "visco-elastic" failure, when the rough zone had no time to form, and the mechanism of failure consisted in the formation and growth of cracks (see Chapter 3). Such a character of failure can be observed in experiments on the testing machine at standard strain rates.

Kase understands under-defects of rubber microcracks, the degree of criticality of which is determined by their dimensions. This suggestion is a theory with great limitations, as the author does not bring any evidence of the existence of cracks in the rubbers before the experiment. If defects in the rubber in a wider sense are investigated, the conclusions of Kase's theory remain unchanged, only instead of cracks one has to examine structural defects of the rubber which are equivalent to them in their degree of criticality.

Kase assumes that the function of the distribution of microcracks in the rubber according to their dimensions looks like this:

$$W(s) = \lambda e^{-cs} \tag{5.3}$$

where s is the area of the cross-section of the microcrack, c and λ are constants of the rubber.

The ultimate tensile stress, calculated on the original cross-section, is $f = f_m(1-\alpha s)$, where f_m is the theoretical strength of flawless rubber and α is an empirical constant. Applying the method of mathematical statistics (in the assumption that the specimen fails in the place of the most critical defect), Kase obtains the double exponential function of the distribution of strength:

$$\varrho(f) = e^y \exp(-e^y) \tag{5.4}$$

where

$$y = \frac{c}{\alpha f_m}(f - f_H) \tag{5.5}$$

The most probable strength f_H is calculated according to the formula

$$f_H = f_m\left[1 - \frac{\alpha}{c}\ln(rV)\right] \tag{5.6}$$

where r is the mean number of defects per unit of volume of the rubber; V is the volume of the specimen or of the stressed part of the specimen.

In eqns. (5.3) to (5.6) c, α, r, f_m are empirical constants.

The obtained function of distribution is unsymmetrical and describes well the experimental data for carbon-black filled rubber. To compare the function of distribution (5.4) with the function of distribution (5.4) with the experimental curve, f_H is determined from the latter, and the value $c/\alpha f_m$ selected and introduced in eqn. (5.5). The experimental curve of strength distribution for filled NK and SKS-30 rubber which was found by Kase, agrees with the equation of distribution (5.4).

5.4. Influence of the Dimensions of the Specimen on the Strength of Rubber

Under the scale effect of strength we understand the influence of the dimensions of the specimen or of any of its parts on its strength. The appearance of the scale effect in rubbers can be distinctly observed on the distribution curves for specimens of different thickness (see Fig. 94). The values of the mean strength of strips of non-filled SKS-30 under tension in dependence on their thickness (width of the specimens 7 mm, stressed length 25 mm) are:

Thickness of the specimen (mm)	0·35	0·50	0·70	1·0	1·5	2·0	2·5
Failure stress (kg/cm²)	200	170	160	140	115	100	86

It can be seen from these figures that the true stress for SKS-30 rubbers is doubled with a decrease of the thickness of the rubber sheet from 2·5 to 0·5 mm,† but the strength grows quicker in thin material. The strength of unfilled rubber changes also with a change of the length of the stressed part and its width (data obtained at a strain rate of 500 mm/min). The influence of the length of the stressed sample (thickness 0·9 mm, width 6·6 mm) on the true

† The apparent failure stress grows within these limits of varying thickness from 19 to 35 kg/cm² but the total strain does not change; the true strength is therefore proportional to the apparent one for all thicknesses investigated.

strength of unfilled SKS-30 rubber is illustrated by the following data:

Length of stressed part (mm)	10	20	40	60	80	100
Failure stress (kg/cm^2)	160	157	143	139	133	130

The influence of the width of the sample (thickness 1 mm, length of stressed part 20 mm) on the true strength of unfilled SKS-30 rubber, is:

Width of sample (mm)	3	5	7	10	15
Failure stress (kg/cm^2)	210	165	150	140	135

These data confirm the existence of a considerable scale or size factor, especially for samples with small dimensions where the strength of the rubber changes very much. One has therefore to consider the decrease of strength of massive technical rubber articles compared with the strength measured on small rubber samples. This facet of assessing the properties of technical rubber articles does usually not get its due attention: owing to this there are no data on the influence of the size factor on the strength of technical rubbers.

5.5. Theory of the Scale Effect of Strength

The different strength values of small and large specimens with the same structure can, in agreement with the statistical theory of strength, be explained by the fact that in large samples the probability of the presence of most critical defects or most critical secondary stresses is greater than in small ones. In very small samples there may be no critical defects at all. The inner secondary stresses may also not appear in very small samples as its structure may happen to be completely homogenous. Consequently, the strength of small samples must be higher than that of big ones.

A connection between strength and scale factor is established in different statistical theories of strength. The best well-known formula for solids is Weibull's (confirmed by Chechulin) which is used for materials which have no anistropic strength properties:

$$\sigma_H = \frac{a}{V^{1/n}} \tag{5.7}$$

where σ_H is the most probable strength; a is a constant, dependent on the material and on the character of the stressed state; V is the volume of the specimen or of its stressed part;† n is a constant of the material which takes into account the character of the distribution of defects.

For elastomers Kase[18] suggested eqn. (5.6) which establishes a connection between the most probable strength and the volume of the specimen. In so far as all coefficients in this formula are empirical ones, it can be simplified:

$$f_H = K_0 - K \log V. \tag{5.8}$$

The applicability of eqns. (5.7) and (5.8) to rubber is confirmed by the data shown on p. 172, as the mean strength values coincide with the most probable ones for unfilled SKS-30 rubber. As the breaking strain of that rubber is almost constant one can understand in eqn. (5.8) under f_H both the arbitrary and the true strength but the coefficients in eqn. (5.8) will have different values. In half-logarithmic coordinates (Fig. 98) the straight line *KL* shows the dependence of the strength on the thickness of the specimen which agrees with eqn. (5.8). In logarithmic coordinates the data lie on a curve, which points to the non-applicability of Weibull's formula to rubber.

On Fig. 98 the values of the influence of the thickness and length of the sample on the strength lie on different straight lines. This indicates the anisotropic scale effect in rubbers which has also been observed in other materials. The explanation for this influence is given below.

The statistical theory of the size effect of strength can, strictly speaking, only be used for the explanation of the influence of the dimension of the sample on the most probable strength assuming an identical structure of the material of both large and small specimens. But in many cases the conditions of producing speci-

† For silicate glass and glass fibres, the strength of which is normally determined by the surface defects, the volume in eqn. (5.7) has to be replaced by the stressed surface (for massive glasses), and the length (for glass fibres).

mens and large or small articles are such that the structure of the material is different in the large mass than in the small one. The structure can differ, in particular, through different degrees of molecular orientation (for instance, a sample with a non-orientated structure, and a fibre drawn from it with an orientated structure). Although cross-sections of the original and of the orientated specimens are different, the observed difference in strength is not caused by the scale effect, as samples of a different structure are compared.

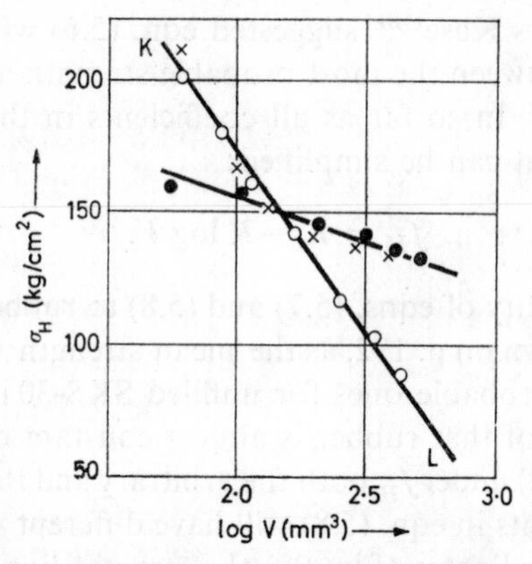

FIG. 98. Dependence between the most probable true strength and the volume of a sample of SKS-30 rubber: ○ — specimens of different thickness; ● —: specimens of different stressed length; × — specimens of different width.

It follows from this that the size effect in its pure form is found in specimens of geometrically similar proportions, obtained, however, from the same original material and by the same manufacturing process; under these conditions the physical similarity of specimens of all dimensions is maintained. If the latter condition is not fulfilled (and that happens often), the same manufacturing process can lead to small and large samples with a different structure and, consequently, also with a different strength. This "scale" effect of strength of non-statistical origin is usually called a "technological" one. It is obvious that this definition is conditional as specimens with different structure are compared.

The complicated effect of the influence of the dimensions on the strength can be observed on glass fibres[19] which are characterized by an anisotropic scale effect (in the longitudinal and transversal directions to the axis of the fibre). A strong dependence of the strength of glass fibres on their diameter can, obviously, be explained not only by the size effect but also by the difference in the structure of thin and thick fibres which were obtained at different drawing speeds. At the same time a slight dependence of the strength on the length of the glass fibres lies completely within the frame of the statistical theory. The dependence of the strength of glass fibres on their length, like that of the strength of solids on the volume, is expressed by Weibull's formula (5.7), where the exponent $1/n$ has in both cases the same value: 0·25.

What has been said above refers to ordinary industrial glass fibres which always have microcracks on the surface layer, i.e. in other words, to defective fibres. Quite recently Izmailova[20] succeeded in obtaining flawless glass fibres which did not show any dependence of the strength on the length and practically none on the diameter. These fibres are characterized by unusually small scatter of experimental results (the coefficient of variation constitutes a few per cent) and by other new properties.

The anisotropic size effect of the strength was also discovered in sheet glasses by Tsepkov.[21] It was found that the strength of the surface of sheet glasses under tests on symmetrical bending depends more on their thickness than on the size of the stressed surface of the sample. This is connected with different conditions of manufacturing of sheet glass and especially with the far greater temperature gradients during the shaping by the method of vertical drawing of thick glasses. As a result the structure and the defects in the glass mass and, especially, on the surface are different for thick glasses and thin ones which are produced under conditions which are nearer to the isothermal distribution of temperature across the thickness of the sheet.

The anisotropy of the scale effect of the strength of rubbers with changing thickness and length of the specimen (see Fig. 98) can, obviously, be explained by similar causes. The scale factor is greater with varying length of the rubber specimen. The structure of thin foils can differ from the structure of thick ones for the same rubber, on account of the more equal heating, the pressure in the moulds which is not always controlled during vulcanization, and other causes which lead to a more compact and homogenous structure of the thin foils.

The investigated scale effect of strength refers to such experiments where the stress in the unfailed parts of the cross-section of the specimen during the break grows continuously (for instance, at failure under static load, under tension with constant rate of deformation or loading, etc.). In these cases the strength is determined by the most critical defects, the development of which leads to a catastrophical parting of the specimen. However, in practice one finds also other kinds of deformation. Of these we have to single out especially the type of fixed (static) deformation by tension or bending, during which the growth of cracks in the sample leads to a gradual unloading of the material. As a result the stress in the remaining section can decrease so much that a further deterioration is halted (see, for instance, § 3.3).

For complete failure, i.e. parting of the specimen, it is imperative that the store of elastic energy in the specimen is not smaller than the expenditure of energy on the process of propagation of cracks across the whole specimen. A small specimen does not possess a store of elastic energy sufficient to provide complete failure. This shows one more scale factor of a non-statistical nature.[22]

Various points of view on the nature of the size effect of the strength of solids during static and dynamic experiments were expressed in discussion.[23]

Literature

1. W. Weibull, *A Statistical Theory of the Strength of Materials*, Stockholm, 1939.
2. T. A. Kontorova, Ya. I. Frenkel, *ZhTF*, **11,** 173 (1941).
3. B. B. Chechulin, *ZhTF*, **24,** 292 (1954); *Masshtabniyi faktor i statisticheskaya priroda prochnosti metallov*, Metallurgizdat, 1963.
4. L. G. Sedrakian, *K. statisticheskoi teorii prochnosti*, Yerevan, 1958.
5. H. Powell, F. Preston, *J. Am. Ceram. Soc.* **28,** 145 (1945).
6. P. Ya. Bonkin, *v sb.* "*Stekloobraznoye sostoyaniye*", Izd. AN SSSR, 1960, p. 415.
7. N. N. Afanas'yev, *ZhTF*, **10,** 1553 (1940).
8. S. D. Volkov, *Statisticheskaya teoriya prochnosti*, Mashgiz, 1960.
9. G. M. Bartenev, *DAN SSSR*, **82,** 49 (1952).
10. P. Flory, N. Rabjohn, M. Shaffer, *J. Polymer Sci.* **4,** 435 (1949).
11. V. I. Romanovskii, *Primeneniye matematicheskoi statistiki v opiytnom dele*, Gostekhizdat, 1947.
12. B. M. Schchigolev, *Matematicheskaya obrabotka nablyudenii*, Fizmatizdat, 1960.

13. A. M. Dlin, *Matematicheskaya statistika v tekhnikye*, "Sovyetskaya nauka", 1958.
14. W. H. Reece, *IRI Trans.* **11,** 312 (1935).
15. O. Davies, S. Horrobin, *IRI Trans.*, **12,** 85 (1936).
16. I. I. Gol'berg, *Kauchuk i rezina*, No. 6, 23 (1957).
17. I. V. Zakharenko, D. L. Feeyukin, J. I. Gol'berg, *v sb. "Fiziko-mekhanicheskiye ispiytaniya kauchuka i reziniy"*, Goskhimizdat, 1960, p. 140 *(Trudiy NIIShP*, No. 7).
18. S. Kase, *J. Polymer Sci.* **11,** 426 (1953).
19. G. M. Bartenev, A. N. Bovkunenko, *ZhTF*, **26,** 2508 (1956).
20. G. M. Bartenev, L. K. Izmailova, *DAN SSSR*, **146,** 1136 (1962).
21. G. M. Bartenev, L. P. Tsepkov, *DAN SSSR*, **121,** 260 (1958).
22. B. A. Drozdovskii, Ya. B. Fridman, *Vliyaniye treshchin na mekhanicheskiye svoistva konstrutsionniykh stalei*, Metallurgizdat, 1960.
23. O vliyanii razmerov obraztsov na ikh mekhanicheskiye svoistva, *Zav. lab.* **26,** 319 (1960); **26,** 1104 (1960).

CHAPTER 6

TIME DEPENDENCE OF THE STRENGTH OF RUBBERS

THE strength of all materials decreases with the increase of the time during which it is stressed. This phenomenon is called static fatigue if the material is under static stress, and dynamic fatigue if it undergoes cyclic loading. This refers also to rubbers. In this chapter we shall investigate the time dependence of strength (static fatigue) of elastomers which is the chief criterion for a fixed period of service of many technical rubber goods.

6.1. Formula of Time Dependence of the Strength of Rubbers

There exists the opinion that the dependence of the durability τ on the tensile stress σ has the same form for solids and for elastomers [see eqn. (1.13)]. According to Zhurkov and Narzullayev[1] this formula is applicable to all elastomers apart from crystallizing ones, the structure of which changes considerably during deformation. However, detailed investigations, carried out by one of the authors and co-workers,[2, 3] led to another, experimentally confirmed equation for the time dependence of the strength of rubber:

$$\tau = B\sigma^{-b} \tag{6.1}$$

where B and b are constants.

Investigations in static fatigue of elastomers were carried out mainly under tensile deformation. SKS-30 rubbers were examined:[2] unfilled ones with 0·3 and 3% sulphur on the rubber (static equilibrium moduli 2 and 12 kg/cm²) and filled ones with 50 parts carbon black and 3% sulphur (nominal static equilibrium modulus 30 kg/cm²). The rubbers were examined under constant loads of different magnitudes at 20°C, practically without any effect of light and ozone. The durability was determined as the mean value of from fifteen to twenty tests. The specimens had a working length of 25 mm, a width of 7 mm, and a thickness of 6 mm. As shown in that paper the time dependence of strength of low-modulus rubbers ($E_\infty = 2$ kg/cm²) in half-logarithmic coordinates is not expressed by a straight line (Fig. 99), whereas the

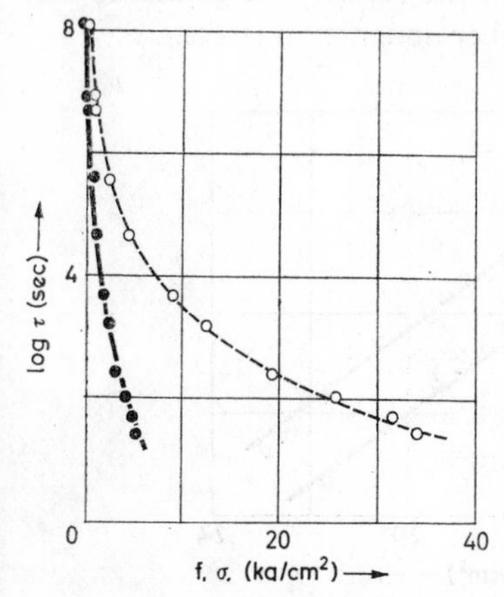

FIG. 99. Time dependence of the strength of unfilled SKS-30 rubber with an equilibrium modulus of 2 kg/cm² at 20°C in half logarithmic scale.

▬▬: apparent stress;
— — — — —: true stress.

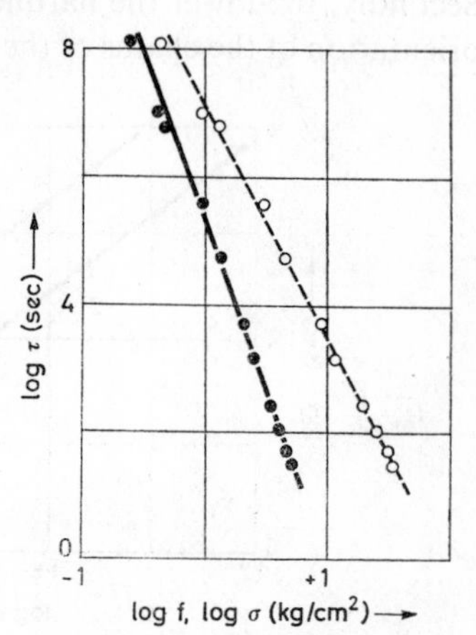

FIG. 100. Time dependence of strength of unfilled SKS-30 rubber with an equilibrium modulus of 2 kg/cm² at 20°C in logarithmic scale.

▬▬: apparent stress;
— — — — —: true stress.

same values in logarithmic coordinates (Fig. 100) lie well on a straight line—which agrees with eqn. (6.1). Rubber with the modulus $E_\infty = 12$ kg/cm² shows the same properties. The scale factor (the thickness of the sample) influences only the constant B in eqn. (6.1), as the incline of the straight lines (as can be seen on Fig. 101), and therefore also to constant b, do not depend on the thickness of the specimen.

The values of the time dependence of the strength of carbon-black filled rubbers which are considerably harder materials than unfilled rubbers, lie almost on straight lines in both coordinate systems. Filled rubbers are, therefore, in view of the strength of the material, intermediate between elastomeric and solid polymers. From this one can draw the conclusion that the less hard the rubber, the clearer appears the departure from the time dependence of strength which is characteristic for solid polymers. There are at least two reasons for this. Firstly, the lower the hardness of the rubber, the more appears the break mechanism which is characteristic for the elastomeric polymers (see Chapter 3). Secondly, the lower the hardness of the rubber, the greater is the orientation of the chains of the rubber under great stress.

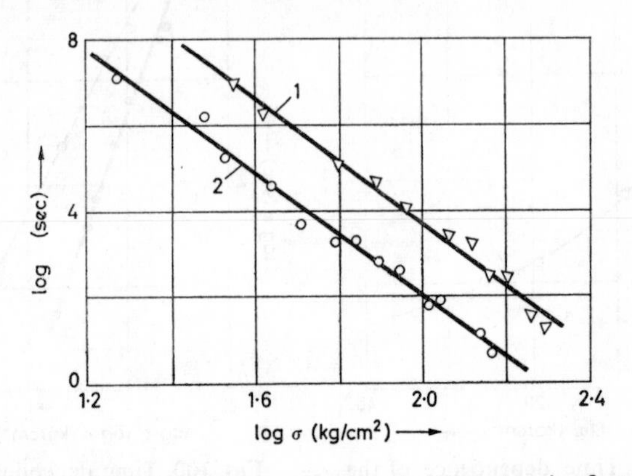

FIG. 101. Time dependence of strength (true stress) of unfilled SKS-30 rubber with an equilibrium modulus of 12 kg/cm^2 at different thicknesses of the specimens. 1. Thickness 1 mm: 2. Thickness 6 mm.

The orientation of the chains at break characterizes the magnitude of the tensile deformation under the effect of a static load. The value of deformation and durability under different sizes of true stress for low modulus rubber (2 kg/cm^2) are shown in Table 8. A considerable increase of deformation (and orientation) with the increase of the static stress leads to a relative deceleration of the speed of growth of the small tears and a departure of the time dependence of strength from a straight line (see Figs. 99 and 102).

The static fatigue which is characterized by the incline of the straight lines of durability, i.e. by the constant b in eqn. (6.1), depends on the hardness of the rubber (see Fig. 102). The infinite value of the constant b corresponds to the straight line which lies parallel to the axis of time, i.e. to the material which possesses a permanent durability. For the various existing rubbers the value

TABLE 8

Deformation and Durability of Low-modulus Rubbers in Dependence on the Stress

σ (kg/cm²)	ε (%)	τ (sec.)	σ (kg/cm²)	ε (%)	τ (sec)
0·4	68	$1{\cdot}44\times10^8$	12·8	380	$1{\cdot}52\times10^3$
0·75	100	$1{\cdot}03\times10^7$	19·4	440	$2{\cdot}9\times10^2$
1·3	170	$6{\cdot}82\times10^6$	25·8	500	$1{\cdot}11\times10^2$
4·6	220	$5{\cdot}38\times10^4$	34·3	540	$0{\cdot}34\times10^2$

of the constant b is, as a rule, within the range 3–12, and grows with increasing hardness.

The true safe stress in rubbers is near zero (see Fig. 99), obviously owing to the effect of two factors which are characteristic for these materials. Firstly, due to the decrease of molecular

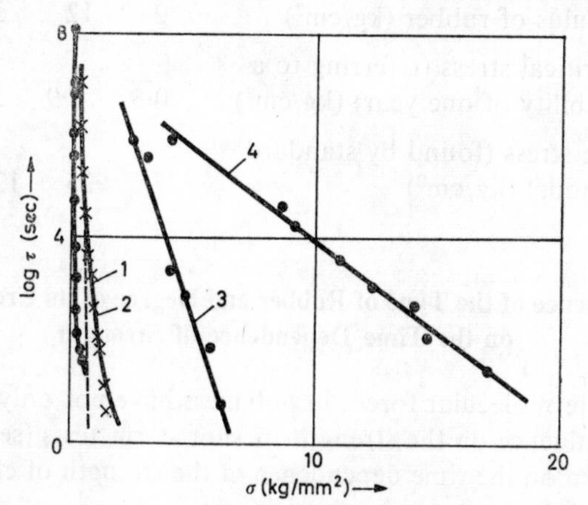

FIG. 102. Time dependence of the strength of rubbers of different hardness. 1. Unfilled SKS-30 rubber (equilibrium modulus 2 kg/cm²). 2. The same with a modulus of 12 kg/cm². 3. The same with a modulus of 30 kg/cm². 4. Technical Buna-S rubber.

orientation with a decrease of load (the safe stress is the lower, the lower the orientation); secondly, due to the process of ageing which weakens the material the more, the longer the load is applied.

In engineering practice one takes as "non-critical" stress that particular failure stress f_0, calculated on the original section, which gives a definite service life of the rubber. This "non-critical" stress can be determined by using logarithmic graphs of the experimentally established time dependences of strength on which they are expressed as straight lines. The range of durability must be sufficiently great (from 10 to 10^5 sec). By extrapolation one can then determine the magnitude of the "non-critical" stress.

The strength, determined by standard methods, has up to now been the main characteristic of rubbers which is used in engineering calculations. However, this property can lead to errors; for the calculation of the resistance of rubber to static loads it is absolutely necessary to start from the "non-critical" stress. This can be seen from the values for rubber with different moduli of elasticity:

Static equilibrium modulus of rubber (kg/cm²)	2	12	30
Non-critical stress (referring to a durability of one year) (kg/cm²)	0·3	7·9	39
Failure stress (found by standard methods) (kg/cm²)	7	25	124

6.2. Influence of the Type of Rubber and Degree of its Cross-linking on the Time Dependence of Strength

The intermolecular forces in polymers have not only a considerable influence on the strength in short-term tests (see Chapter 4) but also on the time dependence of the strength of elastomeric polymers.[3]

The influence of the intermolecular forces was investigated on samples of unfilled non-crystallizing rubbers: poly(butadiene) (SKB, SKBM), butadiene-styrene (SKS-30, SKS-10) and butadiene-nitrile (SKN-18, SKN-26, SKN-40). The durability changed with the conditions of the uniaxial tension under the effect of constant stresses of different magnitude at 20°C. The specimens had the shape of dumb-bells with a length of the stressed part of 25 mm, a width of 6–7 mm, and a thickness of 6 mm.

On Fig. 103 is shown the time dependence of the strength of the mentioned low modulus vulcanized rubbers with approximately the same visco-elastic modulus.† In half-logarithmic coordinates these values do not lie on a straight line [in logarithmic coordinates straight lines result and correspond to eqn. (6. 1)]. The order of distribution of the curves and their shape depends on the intensity of the intermolecular forces: the order of the curves corresponds to the grouping of the rubbers according to their glass transition points, i.e. according to their content of polar groups. Thus, with an increase of nitrile polar groups the

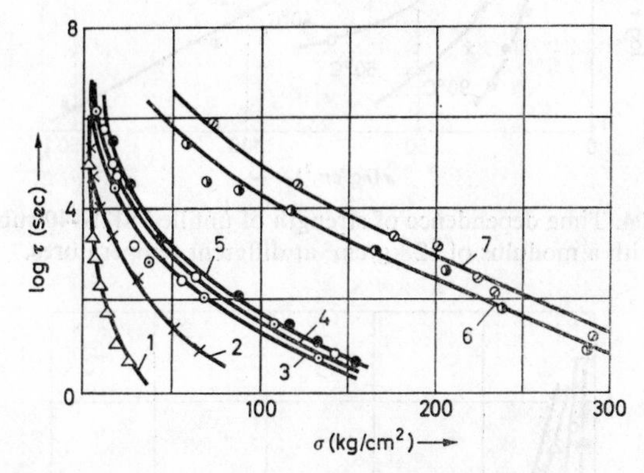

FIG. 103. Time dependence of strength of unfilled low-modulus rubbers. 1. SKS-10; 2. SKB; 3. SKBM; 4. SKS-30; 5. SKN-18; 6. SKN-26; 7. SKN-40.

curves of the time dependence of the strength move more to the right; a part of the curve straightens and becomes longer and after a longer time the curve bends upwards.

Consequently, as the intermolecular forces increase with the transition from non-polar to polar polymers a transition takes place from the laws governing the strength properties of elastomeric polymers to those which are valid in solid polymers. The increasing intensity of the intermolecular forces due to the introduction of an active filler (carbon black) also leads to a transition from one kind of time dependence of the strength to another.

† The modulus E of the vulcanized rubber measured by the method of tensile stress relaxation (24 hr, 20°C) was approximately 3 kg/cm^2.

The investigation of the time dependence of the strength of elastomers from polar SKN-40 rubber in which the intermolecular forces weaken with rising temperature showed that in this case the arrangement of the curves depends on the size of the intermolecular forces (Fig. 104). The time dependence of the

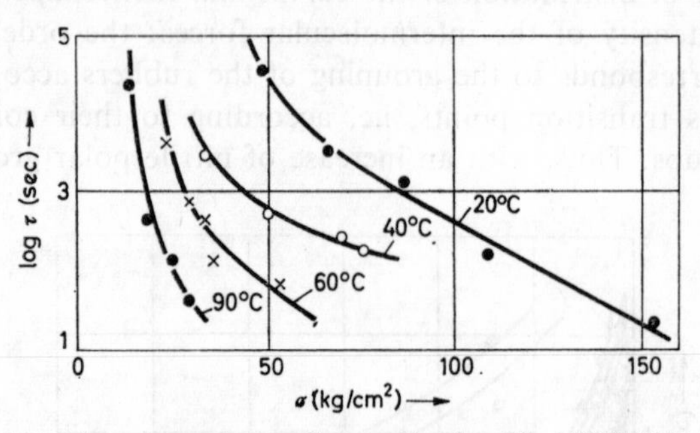

FIG. 104. Time dependence of strength of unfilled SKN-40 rubber with a modulus of 12 kg/cm² at different temperatures.

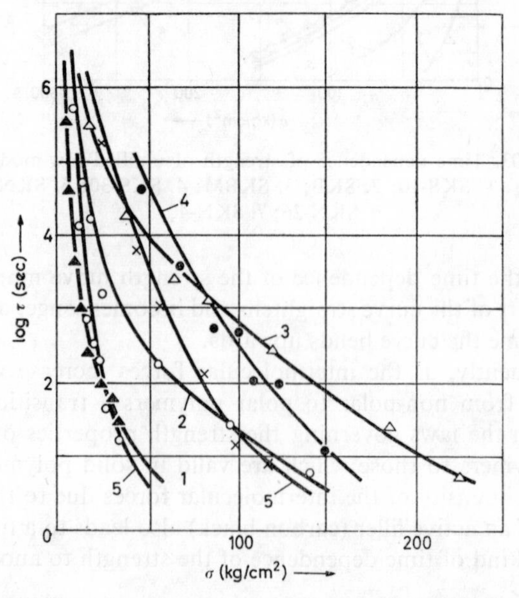

FIG. 105. Time dependence of strength of unfilled SKS-30 rubbers at 20°C with different moduli. 1. 2 kg/cm²; 2. 2·7 kg/cm²; 3. 6 kg/cm²; 4. 10 kg/cm²; 5. 15·6 kg/cm²; 6. 24 kg/cm².

strength of polar SKN-40 rubbers reminds at high temperature of the similar dependence of non-polar SKS-10 rubber at low temperatures (see Fig. 103).

The structure has also an essential influence on the strength of vulcanized rubber (density of the spatial network). Investigations in unfilled SKS-30 rubbers with a different number of cross-links and correspondingly with different values of static equilibrium moduli[3] showed that the time dependence of strength in half-logarithmic coordinates differs clearly from the linear dependences (Fig. 105). *These* values in logarithmic coordinates lie well on a straight line.

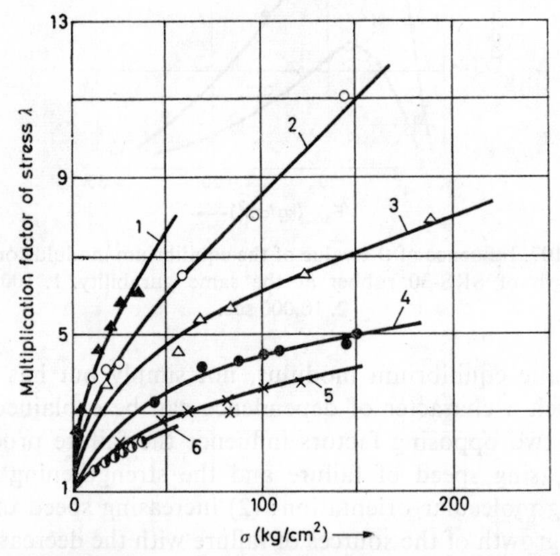

Fig. 106. Dependence between the shift of tension and the true stress of unfilled SKS-30 rubber with different moduli. 1. 2 kg/ /cm²; 2. 2·7 kg/cm²; 3. 6 kg/cm²; 4. 10 kg/cm²; 5. 15·6 kg/cm²; 6. 24 kg/cm².

It follows from the shown data that with an increase of the density of the spatial network the durability at first grows (up to the value of the static equilibrium modulus of 6–8 kg/cm²), and with a further increase of the quantity of cross-links it begins to decrease.

The dependence of the multiplication factor of tension (at a given stress) on the density of the spatial network is shown on Fig. 106.

As can be seen on this diagram, the multiplicator of tension and, consequently, also molecular orientation increases in proportion to the decrease of the number of cross-links. This must lead to an ofcrease of strength and durability. If one compares the strength in SKS-30 rubbers with different static equilibrium moduli at the same durability (such data can be taken from Fig. 105) then, as can be seen on Fig. 107, the strength changes with a decrease

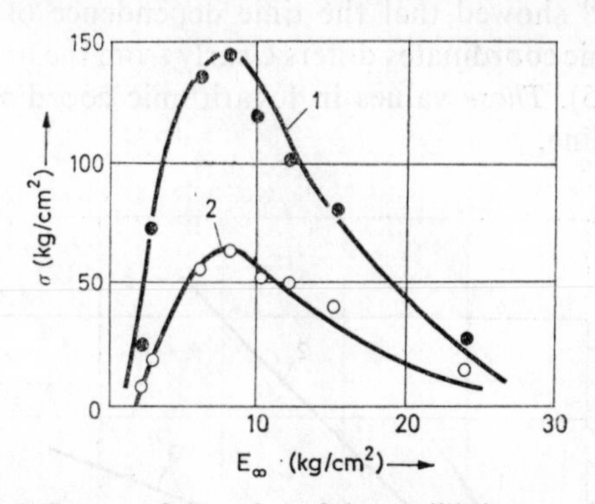

FIG. 107. Influence of the value of the equilibrium modulus on the strength of SKS-30 rubber at the same durability. 1. 100 sec. 2. 10,000 sec.

of the static equilibrium modulus, not simply but has a maximum. Such a character of dependence can be explained by the fact that two opposing factors influence the failure process: (1) the decreasing speed of failure and the strengthening with an increasing molecular orientation; (2) increasing speed of formation and growth of the sources of failure with the decreasing density of the spatial network. The two factors oppose each other, as a result of which the strength values show a maximum.

The time dependence of the strength of rubbers differs therefore from the time dependence of the strength of solids and follows eqn. (6.1). With an increase of the intermolecular forces (polarity, cross-linking and filling) it approaches the dependence which is characteristic for solids.

6.3. Influence of Temperature on the Time Dependence of the Strength of Elastomers

The influence of temperature on the time dependence of the strength of elastomers has been studied in detail[(3)] on unfilled SKS-30 rubbers in the range from 20 to 140°C (Fig. 108).

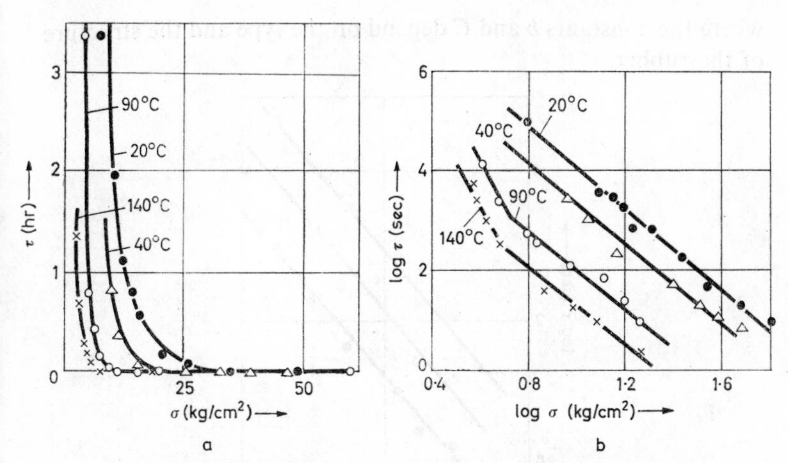

FIG. 108. Time dependence of strength of unfilled SKS rubber (equilibrium modulus of 3 kg/cm²) at different temperatures.

The curves on Fig. 108a give a graphic idea not only of the strong influence of temperature on durability but also of the practical existence of non-critical loads which decrease with rising temperature. In logarithmic coordinates (Fig. 108b) the time dependence of the strength at all temperatures is shown by straight lines in accordance with eqn. (6.1).

At high temperatures (90–140°C), in areas of great durability a deviation of the curves of time dependence of strength from the linear dependence can be observed (see Fig. 108b). It has been established that this is connected with the change of structure in the surface layer of the samples under the effect of the failure process.

The dependence of the logarithm of durability on the inverse temperature (in °K) is expressed by parallel straight lines (Fig. 109). For elastomers these dependences differ from the corresponding dependences for solid polymers where the straight lines radiate fan-like from a pole (see Chapter 1).

The temperature–time dependence of the strength of some elastomeric polymers in a limited range of stresses and temperatures is described by the following equation:

$$\tau = C\sigma^{-b}e^{U/kT} \tag{6.2}$$

where the constants b and C depend on the type and the structure of the rubber.

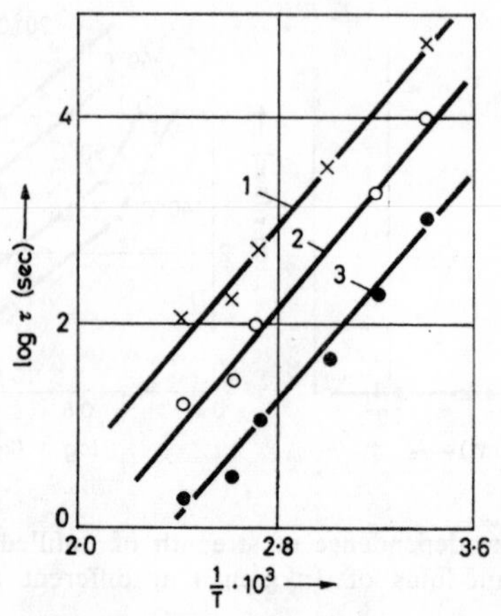

FIG. 109. Temperature dependence of durability of unfilled SKS-30 rubber (equilibrium modulus of 3 kg/cm²) at different stresses. 1. 6·3 kg/cm²; 2. 10 kg/cm²; 3. 17·8 kg/cm².

It follows from eqn. (6.2) that the constant in eqn. (6.1) depends on the temperature. Besides, it follows from it that in contrast to solids, the stress does not influence the activation energy of failure in elastomers but changes the value of the pre-exponential member. The activation energy U, calculated from the inclination of the straight lines on Fig. 109 for SKS-30 rubbers is 13·2 kcal/mol. The low value of the activation energy proves that the kinetic process of failure or rubber is evidently mainly determined by intermolecular bonds.

6.4. Theory of Time and Temperature Dependence of the Strength of Elastomers

The equation of durability

$$\tau = \tau_0 e^{(U_0 - \gamma\sigma)/kT} \tag{6.3}$$

is applicable to many materials (metals, fibres, plastics etc), and independent from the actual failure mechanism. For a number of materials like rubbers, silicate glass, some plastics and ebonite,[4]

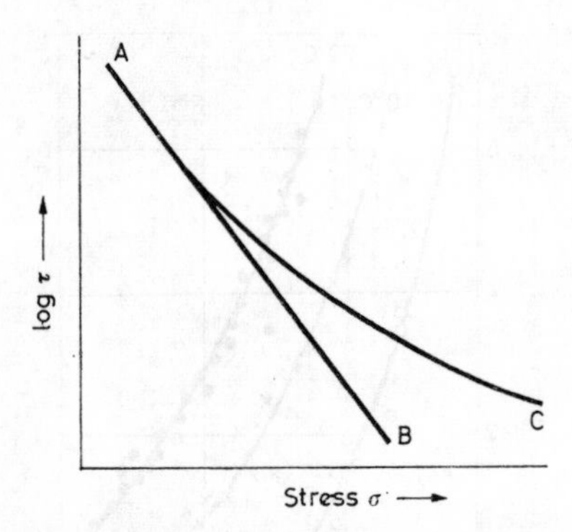

FIG. 110. Normal (*AB*) and anormal (*AC*) time dependence of the strength of the material.

a sharp deviation from eqn. (6.3) can be observed if one considers that the constants which enter this formula do not depend on the stress and the duration of the experiment. Equation (6.3) in half-logarithmic coordinates is expressed by the straight line *AB* (Fig. 110). The actual dependence of the materials mentioned which show an anomalous time dependence of strength is expressed by the curve *AC*. The reasons for this are different for different materials. The dependence of durability on stress for instance for crystallizing elastomers deviates from the straight line owing to crystallization, the extent of which depends on the size of tension. For non-crystallizing rubbers this dependence deviates from the straight line on account of the strengthening of the material with an increasing orientation at great stresses. We note that for ebonite (Fig. 111) and for some plastics a similar deviation occurs due to the change in structure during testing.

Consequently, the common cause of the anomaly of the dependence of strength of some polymers consists in the change of their structure under the effect of stress. This kind of phenomenon can also be observed in other solids. For instance, in silicate glass the fundamental factor which is responsible for the anomaly of the time dependence of the strength, is the surface-active effect of humidity. Under low stresses and small strain rates the molecules of the surface-active substance have time to migrate behind the growing crack. At great strain rates (in regions of great stress)

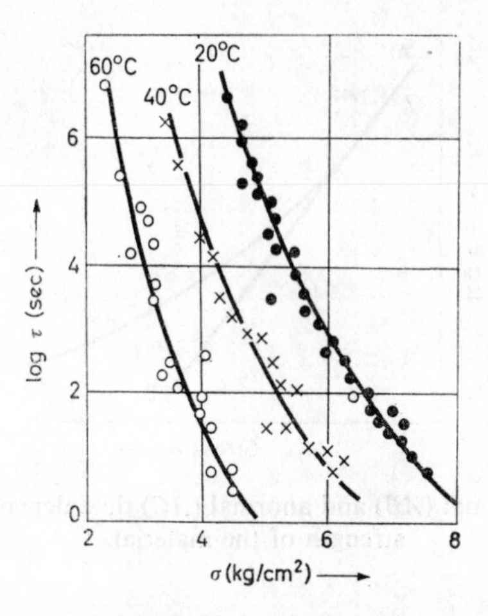

FIG. 111. Time dependence of strength of ebonite at different temperatures.[4]

these molecules have no time to follow the rapidly growing cracks and as a result the strength and durability grows (see curve *AC* on Fig. 110).

According to Zhurkov the equation of the time dependence of strength (6.3) is common to all materials, and apparent deviations from it can mainly be explained by the variability of the structure-sensitive coefficient which depends on the stress, which in its turn causes a change of the structure of the polymer.

The curve of the time dependence of strength *AC* (see Fig. 110) in logarithmic coordinates becomes in certain ranges of durability for some materials a straight line, in particular for elastomers (§ 6.1). This is the reason why time dependence of strength for the mentioned materials, which differ from each other, is expressed by an exponential law — eqn. (6.1).

Besides, the non-dependence of the activation energy of failure on stress [eqn. (6.2)] is characteristic for some rubbers.

One can consider that in prolonged tests during a great part of the time the specimen is in a condition which is near the equilibrium. Using the equation of visco-elastic deformation

$$\sigma = T\varphi(\lambda) \tag{6.4}$$

one can understand why the time dependence of the strength of visco-elastic materials is different from the time dependence of solids and why the activation energy does not depend on the stress.

In fact, we have, from eqns. (6.3) and (6.4):

$$\tau = \tau_0 e^{-\gamma\varphi(\lambda)/k} \cdot e^{U_0/kT}$$

where the coefficient γ depends on the multiplicator of tension (molecular orientation).

Calling the first two factors $\psi(\lambda)$ we get:

$$\tau = \psi(\lambda) e^{U_0/kT}$$

or

$$\tau = F\left(\frac{\sigma}{T}\right) e^{U_0/kT} \tag{6.5}$$

Approximating $[F(\sigma/T)]$, as some value of the function $C \cdot \sigma^{-b}$, we arrive at an equation which is similar to eqn. (6.2).

The independence of the activation energy of failure from stress can be explained, therefore, by the molecular–kinetic nature of the elasticity of visco-elastic materials. This fact connects the process of failure of elastomeric polymers with the process of their viscous flow, as the activation energy of both processes does not only not depend on the stress but that its values coincide in individual cases[5] (the activation energy of the viscous flow of SKS-30 rubber is 13 kcal/mol). This proves the close connection of the failure process and the viscous flow of elastomeric materials[6] and permits the possibility of applying to it the methods of coordinated relations of Ferry (see § 2.6). This connection follows also from the mechanism of slow failure of elastomers which was investigated in Chapter 3. The formation of strands in the stressed visco-elastic material is connected with surmounting of the intermolecular forces as a result of the slipping of individual parts during microstratification of the material. The process of microstratification is probably of the same nature as the viscous flow of polymers.

The formulation of a theory of time dependence of the strength of elastomers is an extremely difficult problem. This explains also the absence of any molecular theory of the strength of elastomers at present.

Literature

1. S. N. Zhurkov, B. N. Narzullayev, *ZhTF*, **23,** 1677 (1953).
2. G. M. Bartenev, S. V. Burov, *ZhTF*, **26.** 2558 (1956).
3. G. M. Bartenev, L. S. Bryukhanova, *ZhTF*, **28,** 287 (1958).
4. M. I. Bessonov, Ye. V. Kuvshinsky, *Plast. massiy*, No. 5, 57 (1961).
5. G. M. Bartenev, *DAN SSSR*, **133,** 88 (1960).
6. V. Ye. Gul', *DAN SSSR*, **96,** 953 (1954).

CHAPTER 7

DEPENDENCE OF STRENGTH OF ELASTOMERS ON STRAIN RATE AND TYPE OF FILLER

IN THIS chapter we examine the strength of elastomers at constant strain rates and the methods which allow the calculation of the durability in experiments with a constant strain rate from the time dependence of strength at static experiments. We shall also examine briefly the influence of active fillers on the strength of rubbers, and the dependence of the strength on the type of stressed state.

7.1. Strength of Elastomers at Constant Strain Rates

Standard tests on visco-elastic materials are carried out at a constant strain rate. In practice, however, one does not come across such conditions. As a rule when used, technical rubber products are either under the effect a static load, or undergo frequent deformations. The system of experiments with a constant strain rate has therefore a purely theoretical importance and is used as a comparative method of determining the qualities of the material.

In a number of works[1-3] attention has been paid to the circumstance that the strength of elastomers depends on the speed of deformation; a strictly fixed speed of deformation is therefore specified for standard experiments. Dorey investigated the strength of crystallizing elastomers on Shopper's testing machine, varying its limits from 150 to 1500% per minute. Noting the negligible change of strength, Dorey arrived at the conclusion that for failure tests the choice of strain rate within the indicated small limits is not of essential importance. However, the works of Dorey and of other investigators do not answer the more important question, to which extent the results of standard experiments correspond to the strength of real products under conditions of use.

In the work of one of the authors[4] the dependence of the strength on the strain rate of unfilled butadiene-styrene rubbers was investigated. The experiments were carried out: at great speeds on Shopper's testing machine, and at low speeds on a Polyan dynamometer. The connection between the true strength σ and the strain rate of deformation ($v = d\varepsilon/dt$) in logarithmic coordinates is expressed by a linear dependence (Fig. 112), whereas in other coordinates the dependence is non-linear.

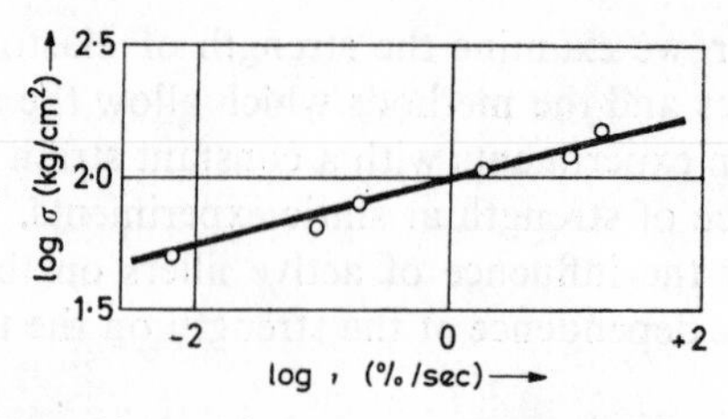

FIG. 112. Dependence of the true strength on the strain rate[4] (strips of SKS-30 rubber, 1 mm thick).

For the comparison of strength at static experiments and experiments with constant speed of deformation, samples of the same elastomer were tested on Shopper's machine under standard conditions, and under a static load for 2 hr. In the first case the nominal strength turned out to be 26 kg/cm^2, and in the second about 8 kg/cm^2. Consequently, the strength of elastomers in a static loaded state is far smaller than under standard experiments, and the difference is the more pronounced, the smaller the static load. It is therefore absolutely necessary to carry out tests on static fatigue at the same time as the standard experiments.

The calculation of the part played by time in the failure process of rubbers must lead to the conclusion that the strength grows with increasing rate of loading. However, at great strain rates anomalous deviations from this principle can be observed, as was discovered by Zhurkov and co-workers[5] for natural and synthetic rubbers. The experiments were carried out with a wide range of strain rates—from 0·2 cm/sec to 31 m/sec, which corresponds to deformation speeds from 7% to $1{\cdot}3\times10^5$% per second. In some experiments the deformation speed went down to $10^{-5}\%$ per second. With a rising deformation speed the modulus of visco-elasticity rises, and the strength of the rubber changes.

For unfilled non-crystallizing rubbers the strength grows, as a rule, with an increase of the strain rate; for crystallizing rubbers the strength is great in some ranges of strain rates, then falls with increasing strain rates, passes a minimum, and increases again.[5] A similar character of change of the strength of elastomers was observed in Chapter 6.

The anomalous reduction of strength of crystallizing elastomers in some ranges of strain rates can be explained by the retardation of the process of crystallization at rapid deformation (for an almost complete crystallization of NK rubber during tension a time of the order of 1–2 sec is required). At rapid deformation the process of crystallization has no time to develop sufficiently, and the elastomer has less strength than under slow tensioning. In ranges of small speeds of deformation the process of crystallization under stress has time to develop fully. The decrease of strength at further lowering of the deformation rate can therefore be explained by the usual time dependence of strength.

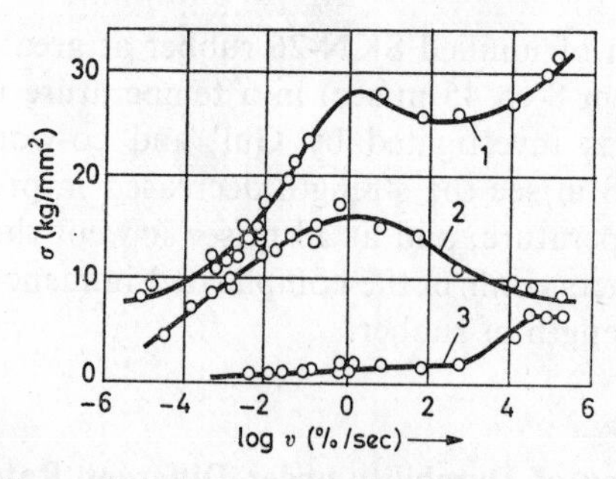

FIG. 113. Dependence of true strength of crystalline and non-crystalline rubbers on the strain rate.[5] 1. NK rubber. 2. SKS-30 rubber carbon black. 3. SKS-30 rubber without carbon black.

And so the strength of crystallizing rubbers grows at first with decreasing strain rates and then falls (curve 1, Fig. 113). The maximum strength is found in NK rubber and poly(chloroprene) rubbers in the range of strain rates from 0·1 to $10^3\%$ per second. The whole curve of strength of crystallizing rubber can be divided in three regions: (1) the region of slow tension, where the increase of strength is caused by the time dependence of strength of the rubber which had started to crystallize during tensioning; (2) the transition region where an anomalous character of the change of strength can be observed, due to insufficiency of time for full crystallization; (3) the region of high strain rates, where the strength characteristics are determined by the time dependence of strength of non-crystallizing rubbers. In crystallizing rubbers at great strain rates a deviation of the growth of strength from the general principles with increasing strain rates can thus be observed.

The strength of unfilled non-crystallizing SKS-30 rubber grows throughout the whole range of strain rates (curve 3, Fig. 113). At great speeds its strength approaches the strength of filled elastomers from the same rubber. On the strength curve of carbon-black filled SKS-30 (curve 2, Fig. 113) there is a maximum like with crystallizing rubbers. Besides, under slow tension the strength of filled rubber is considerably higher than of unfilled which, anyhow, is well known from the usual tests on the testing machine. These results show that the interaction of filler and rubber is a complicated process which depends on the speed of deformation.

The strength of unfilled SKN-26 rubber at great speeds of deformation (from 8 to 45 m/sec) in a temperature range of -20 to $+100°C$ was investigated by Gul' and co-workers.[7] At a strain rate of 8 m/sec the strength decreased in proportion with the rising temperature, and at 28 m/sec it went through a minimum. These facts confirm the complicated influence of the strain rate on the strength of rubber.

7.2. Calculation of Durability under Different Rates of Loading

The behaviour of the same material under different rates of loading can be sharply different. However, in many cases this behaviour reflects the same principles of failure under load, but under different conditions.† Generally, if one knows these principles one can predict the behaviour of the material under any

† No change of the structure of the original material must take place during the experiments.

strain rate. At the same time it is indispensible to keep in view the difficulties of developing methods of calculations of the strength for any condition of stress on account of the insufficient knowledge of the behaviour of materials under complicated experimental conditions as well as the lack of development of the theory of failure of such materials as plastics and elastomers.

One of the most important strength characteristics of the material is its durability (the time which elapses from the moment of the application of the load to the complete failure of the specimen). Usually, the durability of the material under different conditions which imitate those of actual use is determined directly from the experiment, and is not calculated.

An exceptional interest presents the simple system of constant tensile loads which has been completely determined and clarified (see Chapters 1 and 6). The question crops up whether one can, knowing the time dependence of strength with this type of experiment, determine mathematically the durability of the material under any other type. The answer to this question has an enormous practical importance.

For the development of mathematical methods it is first of all important to know whether the material under test has a mechanically reversible or non-reversible process of failure. As, for instance, Haward's,[8] Zhurkov's and Narzullaev's experiments showed (see Chapter 1) the accumulation of flaws in plastics is a non-reversible process. If one defines with τ the time of failure of the specimen at a given constant stress σ with t_1, the time of loading an identical specimen with the same stress ($t_1 < \tau$), after which the load will be removed, and with t_2 the time during which this specimen, which had been unloaded, breaks under the same stress σ, then it is found that for plastics the following law of additive durability is observed:

$$\tau = t_1 + t_2 \tag{7.1}$$

This means that the number of local flaws which had accumulated in the material before the relaxation (microcracks, silver cracks, etc.) do not disappear during the period of relaxation. The causes for the irreversibility of the failure process in plastics are clear from the mechanism of their failure (see Chapter 3).

Let us examine a material for which failure is an irreversible process (first hypothesis); let us also examine the speed of failure (under which we understand the speed of growth of cracks or small tears) which depends only on the nominal stress σ, and not on the amount of already existing flaws (second hypothesis). For such a material the postulate of Bailey[9] will be correct, which is inferred from the following considerations.

If a variable stress $\sigma(t)$ is divided into a number of constant stresses σ_i which act during consecutive and extremely small intervals of time Δt_i, then the change of σ in that time interval can be disregarded, and the stress can be regarded as σ_i. It follows now from the first and second hypothesis that in the time Δt_i the arising local failures of the specimen amount to the fraction $\Delta t_i/\tau_i$, where τ_i is the durability under constant stress σ_i. When the sum $\Sigma \Delta t_i/\tau_i$ equals 1, a complete break of the specimen takes place (rupture). At $\Delta t_i \to 0$ we arrive at Bailey's postulate of failure:

$$\int_0^{\tau'} \frac{dt}{\tau(\sigma)} = 1 \tag{7.2}$$

where τ' is the durability of the specimen at any given type of experiment; dt is the infinitely small time interval during which the changing stress σ can be regarded as constant; $\tau(\sigma)$ is the durability under constant tensile stress, determined by the known time dependence of strength.

The two hypotheses are not valid for all materials and conditions of experiments. Actual failure of brittle solids, and also of elastomers agree with the third hypothesis of Alfrei[10] which, strictly speaking, is not compatible with the condition (7.2), as according to this hypothesis the speed of failure depends on the amount of the already existing local flaws in the material. Therefore eqn. (7.2) is only approximately correct under not very low strain rates for brittle bodies and elastomers, when the mirror zone (for elastomers the rough zone) occupies only a small part of the surface of the break. In this case one can practically consider that the speed of the process of failure is determined by the nominal stress σ, which is not very different from stress σ', and is calculated on the sound cross-section of the sample.

For plastics at low strain rates, when the process of cracking (silver) proceeds throughout the experiment, Bailey's postulates are almost exactly fulfilled as the speed of growth of silver cracks, in contrast to ordinary failure cracks, is in practice determined by the nominal stress (see Chapter 3).

If we examine the problem quite strictly, then Bailey's criterion demands the fulfilment of one more additional assumption which is usually not mentioned. This is the assumption of the independence of the critical length of crack l_K from the nominal stress σ, during which the transition from slow to rapid failure takes place. Under the failure fraction $\Delta t_i/\tau_i$ we understand the expression $\Delta l_i/l_K$ through which the growing crack went in the time Δt_i at $\sigma = \sigma_i$. Assuming the validity of the first two hypotheses and the correctness of the additional assumption, the crack during the time Δt_i under the nominal stress σ_i grows by the length $\Delta l_i = l_K (\Delta t_i/\tau_i)$. In agreement with the second hypothesis the crack grows uniformly. As $l_K =$ const., one can lift l_K from the sum in the expression: $l = l_K \Sigma(\Delta t_i/\tau)$, where l is the length of the crack which has grown during the time $t = \Sigma \Delta t_i$. Allowing that Bailey's postulates are fulfilled, i.e. the sum $\Sigma(\Delta t_i/\tau_i) = 1$, we obtain $l = l_K$, i.e. the specimen fails in agreement with Bailey's formula.

The additional assumption, as well as the two hypotheses, approximates conditions of brittle failure, as l_K depends essentially on the nominal stress (see Chapter 1).

One can show that in general the condition of failure for brittle fracture at any strain rate looks like this.†

$$\int_0^{\tau'} \frac{dt}{\tau(\sigma)} = 1 - \frac{\alpha}{DL^2} \int_0^{l_K} (L-l)l \exp\left[-\frac{\alpha\sigma(L+l)}{L-l}\right] \frac{d\sigma}{dt}\, dl$$

where $\alpha = \gamma/kT$; $D = \lambda\gamma_0 \exp(-U_0/kT) \exp(a/k)$; L is the width of the sample.

We apply eqn. (7.2) to the calculation of durability of elastomers under constant strain rate, considering that the time dependence of strength for unfilled elastomers looks like this: $\tau = B \cdot \sigma^{-b}$, and that the sample is stretched with the constant speed v. This means that $\varepsilon = \varepsilon_0 + vt$, where ε is the deformation of the specimen at the moment of time t, ε_0 is its original static deformation at $t = 0$.

For unfilled SKS-30 rubbers at equilibrium or near equilibrium slow tensioning,[11] a straight proportionality exists between true

† For explanation of the symbols in this equation see pp. 48 and 49.

stress and deformation, practically right up to an elongation of 400%, i.e. $\sigma = E_\infty \cdot \varepsilon$, where E_∞ is the static or the apparent static equilibrium modulus. For unfilled rubbers, as schematically illustrated on Fig. 114, the region of small strain rates corresponds to the part of the curve *AB*, where the visco-elastic modulus differs in practice little from the static equilibrium modulus. In the region of medium strain rates the modulus *E* grows

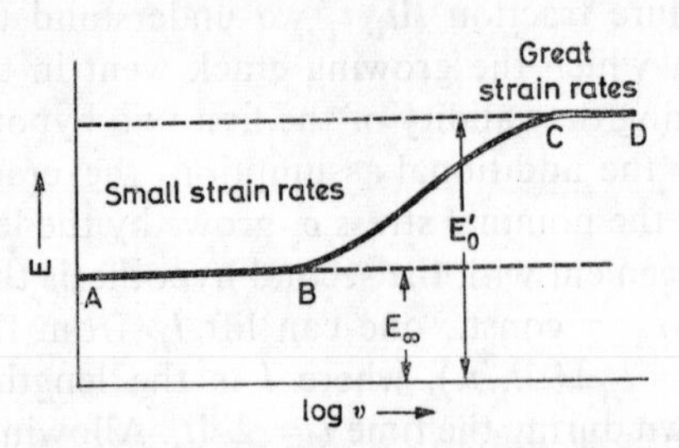

FIG. 114. Influence of strain rate on the visco-elastic modulus of non-crystalline unfilled rubber (schematic).

with increasing strain rates. At very great strain rates, near the speed of sound in rubber, the visco-elastic modulus reaches the value E_0' and does also virtually not depend on the strain rate. This modulus was called the primary visco-elastic modulus.

For unfilled SKS-30 rubbers[12] the height of the "threshold" $E_0' - E_\infty$ (see Fig. 114) at 20°C is 11·5 kg/cm². The speed of elastic waves at 20°C in SKS-30 rubber is, from the same sources, 65 m/sec. The great strain rates in Zhurkov and co-workers'[5] experiments (between 10 and 30 m/sec) approach this order of magnitude.

It follows from these figures that in the region of great strain rates unfilled SKS-30 rubber obeys the law $\sigma = E_0'\varepsilon$. If we assume that for this rubber one can in the transitional region of strain rates approximate the stress curve by a straight line, we can write the law of deformation at all strain rates

$$\sigma = E\varepsilon \tag{7.3}$$

where σ is the true stress; $E = E(v)$ is the visco-elastic modulus which depends on the strain rate.

Under small strain rates the failure process of elastomers is caused by the growth of small tears, i.e. irreversible. The equation of failure (7.2) can be applied with a sufficient degree of accuracy only if the rough zone of the failure surface is small. In view of the slow development of the rough zone this condition is applicable in a wide range of strain rates, and the more so at great speeds. The time dependence of the strength of rubbers which takes into account the law of deformation (7.3) can be written down in this form:

$$\tau = BE^{-b}\varepsilon^{-b}.$$

The substitution of τ in the equation of failure (7.2) gives

$$\frac{E^b}{B}\int_0^{\tau'} (\varepsilon_0+vt)^b\, dt = 1.$$

From this, taking into account that the failure stress $\sigma = \sigma_0 + Ev\tau'$ (where σ_0 is the stress which corresponds to the initial static deformation ε_0), we obtain the following dependence of the true failure stress σ on the strain rate v:

$$\sigma = [BE(1+b)v]^{1/(1+b)} + \sigma_0.$$

In practice one meets mostly with $\varepsilon_0 = 0$ and $\sigma_0 = 0$. We obtain therefore finally the following equation:

$$\log \sigma = C_0 + C_1 \log v \tag{7.4}$$

where the constants C_0 and C_1 are

$$C_0 = \frac{1}{1+b}\log[(1+b)EB]; \qquad C_1 = \frac{1}{1+b} \tag{7.5}$$

This result agrees with the experimental data (see Fig. 112) for unfilled SKS-30 rubber, according to which a linear dependence between the logarithm of true strength and the logarithm of strain rate is observed in regions of small deformation speeds. The values of Fig. 112 correspond to the comparatively narrow range of strain rates from 0·01 to 20% per second.

At an infinite decrease of the strain rate the rough zone in the moment of break becomes greater and greater and eqn. (7.2) less and less applicable. However, qualitatively one can predict that the strength must decrease sharper with the decrease of the deformation speed than on part *AB* (Fig. 115). This follows from the fact that at low strain rates the failure speed of the rubber grows relatively quicker. To the same small stress σ corresponds therefore a lesser durability or a greater strain rate than it would follow from Bailey's equation (7.2). As a result, the experimental data, instead of lying on the chain-line *A'A*, move to the right and lie on the solid curve which turns swerving downwards.

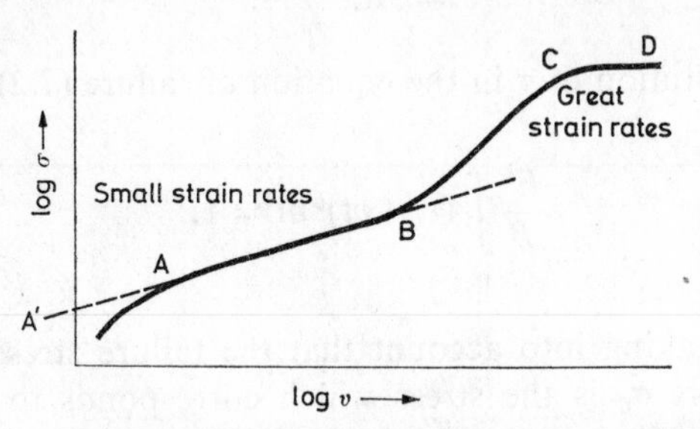

FIG. 115. Influence of strain rates on the true strength of non-crystalline unfilled rubber (schematic).

The region *BC* on Fig. 115 corresponds to the region of appreciable growth of the visco-elastic modulus from E_∞ to E_0' on Fig. 114. In this region the rates of loading grow quicker than the speeds of deformation which leads to a more pronounced growth of strength. Equation (7.4) describes this curve, where in the region *BC* the constant C_0 grows together with the modulus *E*. The latter leads to a rise of the curve until the modulus becomes E_0'. After this the dependence acquires its former linearity. Of course, if the assumption is correct, at rapid failures the course of the time dependence of strength and the value of the constant *b* remains as before.

7.3. The Influence of Fillers on the Strength of Rubbers

The basic problem of rubber technology is the production, from rubber, fillers and other ingredients — of a material which possesses maximum mechanical strength and elasticity and also the capacity to preserve its properties during long periods of use.

According to their influence on the strength of rubbers fillers can be divided in two groups: active fillers which increase the strength of the rubber, and inert fillers which practically do not change its strength. The principal active filler for rubbers is carbon black. The ideas about the part played by fillers and the mechanism of their effect, as they existed before 1941, are found in Margaritov's book.[13] According to Boonstra,[14] the reinforcing effect of fillers consists not in the increase of the absolute value of strength but in the change of its temperature dependence. The introduction of carbon black shifts the maximum temperature dependence of strength into the region of higher temperatures (Fig. 116). The

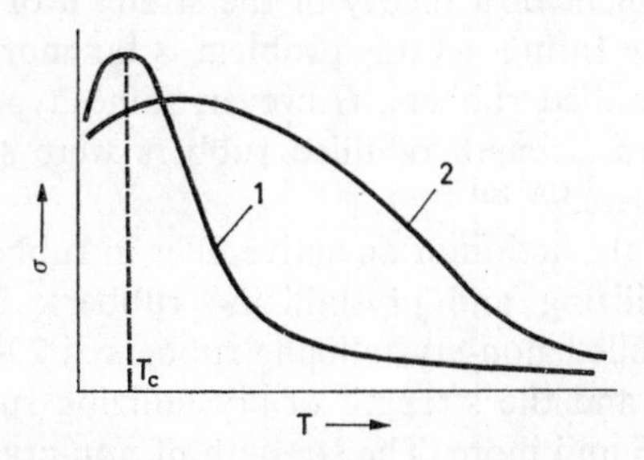

FIG. 116. Influence of filler on the time dependence of strength of rubber under tension (according to Boonstra). 1. Unfilled rubber. 2. Filled rubber.

active filler makes the temperature dependence of strength smoother, lowering the strength a little near the glass transition point but then raising it over a considerable temperature range which is of interest to the users of rubbers. In the same temperature region the filler increases also the visco-elastic modulus of the rubber.

The reasons for these phenomena are discussed in different theories on the strengthening of rubber. In the majority of these the main point of investigation is the influence of the fillers on the deformation — and relaxation properties of rubbers from the point of view of the nature of the bonds which exist between the filler particles and the macromolecules of the rubber. The strength of the materials as such is not examined in these theories but the strength of the structure, for instance the strength of the rubber-filler bonds and its influence on the deformation properties and on the flow of elastomeric polymers.[15-25]

The particles of the active filler (carbon black) change the structure of the rubber and enlarge the number of types of bonds.

Besides chemical cross-links and secondary local intermolecular links in filled rubbers there appear bonds between filler and rubber which are mainly adsorption bonds. These bonds of carbon black–rubber have a strength which is intermediate between the strength of chemical and intermolecular links between the rubber chains which leads to the appearance of clearly expressed thixotropic properties at deformations of carbon black filled rubbers (Patrikev–Mullins effect).[26–28] The thixotropic properties can also be observed in unfilled crystallizing rubbers.[29]

The development of a theory of the strength of filled rubbers is a thing of the future, as this problem is far more complicated than that of unfilled rubbers. However, some hypotheses which explain the high strength of filled rubbers were suggested in a number of papers.[16, 30]

The effect of the action of an active filler in rubbers is different for non-crystallizing and crystallizing rubbers. The apparent strength of unfilled non-crystallizing rubbers is 20–30 kg/cm^2 or a little higher, and the strength of crystallizing rubbers reaches 200–500 kg/cm^2 and more. The strength of non-crystallizing rubbers can be increased up to the strength of crystallizing rubbers by the introduction of active fillers, whereas their introduction into crystallizing rubbers does not give any noticeable increase of strength (Table 9).

TABLE 9

Influence of Filler on Tensile Strength of Elastomers from Different Rubbers

Rubbers	Apparent strength (kg/cm^2)		Coefficient of strengthening
	Unfilled elastomer	Filled elastomer	
Non-crystallizing rubbers			
sodium butadiene	15–20	120–180	5–12
butadiene-styrene	20–30	150–200	5–10
isoprene	20	80–120	4– 6
Crystallizing rubbers			
natural	200–300	300–330	1–1·6
Chloroprene	150–200	200–250	1–1·7
butyl rubber	250–300	300–350	1–1·4

The figures on Table 9 refer to tests on rubbers under great tension. If one compares the strength of non-crystallizing and crystallizing rubbers under conditions where failure takes place with small deformations (for instance under compression, shear, abrasion), and also when testing with repeated deformation the value of the strength of both types of rubbers differs only little.

The reason for the high strength of filled rubbers can be explained by two fundamental hypotheses: Alexandrov's and Lazurkin's relaxation hypothesis,[30] and Dogadkin's orientation hypothesis.

According to Alexandrov and Lazurkin the mechanism of strengthening through active fillers consists mainly in the fact that the filler promotes the equalization of the overstress in the material. The construction of the spatial network of the rubber is irregular; therefore a considerable overstress can appear in some places of the network under deformation which leads to the formation of nuclei of break, whilst the greater part of the material is hardly stressed. In rubber with active fillers a considerable part of the chains of the spatial network is adsorbed on the surface of the filler particles. The bonds of the polymer chains with the carbon black particles are stronger than those between the chains but weaker than the individual chains and the spatial network as a whole. Due to this the overstress in the network cannot exceed the forces of the interaction of the filler particles with the rubber chains. If the overstress in some part becomes greater than the force of the linkage of the rubber with the filler, the stressed chain will break off (desorb) and will unload instead of rupturing. At the same time the low stressed chains do not break off from the filler particles and take a part of the load on themselves. In this way the stress is distributed more evenly in the filled rubber between the elements of its spatial network. Rupture sets in when the process of the levelling of the stresses ceases due to the desorption of the chains.

The relaxation character of this strength mechanism of filled rubbers is shown by the fact that with rising temperature (and decrease of strain rate) the probability W of the breaking off of the chain from the filler particle grows under the same stress, and the mean relaxation time of the desorption process τ_D (magnitude, inverse probability W) decreases. If the testing time is considerably above τ_D the relaxation mechanism of the action of the filler does not appear, and no strengthening effect is observed.

If the length of the experiment is much less than τ_D, and this is possible at low temperatures and at high strain rates, the elastomer will break sooner than the desorption mechanism will come into force. This is the reason for the appearance of the peak on the curve of the dependence of strength on the strain rate for filled SKS-30 rubber (see Fig. 113, curve 2), and also the peak on the curve of the temperature dependence of strength (see Fig. 116).

When tensioning unfilled crystallizing rubbers the appearing crystallites play much the same part as the filler particles but the surface sorption is here replaced by the ordered packing of the polymer chains from the amorphous phase on the crystallite surfaces. The small influence of active fillers on the strength of crystallizing rubbers (see Table 9) is caused by the circumstance that during tension a great number of crystallites appear in the rubber, nearly the optimum number of active filler particles. Therefore, the introduction of an active filler raises the strength only insignificantly, and in great quantities can even lead to a reduction due to the lowering of the capacity for crystallization of the rubber.

A sharp decrease in the strength of crystallizing rubber near 100°C can be explained by the melting of the crystalline phase.

The filler can take part in the formation of two types of structures: (1) particles of filler or their aggregates are distributed at random in the rubber mass, and basically isolated from each other by layers of rubber; (2) the filler particles form a spatial network. The character of the forming structure depends on the quantity of introduced filler, its degree of dispersion, and also on the relation of the strength of the filler–rubber, and filler–filler bonds. If the bonds of filler with rubber are stronger, mainly a structure of the first type is formed (such a structure is formed by non-active and little-active fillers). If, however, the filler–filler bonds are the stronger ones, then chain structures are formed which serve as a matrix on which the rubber molecules are packed and orientated. Such structures are formed by active fillers.

The influence of the structures of active fillers in the rubbers on their strength is very considerable.[16] For instance, carbon black structures, from a certain degree of filling onwards are a spatial network which consists of carbon black chains, the number of which grows in proportion to the increase of filling, and reaches its highest value at approximately 20–30% of filler to rubber. The strength has a maximum in the region of those values, at which the formation of the carbon black structures comes to an end.

According to Dogadkin[31, 32] the elastomer chains in the rubber are distributed mainly along the short carbon black chains, forming regions with an orientated phase. These orientated regions are obviously blocks of macromolecules.

Thus, with the introduction of an active filler into the elastomer, a transition from the random structure of the polymer chains to the block structure of the rubber takes place with an increase of its quantity and the formation of short chains. This structural peculiarity of filled rubbers is probably the reason for its high strength.

It is known that there is an optimal content of active filler at which the true strength reaches a maximum.†

As can be seen on Fig. 117, during vulcanization the strength passes beyond a maximum not only for unfilled but also for filled rubbers.

The influence of the type and quantity of filler on the strength of non-crystallizing SKS-30 rubbers during tests on the testing machine was examined in detail by Bartenev and Byelostotzka. As fillers were employed: carbon black (active filler), graphite (semi-active), and ground glass (non-active filler).

When investigating the strength of rubbers in dependence on the content of different fillers (Fig. 118) it was found that the non-active filler lowers the strength of the elastomer with increasing quantity at first rapidly, and then slower. Semi-active filler has in small quantities practically no influence on the strength of the elastomer, and with increasing quantity causes a small increase of strength.

Active fillers cause already in small quantities an increase of strength. With approximately 12 vol% of carbon black the true strength reaches a maximum, and with a further increase of carbon black it falls. At the same time it is known that active fillers at a concentration of 10–15 vol% form a continuous spatial structure which causes, according to Dogadkin, the emergence of an orientated rubber phase. Consequently, both processes (the strengthening of the rubber and the formation of carbon black

† In engineering the strength is usually calculated on the initial cross-section of the specimen. By this method of establishing the experimental data the picture of the influence of filler on strength becomes blurred.

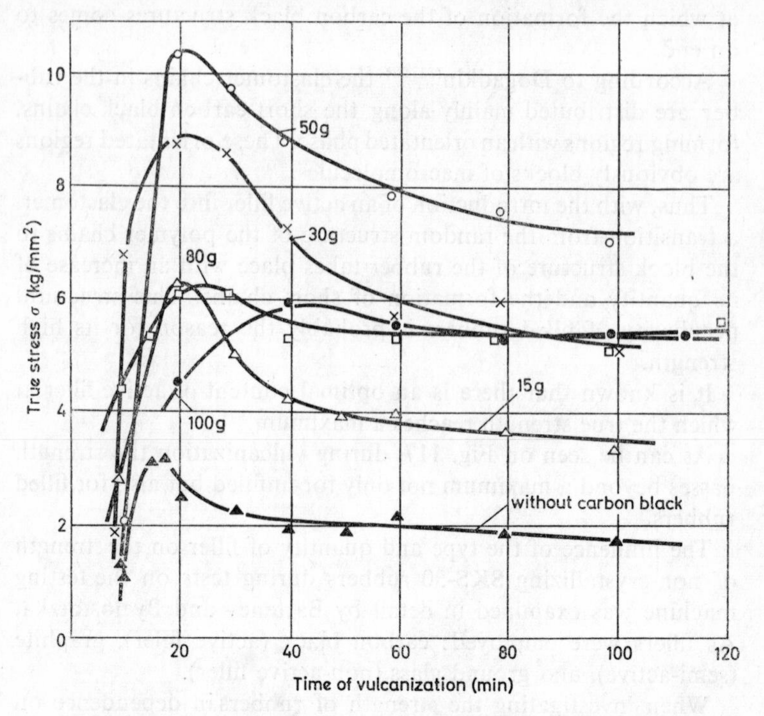

FIG. 117. Influence of the time of vulcanization at 143°C on the true strength of SKS-30 rubber (100 g rubber, 3 g sulphur, 1 g Captax, 2 g stearic acid, 5 g ZnO) with different contents of lamp black. (according to Bartenev's and Belostotzka's data).

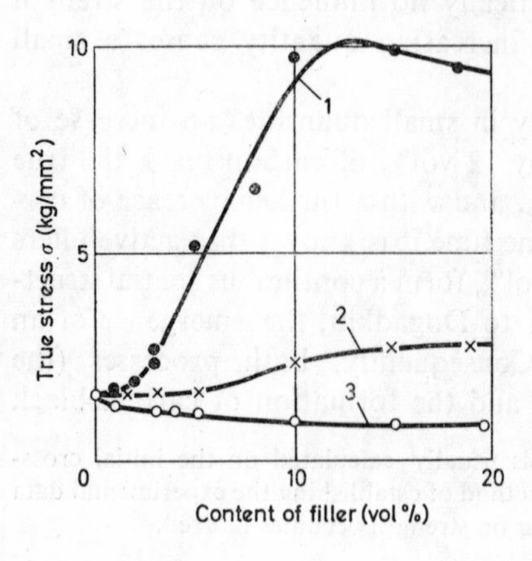

FIG. 118. Influence of filler on the true strength of SKS-30 rubber. 1. Carbon black. 2. Graphite. 3. Ground glass.

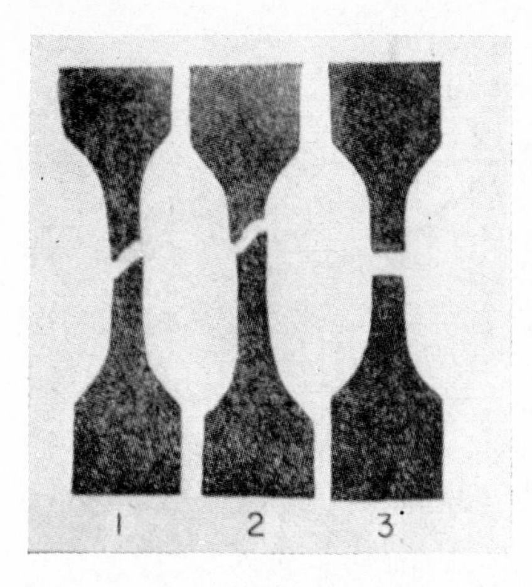

FIG. 119. Types of failure of technical rubber under tension. 1. Shear. 2. Shear and rupture. 3. Rupture.

FIG. 120. Failure of rubber sample under compression (chief tear under an angle of 45°).[33]

FIG. 121. Failure of rubber sample under compression (tear on the lateral surface of the rubber packing).[33]

structures) are interrelated which is the confirmation of the second hypothesis of strengthening.

In conclusion we note that the high strength of filled rubbers can partly be explained by the inhibition of the growth of cracks and other defects in the presence of a filler which creates steric hindrances for growing cracks.

7.4. Influence of the Type of Stress on the Failure of Rubbers and Elastomers

The most critical type of stressed state for elastomers is the tension under which failure is observed: through rupture, through shear and through a combination of rupture and shear (Fig. 119). A complicated failure is observed in the specimen when in neighbouring levels two small tears grow towards each other and then join by local shear.

Under compressive loads the elastomer fails usually by shear or by rupture if the compression proceeds without slipping on bearing surfaces. Such types of failure are frequently met with in elastomeric strips[33] which are used to make sealing rings for stuffing boxes. Under great and prolonged compression there appear, due to fatigue, slight tears in the packing which are either under an angle of 45° to the direction of the compression (Fig. 120) or parallel to the contacting surfaces (Fig. 121), where the side of the deformated packing bulges most. The rectangular cross-section of the packing takes on the shape of a "barrel" (Fig. 122), whereby in the points *A* and *B* under great compression a tension appears. The greater the compression of the packing, the

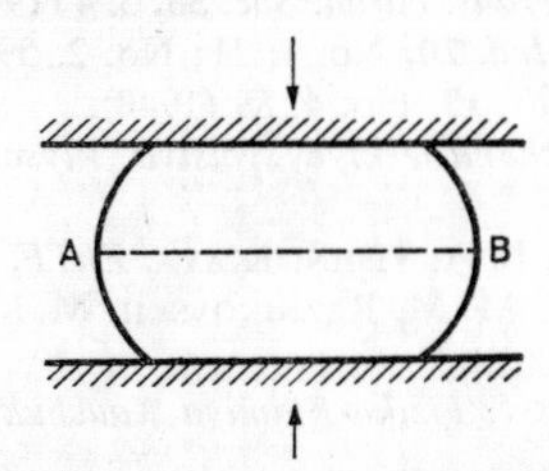

FIG. 122. Schematic deformation of rubber specimen under compression without slipping along the supporting surfaces (formation of "barrel").

more the free side surfaces bulge. Along the line *AB* of that "barrel" a growth of small tears takes place from the surface into the depth of the material.

Under biaxial tension which rubber sheets, balloons, valves, etc., undergo in use, the failure takes place through rupture. The line of such a rupture can travel along the sheet in the most fantastic way, as its path is determined both by the highest stressed and the weakest places of the sheet which emerge due to the irregularity of the thickness of the sheeting, the heterogeneity of the material, etc. As a result of this the line of the break can close up. For instance, at failure of rubber sheets and coatings "flakes" may break off.

On a completely homogenous material with constant sheet thickness the probability of failure is the same in every direction, and it should take place in small areas all over the sheet.

It should be noted in conclusion that very little work has been done on the failure of polymers and glasses under different types of stress.[34] (See also § 1.14).

Literature

1. W. Reece, *IRI Trans.* **11,** 312 (1935).
2. O. Davies, R. Horrobin, *Rubb. Chem. Technol.* **10,** 180 (1937).
3. R. Dorey, *IRI Trans.* **7,** 158 (1931).
4. G. M. Bartenev, *DAN SSSR*, **82,** 49 (1952).
5. S. N. Zhurkov, T. P. Sanfirova, E. Ye. Tomashevskii, *Viysokomol. soyed.* **4,** 196 (1962).
6. D. S. Villars, *J. Appl. Phys.* **21,** 565 (1950).
7. V. Ye. Gul', C. V. Kovriga, Ye. G. Yeremina, *Viysokomol. soyed.* **2,** 160 (1960).
8. R. N. Haward, *Trans. Farad. Soc.* **38,** 394 (1942).
9. J. Bailey, *Glass Ind.* **20,** No. 1, 21; No. 2, 59; No. 3, 95; No. 4, 143 (1939); *Ceram Abs.* **19,** No. 4, 89 (1940).
10. T. Alfrei, *Mekhanicheskiye svoistva Viysokopolimerov*, Izdatinlit, 1952.
11. G. M. Bartenev, L. A. Vishnitskaya, *ZhTF*, **20,** 859 (1950).
12. G. M. Bartenev, M. M. Reznikovskii, M. K. Khromov, *Kolloid zh.* **18,** 395 (1956).
13. V. B. Margaritov, *Fiziko-Khimiya Kauchuka i reziniy*, Goskhimizdat, 1941, p. 305.
14. B. S. T. T. Boonstra, *Rubb. Chem. Technol.* **23,** 338 (1950).
15. V. A. Kargin, G. L. Slonimskii, Ye. V. Reztsov, *DAN SSSR*, **105,** 1007 (1955).

16. B. A. Dogadkin, *i dr.*, *Kolloid. zh.* **8,** 31 (1946); **10,** 357 (1948); **22,** 663 (1960); *Trudiy III konferentsii po kolloidnoi khimii*, Izd. AN SSSR, 1955, p. 363.
17. Ya. B. Aron, P. A. Rebinder, *DAN SSSR*, **52,** 235 (1946).
18. A. F. Blanchard, D. Parkinson, *Ind. Eng. Chem.* **44,** 796 (1952); *J. Polymer Sci.* **14,** 355 (1954); *IRI Trans.* **32,** 124 (1956).
19. Bueche, *J. Appl. Phys.* **23,** 154 (1952).
20. H. K. de Decker, *Rubb. Abs.* **34,** No. 11, 507 (1956).
21. A. J. Carmichael, H. W. Holdaway, *J. Appl. Phys.* **32,** 159 (1961).
22. V. A. Garten, *Nature*, **173,** 997 (1954).
23. J. W. Watson, *IRI Trans.* **32,** 204 (1956).
24. G. M. Bartenev, L. A. Vishnitskaya, *Kolloid. zh.* **18,** 135 (1956).
25. G. M. Bartenev, N. V. Zakharenko, *Kolloid. zh.* **24** (1962).
26. L. Mullins, *J. Phys. Coll. Chem.* **54,** 299 (1950); *J. Rubb. Res. Inst. Malaya*, **16,** 275 (1947).
27. *Obshchaya Khimicheskaya tekhnologiya*, edited by S. I. Vol'fkovicha, vol. II, ch. 12, Goskhimizdat, 1946.
28. M. M. Reznikovskii, L. S. Priss, B. A. Dogadkin, *Kolloid, zh.* **16,** 24 (1954).
29. G. M. Bartenev, L. A. Vishnitskaya, *ZhTF*, **22,** 416 (1952).
30. A. P. Aleksandrov, Yu. S. Lazurkin, *DAN SSSR*, **45,** 308 (1944).
31. B. A. Dogadkin, *Trudiy III Vsesoyuznoi Konferentsii po kolloidnoi khimii*, Izd. AN SSSR, 1955, p. 136.
32. B. A. Dogadkin, D. L. Fedyukin, V. Ye. Gul', *Kolloid. zh.* **19,** 287 (1957).
33. G. M. Bartenev, *Khim. prom.* No. 8, 15 (1955).
34. R. N. Haward, *The Strength of Plastics and Glass*, New York, 1949.

CHAPTER 8

STRENGTH AND FATIGUE OF ELASTOMERS UNDER CYCLIC LOADINGS

THE strength properties of elastomers under single and repeated deformations differ greatly from each other, as at cyclic loadings specific physico-chemical processes are added to the physical process of the failure of the material.

The smaller the maximum stress during the cycle of deformation, the greater is the part played in the fatigue of the rubber by the chemical processes which are activated by stress (mechanochemical processes). It follows from this that the slow failure of elastomers under cyclic loading is a more complicated process than that of solids.†

8.1. The Principles of the Dynamic Fatigue of Elastomers

Under the phenomenon of dynamic fatigue of elastomers we understand the decrease of the strength of the material under the effect of frequent periodic loads or deformations.

A summary of the data on dynamic fatigue of polymers, in particular of elastomers from papers of foreign investigators before 1950, can be found in Dillon's article.[1] A classification of the dynamic experiments is also presented.

According to a widely spread opinion[2-5] the dynamic fati-

† The fatigue of the material is the result of the time dependence of the strength under static and dynamic loads. However, the conception of the processes which take place in the stressed elastomers is not exhausted by this, as in elastomers, especially at frequent deformation, an accelerated irreversible change of structure takes place which influences the strength, the durability and other properties of the elastomer.

gue in elastomers as opposed to dynamic fatigue of metals is basically the result of chemical, oxidizing processes, and the failure of elastomers under repeated deformations takes place through a break of the molecular chains in the whole volume of the specimen as a consequence of mechanically activated chemical processes. There are, however, papers[1, 6–8] in which it is shown that physical factors also have an essential and direct influence on the dynamic fatigue of visco-elastic materials.

Many rubber products are used under conditions of frequently repeated deformations. In some cases the process of deformation is such that the maximum arising during the cycle of deformation of compression, tension or bending is predetermined, and the maximum load decreases as a result of stress relaxation. In other cases the maximum value of the deforming load remains constant, and the size of the maximum deformation grows in time due to creep.† To these types of using rubber products correspond two methods of testing rubber samples for dynamic fatigue under repeated tensioning.

1. The method of constant maximum elongation or deformation $\varepsilon = \text{const.}$

2. The method of constant maximum load or apparent stress $f = \text{const.}$

In the first method the span between the clamps of the testing machine is set, and during the experiments there accumulate in the specimen "residual" deformations the size of which depends on the properties of the elastomer under investigation, on the duration of the experiments, the predetermined deformation of the specimen, and also on the frequency of deformation and temperature. The maximum stress during the cycling decreases in this method, and in time is reduced to a certain limit. To achieve the second method a device is employed which permits to increase the amplitude of deformation after each cycle in such a way that during each cycle the range of loading from 0 to f is kept constant (where f is the maximum apparent stress).

During experiments one usually determines either the number of cycles before failure N, or the corresponding testing time before the break (or durability $\tau' = N/r$ where r is the frequency of

† Besides these two basic systems there are many others, the investigation of which does not add anything principally new to the laws which govern the dynamic fatigue of rubber.

deformation). In the first method of testing the maximum deformation ε is constant, and the load, preset on the apparatus, gradually decreases due to the process of "dynamic" relaxation, and is found to be lower than the maximum at the beginning of the cycling. In the second method the maximum load equals the initial one during the whole experiment, and the deformation ε grows. The failure load, referring to the initial or broken cross-section of the specimen, gives respectively the apparent (f) and true (σ) dynamic strength of the specimen in these experiments.

According to Galil–Ogly's[7] data for both dynamic methods the dependence of the number of cycles before failure on the magnitude of the maximum deformation, the number of cycles and the temperature† are analogous. On Fig. 123 the dependence of the number of cycles on the maximum deformation under tension is shown for unfilled SKS-30A rubber for both methods at 30°C. For the first method the initial maximum deformation ε is at the same time the deformation at rupture; for the second method the maximum breaking deformation ε is greater than the initial one owing to the elongation of the sample under the effect of the periodically acting load. Correspondingly, two characteristic lines are shown on Fig. 123 for $f =$ const.: one for the initial and one for the failure deformation. Comparing the two methods of testing with identical initial maximum deformations, we see that the second method is the "harder" one; with identical failure deformations, on the contrary, the first method of testing is found to be the "harder" one (see Fig. 123). The two testing methods are not equivalent; this can be explained by the relaxation properties of the elastomer.

The following relation exists between the number of cycles before failure N and the maximum tensile deformation ε during the cycle for both methods:‡

$$N\varepsilon^m = C \tag{8.1}$$

where ε is either the predetermined (maximum) deformation with $\varepsilon =$ const. or the failure deformation (maximum during cycle) with $f =$ const. In the investigated interval (see Fig. 123)

† The temperature of the rubber specimens themselves was measured and not, as is usually done, that of the surrounding air chamber.

‡ This relation can also be observed for the symmetrical loading cycle which is achieved in laboratory tests by torsional bending.[8]

both constants m and C do not depend on the number of cycles and m, besides, neither depends on temperature nor on the method of testing.

During the cycle the maximum true stress σ changes in both methods of testing due to the relaxation properties of the elastomer, approaching a certain value which is also the true breaking stress. Between the latter and the predetermined deformation with ε = const. or the maximum deformation with f = const., one finds a connection $\sigma = E\varepsilon$, where E is the usual dynamic modulus

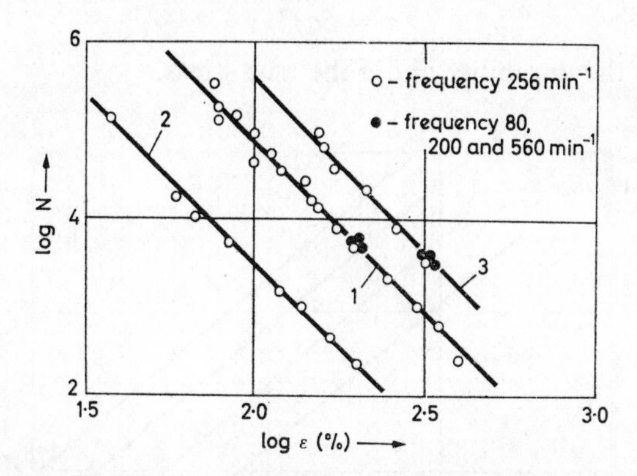

FIG. 123. Number of cycles before failure in dependence on the maximum deformation[7] (unfilled SKS-30-A rubber, 1 mm thick). 1. Number of cycles during tensioning from 0 to ε = const. 2. Number of cycles during tensioning from 0 to f = const. in dependence on the starting maximum deformation during the cycle. 3. The same, in dependence on the ultimate (rupture) maximum deformation.

of the elastomer under loading. If we take this law of deformation into account which is correct over the whole range of the maximum deformation and stresses used in the tests, we obtain, instead of relation (8.1):

$$N\sigma^m = C' \tag{8.2}$$

where σ is the true breaking stress; $C' = CE^m$, where the modulus E virtually does not depend on the number of cycles in the investigated range of frequencies (50–500 min^{-1}).

The relation (8.2) is valid also for metals. This circumstance points to a similarity of features in the phenomena of fatigue of elastomers and metals, in spite of the difference in the specific relaxation properties and failure mechanisms of visco-elastic and solid materials.

The dynamic endurance or "life" N, and the dynamic strength of an elastomer which is characterized by the breaking stress σ, have the same statistical character as the strength of the elastomer under static tests. This follows from the scatter of the results of experiments on dynamic endurance.

In agreement with the data mentioned in Chapter 6 the static fatigue of elastomers is expressed by

$$\tau = B\sigma^{-b} \tag{8.3}$$

where τ is the durability and σ the true stress.

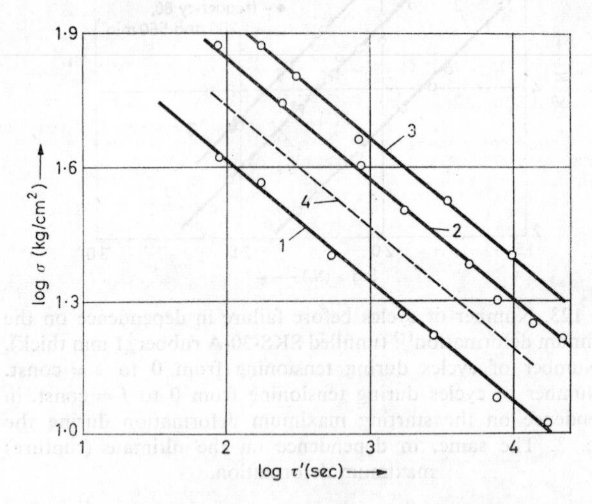

FIG. 124. Relation between durability τ' and maximum true stress in the cycle σ under different methods of experimenting (unfilled SKS-30A rubber).[7] 1. Dynamic method of straining from 0 to ε const. 2. Dynamic method from 0 to $f =$ const. 3. Static method $f =$ const. (time dependence of the strength of rubber). 4. Dynamic method $\sigma =$ const.

An analogous formula is also observed for the dynamic fatigue of elastomers. The static and dynamic fatigue (Fig. 124) is characterized by the same value of the constant b, but by different values for B and B', so that one can write

$$\tau' = B'\sigma^{-b} \tag{8.4}$$

The constant b does not depend on the temperature during the test nor on the method of loading the elastomer nor, consequently, on the frequency of loading ν. If one takes into account that $N = \nu\tau'$ and further assumes $m = b$ and $C' = \nu B'$ it is easy to see that eqns. (8.2) and (8.4) express the same law of dynamic fatigue of elastomers. As the constant C does not depend and the dynamic modulus only slightly depends on the frequency in the frequency's range under investigation (50–500 min^{-1}), the constant $B' = CE^{b/v}$, in the same range of frequencies is inverse proportional to the frequency. At the transition to the static method, when $\nu \rightarrow 0$† and $E \rightarrow E_\infty$ this relation loses its sense, as the constant B' must approach not infinity but the constant B which enters eqn. (8.3).

As $b \gg 1$, great changes of durability τ' correspond to small changes of the strength σ. As a consequence the durability is a more sensitive characteristic of the fatigue of elastomers than the strength. That is precisely why technologists use the "life" N and not the strength as the characteristic of dynamic fatigue.

It follows that the principles of dynamic and static fatigue of elastomers are identical but the static method of testing is the "softer" one compared with the dynamic one. In spite of the fact that under static conditions the elastomer is all the time in a stressed state, its failure takes place considerably later than in dynamic experiments, where the elastomer is stressed only for part of the time. This can be explained first by the circumstance that under periodic loads the overstress on the microdefects has no time to relax during each cycle of loading, whereas during static load they even out in time and approach an equilibrium value.[7] Secondly, the failure of polymers at cyclic loading is mechanically accelerated by activating chemical processes[2, 3, 6]

The above mentioned principles are valid for unfilled rubbers. Filled rubbers[8] are characterized by more complicated principles of dynamic fatigue (Fig. 125). The dependence of dynamic fatigue of rubber on the magnitude of the static tensile deformation has been investigated. The elastomer was prestressed up to ε_{st}

† Strictly speaking ν is not zero for the static method but has some small value. This is connected with the fact that one can present the static method as the first half-period of a cyclic loading with rectangular cycles. Then τ, the time of effect of the constant load (or durability), is the half-period of the cycle, to which corresponds the frequency $\nu = \frac{1}{2}\tau$.

a deformation and then underwent prolonged cyclic loading. It was found that with such a method of testing the number of cycles before failure cannot only decrease proportionately but can for some rubbers change even in a more complicated way.

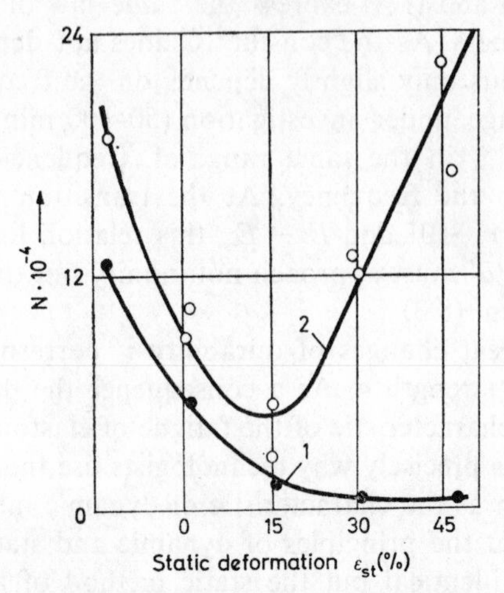

FIG. 125. Dependence of the number of cycles before failure of unfilled rubber on the static component of deformation under stretching[8] (constant amplitude of deformation 30%, frequency 50 min^{-1}). 1. SKB+40 g channel black, 115°C. 2. NK+40 g channel black, 100°C.

The number of cycles before failure depends also on the temperature; with increasing temperature it decreases at first rapidly and then slowly.

8.2. Calculation of Durability of Plastics and Elastomers under Cyclic Loading

The general method of calculating the strength and durability of materials under different methods of deformation, and the calculation of the strength of elastomers at constant strain rates has been examined in Chapter 7. In this section we bring similar calculations for the method of cyclic loading. The basic calculation is, as before, the condition of failure by Bailey (see p. 182). Zhurkov and Tomashevsky (see § 2.5) applied this method to calculations of the durability of plastics at constant strain rates and under cyclic loading with cycles of rectangular shape, Panshin and others[9] at cycling loading with cycles of serrated form, Regel and Leksovsky[10] under tension with cycles of sinoidal shape.

The durability under cyclic loading with equal cycles of symmetrical serrated form is[9]

$$\tau' = \alpha \frac{(1-1/K)\sigma_2}{1-\exp[-\alpha(1-1/K)\sigma_2]}\tau_2 \tag{8.5}$$

where σ_2 is the upper, σ_1 the lower limit of stress in the cycle; $K = \sigma_2/\sigma_1$, is the coefficient of asymmetry of the cycle; $\tau_2 = A \exp(-\alpha\sigma_2)$ is the durability under constant stress σ_2.

The method of testing with constant strain rate (see § 2.5) is frequently found to correspond to serrated cyclic loading, when failure sets in at the end of the first half-period. This case can be observed if one chooses as σ_2 the strength of the material at a predetermined strain rate.

Equation (8.5) is valid for stresses σ_2, which are smaller than the strength, when the number of cycles before failure ($N = \tau'/\theta$, where θ is the period of the cycle) is sufficiently great—say, above 100. It follows from this equation that the durability does not depend on the period of the cycle.

It was found that the durability of poly(methylmethacrylate)[9] under cyclic testing agrees with the calculated values only in the region of small durabilities, or in the region of great maximum stresses σ_2. In the region of small σ_2 which for the practical use is of the greatest interest, the durability is lower by one or two orders of magnitude than the calculated value. This result shows that the applied method is not always advantageous for the calculation of the durability of plastics under cyclic loading, and it is indispensible also to calculate the specific phenomena which are introduced by the cyclic loading itself. Some of them are analysed in papers[9] and will now be examined in detail.

One of the reasons of the increase of speed of failure in polymers can be the heating of the material in places of overstress and at the roots of microcracks. During this the rise in temperature at the roots of cracks can considerably exceed the heating of the specimen as a whole. With rising temperatures in places of stress concentration the speed of formation and propagation of microcracks increases, and the durability decreases. Under simple tension the mechanical losses are small and do not cause any essential effect. With a small number of cycles the local rise of temperature is also negligible, and the actual and calculated durabilities virtually coincide. With an increase of the number of cycles the temperature rises noticeably in places of stress concentration and approaches a certain ultimate value, at which a thermal balance is established: the quantity of heat which is produced during the cycle equals the quantity of heat which is dissipated due to the thermal conductivity of the material. Therefore, with great numbers of cycles the thermal effects are greatest and the durability decreases to values which correspond to the durability at elevated temperatures. In order to explain the observed discrepancy of the durability of poly(methylmethacrylate) from the calculated results it is sufficient to assume that in places of stress concentration the temperature rises by 30–50°C. A local heating occurs also in elastomers under frequent deformations.

Other reasons for the decrease of durability are also possible. For instance, the specimens under cyclic loading may not have time, because of their molecular orientation, to get strengthened as much as under static[10] loading and, besides, under prolonged cyclic loading the processes of ageing of the material may be activated.

A disagreement of the experimental results with the calculated ones is also possible as the employed conditions of failure by Bailey are not strict; for instance the speed of failure depends not only on the stress but also on the extent of the already existing degradation. If it follows from Bailey's condition that under cyclic loading (with identical cycles) all cycles decrease the strength evenly, each following cycle is in fact more critical than the previous one as the dimensions of the growing cracks, and therefore also the stress at their roots, increase continuously. In the development of his condition of failure Bailey based himself on Petersen's and De Forest's[11] experiments on cyclic testing of metals. However, Serensen and co-workers[12] showed that the equation $\sum_i (n_i/N_i) = 1$ which follows from Bailey's condition (where n_i is the number of identical cycles of loading of a predetermined form, N_i is the number of cycles of that form which lead to failure) is neither fully valid for metals.

Let us now examine the durability of elastomers as calculated with Bailey's method under cyclic loading with a constant maximum true stress σ (cycles from 0 to σ). The method $\sigma =$ const. is difficult to achieve with accuracy under laboratory conditions, and even more so in conditions of practical use. However, this method is of interest as it is easily calculated, and is intermediate between the "soft" method $\varepsilon =$ const. and the "hard" method $f =$ const.

The maximum true stress during the cycle can be calculated with the formula $\sigma = (1+\varepsilon)f$. With $f =$ const. the maximum tension ε grows. To bring about $\sigma =$ const. can therefore only be done by a partial loading of the sample which corresponds to the softer method of testing. On the other hand, with $\varepsilon =$ const. the true stress falls owing to the relaxation process, and it is indispensible to increase ε to maintain the constant value σ, i.e. make the method "harder". Together with this, the methods $f =$ const. and $\varepsilon =$ const. correspond in prolonged experiments in practice to the method of constant maximum true stress (Fig. 126). This can

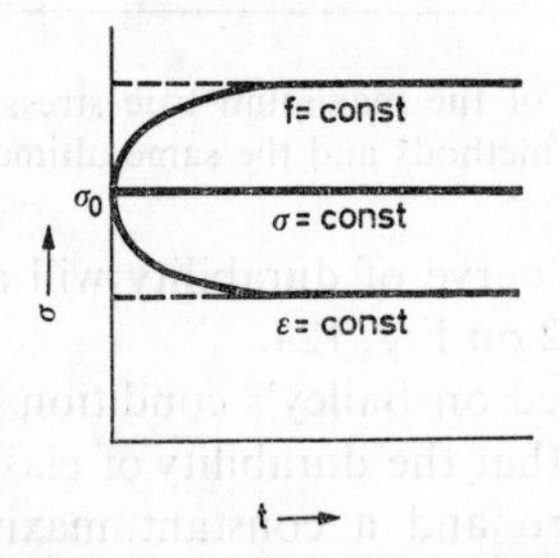

FIG. 126. Change of the maximum true stress during the cycle whilst testing with different methods and the same initial true stress σ_0.

be explained by the comparatively rapid completion of the relaxation process (curve $\varepsilon =$ const.) and the visco-elastic after-effect (curve $f =$ const.). If the number of cycles before failure and the durability is sufficiently great, one can in practice reckon that the true stress is constant during the whole experiment. The size of the true failure stress will differ from the stress at the beginning: it will be greater for $f =$ const., and smaller for $\varepsilon =$ const. The relation between the durability and the true breaking stress with $\sigma =$ const. lies also between the respective relations of the two methods of testing ($f =$ const. and $\varepsilon =$ const.), as is shown by the dotted line on Fig. 124 (see p. 216).

The higher the modulus of the elastomer, the smaller the disagreement between the different methods. For elastomers with an equilibrium modulus above 15–20 kg/cm² there is practically no difference in durability, and the straight lines 1 and 2 on Fig. 124 virtually coincide.

If one compares all three methods of testing under the same ultimate true stress, the method $\varepsilon =$ const. is again found to be the "hardest" (Fig. 127) as was noted earlier on p. 221. However,

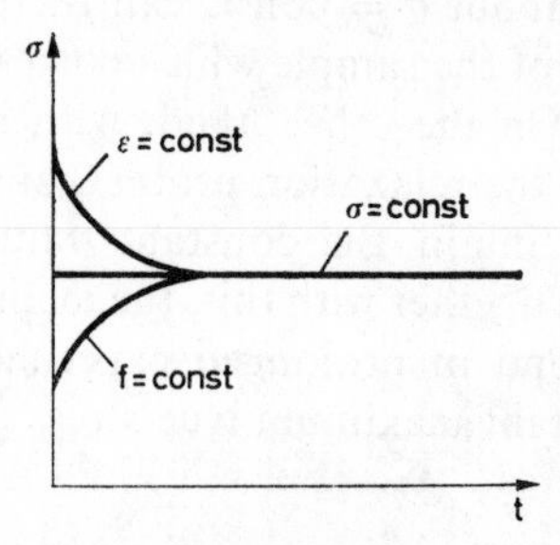

FIG. 127. Change of the maximum true stress during the cycle with different methods and the same ultimate true stress.

with $\sigma =$ const. the curve of durability will also lie between the straight lines 1 and 2 on Fig. 124.

A calculation based on Bailey's condition (see Chapter 7) and on eqn. (8.3) shows that the durability of elastomers under cyclic loading between zero and a constant maximum true stress σ with cycles of rectangular shape is

$$\tau' = \frac{B}{r}\sigma^{-b} \tag{8.6}$$

and for cycles of serrated form

$$\tau' = B(b+1)\sigma^{-b} \tag{8.7}$$

The respective number of cycles before failure is

$$N = \nu\frac{B}{r}\sigma^{-b} \quad \text{and} \quad N = \nu B(b+1)\sigma^{-b} \tag{8.8}$$

where $\nu = 1/\theta$ is the frequency of the cyclic loading; B and b are the constants in the equation of time dependence of strength

(8.3); r is the fraction of the period θ of the rectangular cycle of deformations, when the elastomer is loaded.

The independence of the durability and the proportionality of the number of cycles before failure to the frequency is common to both methods.

Besides, the character of the dependence of the durability on the stress under cyclic loading coincides with the time dependence of strength under static loads. One has, therefore, to expect that eqn. (8.4) is common for all methods, where $B' \leqslant B$, and the constant b is the same for all methods (identical slope of straight lines on Fig. 124). Besides, these common properties of the durability of elastomers do not depend on the form of the loading cycle and are valid, in particular, for sinoidal cycles.

The theoretical straight line of durability which corresponds to the method of constant maximum true stress lies between the straight lines of durability with $f =$ const. and $\varepsilon =$ const.

Not everywhere, however, can an agreement of calculation with experiment be observed. In agreement with eqns. (8.6) and (8.7) the value of the constant B' is greater than B, whereas experiments give an opposite result. Further, it is known[1, 7] that the number of cycles before failure does not depend on the frequency of deformation, at least in a limited range of frequencies. Still, we have, in agreement with eqn. (8.8): $N \sim \nu$. The possible reasons for these discrepancies have already been examined above when discussing Bailey's method of calculation. The basic reason for the discrepancy of calculation and experimental data consists, however, in the mechanical-chemical processes which lower the strength of elastomers under cyclic loadings.

8.3. Phenomenon of Fatigue of Elastomers under Cyclic Loading

The conclusions at which we have arrived in the preceding section are some of the proofs that the failure of elastomers under cyclic loading can be explained not only by physical processes. In an unstressed state slow ageing processes proceed in elastomers—chemical processes under the influence of oxygen, heat and light, which lead to a change in the structure of the elastomer and to a deterioration of its service properties. In stressed elastomers these processes are accelerated, especially under the effect of alternating stresses. In other words, the chemical processes in elastomers are activated by mechanical stresses which is expressed, in particular, in the decrease of the activation energy of these processes.[2, 3, 6, 13, 14]

One of the important mechanisms of elastomer fatigue is the mechanically activated oxidation of rubbers.[2, 3, 14] The fatigue of polymers is connected both with oxidation processes and with the direct failure of the polymer under the influence of mechanical forces.[6]

The part played by mechanical-chemical processes in the fatigue of rubbers is confirmed by numerous facts. The most direct proof may be the following phenomenon, discovered and mentioned in the works of Bartenev and Galil-Ogly.[7] Whilst for well-protected rubbers the durability grows with decreasing thickness of the sample—i.e. the usual scale effect of strength† can be observed—the rubber which is only slightly protected from ageing processes gives an anomalous dependence of durability on the thickness of the samples (Fig. 128). With the decrease of the

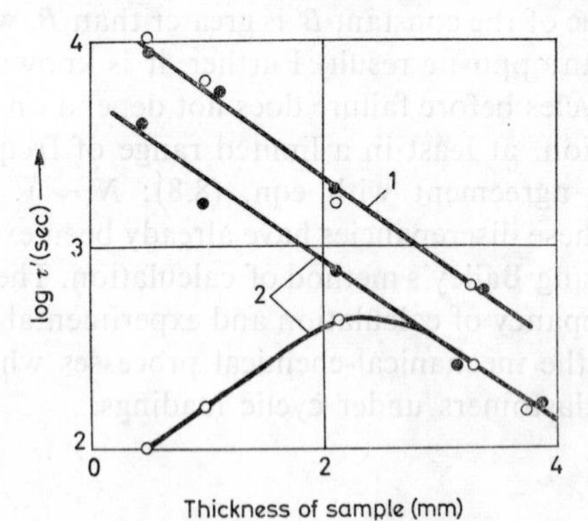

FIG. 128. Influence of the thickness of the sample on the durability of SKS-30A rubbers[7]. 1. Static method. 2. Dynamic method. ○ — badly protected rubber. ● — well-protected rubber.

maximum deformation the "fork" on curve 2 of Fig. 128 moves in the direction of increasing thickness. In static experiments the

† The scale effect of dynamic strength was observed free from any extraneous influence,[7] as the temperature of both thin and thick samples was carefully controlled at 30°C. The influence of higher temperatures on the fatigue of thick samples was excluded from these experiments.

durability of both thick and thin elastomers shows the normal dependence on the scale factor. The durability in static and dynamic methods of testing of samples thicker than 2·5 mm shows the same trend; this can partly be explained by the circumstance that the testing of thick specimens to failure takes only 3–10 min and the oxidizing processes have no time to develop. Evidently, the basic reason is that the ageing process proceeds not only due to the oxygen, dissolved in the elastomer before the experiment, but also due to the oxygen which diffuses from the atmosphere. In thin samples the speed of diffusion is sufficient not to hinder the ageing process, and as a result the strength decreases rapidly.

The anomalous dependence of the dynamic fatigue on the thickness of the specimen can partly or fully be eliminated by the introduction of special chemical substances—antioxidants—into the rubber. The dependence of the durability of SKS-30A rubber on the thickness of the sample is completely different if different antioxidants have been introduced in equal quantities during the mixing (1 g on 100 g rubber) (Fig. 129). With increas-

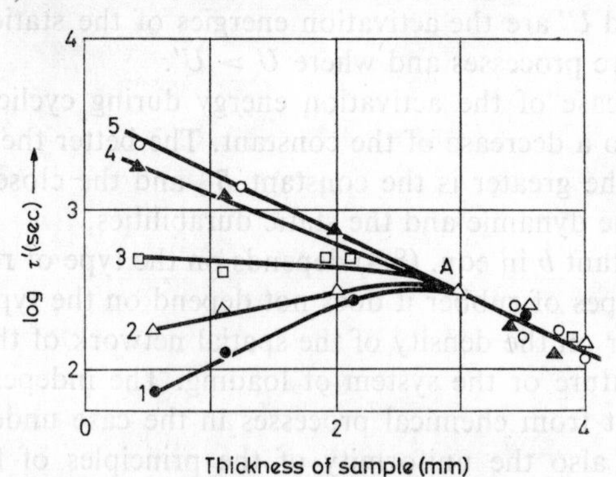

FIG. 129. Influence of the thickness of the sample on the durability of SKS-30A rubber with different antioxidants.[7] 1. Without antioxidant. 2. Aldol-α-naphthylamine. 3. *p*-Anisidine. 4. 1,4-Diphenylphenylenediamine. 5. Ketamin D.

ing thickness of the sample the difference in the influence of the antioxidant decreases, and above 2·5 mm the dynamic resistance of the different samples is the same and independent from the type of the employed antioxidant. For rubbers which are protected by a good antioxidant (e.g. 1,4-diphenyl-phenylenediamine) the character of the change of durability which depends on the thickness of the material is the same in dynamic and static experiments. However, these dependences do not tally, as more favourable conditions are created in static experiments for the relaxation of local overstresses which appear with deformations and are possible centres of the emergence of cracks and tears in the rubber.

The constant B' in eqn. (8.4) is smaller with cyclic than with static loadings. The direct reason for this is the decrease of the activation energy of the failure process during dynamic loading. This becomes evident if one considers the meaning of the constant B, remembering eqn. (6.2) which expresses the temperature-time dependence of the strength of elastomers. Assuming an analogous dependence also for the constant B' in the dynamic method, we have in both cases:

$$B = Ce^{\frac{U}{kT}}; \qquad B' = C'e^{\frac{U'}{kT}} \tag{8.9}$$

where U and U' are the activation energies of the static and dynamic failure processes and where $U > U'$.

The decrease of the activation energy during cyclic loadings leads also to a decrease of the constant. The better the rubber is protected, the greater is the constant B', and the closer to each other are the dynamic and the static durabilities.

The constant b in eqn. (8.4) depends on the type of rubber but for some types of rubber it does not depend on the type of antioxidant, nor on the density of the spatial network of the rubber, the temperature or the system of loading. The independence of the constant from chemical processes in the case under investigation and also the uniformity of the principles of fatigue of rubber in static and dynamic experiments indicates that the elementary acts of the formation of centres of failure have in both cases a common mechanism. Chemical processes play an essential part only in the acceleration of the processes of formation of centres of failure. The role played by these mechanical-chemical processes in dynamic experiments is so great that in many cases it is precisely they which determine the life of rubber articles.

The fatigue of rubbers is thus a complex of simultaneously proceeding interacting physical and chemical processes. A big part is played in it by the heterogeneity of the microstresses and the heterogeneity of the distribution of oxygen in the rubber and by inhibitors and other ingredients. All this leads to an uneven speed of oxidization and of the processes of fatigue which differ in character in the different parts of the specimen; this accelerates the appearance of separate centres of failure, where the material has undergone, up to the moment of examination, the greatest structural changes with comparatively small changes of properties in the basic rubber mass.

8.4. The Part Played by Mechanical Losses in the Fatigue of Rubbers

The actual process of deformation of rubber proceeds always with a finite speed and is therefore thermodynamically irreversible. As a result of inner friction in each cycle of deformation some part of the work turns into heat (phenomenon of hysteresis). The work of the external forces can be presented as the sum of two components: the work which is expended in surmounting the elastic forces, and the work which is expended in surmounting the forces of inner friction. The first one is not accompanied by mechanical losses, and does not lead to the development of heat. The second one turns entirely into heat. Under cyclic loading of the rubber the development of heat due to hysteresis leads to a considerable heating of the material. The more heat is produced per time unit, and the less it passes into the surroundings through heat conductivity and radiation, the greater is the temperature the rubber will attain. The size of temperature in cyclic loading lowers sharply the fatigue strength.

The inner friction plays, therefore, always (with exception of shock absorbers) an unfavourable part in the use of rubbers. Under frequent and shock deformations the inner friction leads to mechanical losses and to a considerable build-up of heat which has a damaging influence not only on the fatigue strength but also on the wear and tear of the rubber, and on the strength of the adhesion between the elements of laminated constructions. As the heat conductivity of rubbers is small, the accumulation of heat leads to a rapid rise of temperature of the rubber article. This is especially critical in connection with the circumstance that the strength of the rubber falls sharply with a rising temperature, and the speed of the ageing processes increases. Because of the slower heat removal the heating-up of big articles proceeds quicker and they fail more rapidly (influence of thermal scale factor in the fatigue of rubbers).

As experience shows, the dynamic visco-elastic modulus consists of two parts:

$$E = E_{\infty}+E_1$$

where E_{∞} is the equilibrium modulus, and E_1 is the non-equilibrium part of the dynamic modulus, and corresponds to the forces of the inner friction.

The equilibrium modulus depends mainly on the extent of the cross-linking (vulcanization). The size of the non-equilibrium part of the dynamic modulus does practically not depend on the extent of vulcanization.[15] One can, therefore, change the size of the dynamic modulus through vulcanization, without changing the inner friction of the rubber. The non-equilibrium part of the modulus, as well as the inner friction, depends essentially on the number of polar groups in the rubber chain, and on the quantity of active filler, i.e. on the character and intensity of the intermolecular forces. The influence of the filler on the dynamic modulus is shown in the change of E_1, whereas E_{∞} is practically unchanged.

In the dynamic methods of testing there exist a few characteristics of mechanical losses which are connected with each other.[16] The coefficient of the mechanical losses (or the relative hysteresis) $\varkappa$ we call the relation of the included area of the hysteresis curves to the area which is included between the stress curve and the abscissa which denotes the deformation. From this definition follows that

$$\Delta W = \varkappa W \tag{8.10}$$

where ΔW is the mechanical loss during the cycle; W is the energy which is transmitted to the specimen during each cycle by the external forces. These are

$$W = \frac{1}{2}E\varepsilon^2 \tag{8.11}$$

where E is the dynamic modulus of the rubber. ε is the deformation.

If one introduces another characteristic of the mechanical losses $K = \varkappa E$, then in agreement with eqns. (8.10) and (8.11) the losses during the cycles are expressed:

$$\Delta W = \frac{1}{2} K\varepsilon^2 \tag{8.12}$$

where K is the "modulus" of the inner friction, so called because it has the dimensions of the modulus of elasticity.[17]

Equation (8.11) and, following from it, (8.12) is valid for small deformations (10–20%). For great deformations (up to 100–200% elongation)

$$W \approx \frac{1}{2} E\varphi(\varepsilon)$$

where $\varphi(\varepsilon)$ is a function. Therefore

$$\Delta W \approx \frac{1}{2} K\varphi(\varepsilon).$$

It is accepted that the mechanical losses in cyclic loading are also defined by $\tan \delta$, the angle of the mechanical losses in connection with the coefficient $\varkappa$ by the equation

$$\varkappa = \frac{2\pi \tan \delta}{4\sqrt{1+\tan^2 \delta} + \pi \tan \delta} \tag{8.13}$$

The principles which have been established for the mechanical losses in cyclic loading are found to be valid also in single impact tests.[18] In particular eqn. (8.11) is applicable up to 20–30% deformation of impact compression.

The mechanical losses are determined by different methods. In Table 10 data are shown for different unfilled rubbers, obtained by the method of free contraction; they illustrate the influence of the intermolecular forces on the inner friction of rubbers.

The product $\varkappa E$ is ten times greater for polar SKN-40 rubber than for rubbers based on non-polar natural rubber. According to data[19] the equation $\varkappa E/E_i \approx \frac{1}{2}$ does, practically, not depend on the type of rubber, nor on temperature, filler, plastification and vulcanization. With a decrease of temperature, increase of frequency and strain rate the non-equilibrium part of the modulus E_1, and the coefficient of the mechanical losses $\varkappa$ grows, reaching a maximum at the transition of the rubber to the glassy state.[16]

The greater $\varkappa$ of the rubber at usual temperatures, the higher the temperature of its glass transition point. At temperatures above T_g the value of $\varkappa$ is approximately proportional to the frequency of cycling. The dependence of $\varkappa$ on the temperature is more pronounced than on the frequency.

The mechanical losses of the rubber are thus extremely important and depend to a great extent on the type of rubber and its composition. Filled rubbers are characterized by greater losses than unfilled ones. The mechanical losses of all rubbers depend on the frequency and strain rate of deformation. At free contraction the mechanical losses reach their greatest values, as shown in Table 10.

TABLE 10

Characteristics of Unfilled Rubbers with Equilibrium Modulus $E_\infty = 9$ kg/cm² at 20°C and Great Strain Rates

Rubbers	T_g (°C)	E (kg/cm²)	$\varkappa$
NK	−50	12·7	0·16
SKB	−33	17·0	0·26
SKS–30	−43	20·5	0·33
SKN–18	−27	26·0	0·31
SKN–26	−20	32·0	0·34
SKN–40	−10	48·0	0·42

From this follows that the resistance of technical rubber products and tyre fatigue must depend to a considerable degree on the mechanical losses.[19–22]

8.5. Influence of Molecular Weight of Rubber, Filler and Temperature on Fatigue Strength of Rubbers

The type of rubber and, to a smaller degree, the composition of the rubber (the type of the vulcanizing agent and the filler)[23] has a great influence on the fatigue of rubber under cyclic loading. The filling with carbon black, which leads usually to a noticeable increase of such characteristics as strength, resistance to tearing and ageing, has a comparatively small influence on the fatigue strength. Therefore the type of rubber determines to a considerable extent its fatigue properties. Together with this, the comparison of the fatigue properties of rubbers from different elastomers during the transition from one method of experiments to the other may give ambiguous results; it is absolutely necessary to keep this in mind when compounding rubbers for some particular condition of practical use.

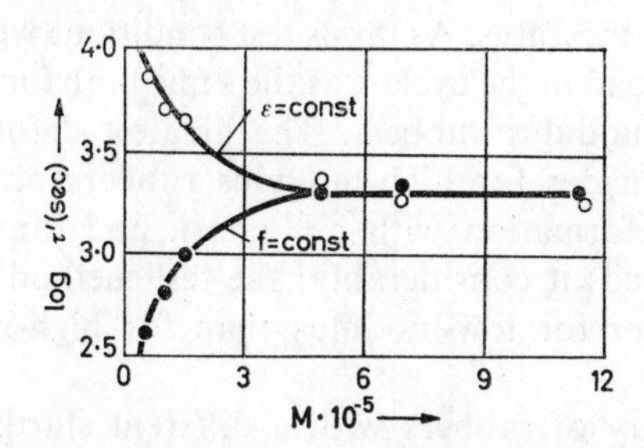

FIG. 130. Influence of molecular weight of SKS-30A on the durability of vulcanized rubber with the same quantity of bonded sulphur (2·05% sulphur, $\varepsilon = 100\%$, $f = 7{\cdot}6$ kg/cm², frequency 256 min^{-1}, temperature of specimens 30°C.)[24]

The influence of the molecular weight of the basic elastomer on the fatigue properties of rubbers was studied on vulcanized SKS-30A rubber,[24] under the methods: $\varepsilon =$ const. and $f =$ const. It was found that this produces on vulcanized rubbers with an equal quantity of chemically bonded sulphur but different densities of the spatial network† considerable differences with the two methods (Fig. 130). If, with $\varepsilon =$ const. the durability τ grows, it decreases with $f =$ const. with decreasing molecular weight of the elastomer base. For high-molecular rubbers the data on Fig. 130 correspond to the conditions under which the maximum load with $f =$ const. in the loading cycle corresponds to the maximum deformation with $\varepsilon =$ const. As practically no plastic deformations develop during the life of these rubbers with a high static equilibrium modulus (E_∞ = 15–17 kg/cm²), both methods give for them

† Vulcanized rubbers of low-molecular types have a sparser spatial network and a lower static equilibrium modulus than vulcanized rubbers of a high-molecular type.

practically the same durabilities. The high value of the durability τ' of vulcanized rubbers with a low starting molecular weight with $\varepsilon =$ const. can be explained by the fact that low-molecular rubbers have a low equilibrium modulus and a sparser spatial network. This leads to the development of great plastic deformations during experiments, which lowers the stress and softens the test conditions.

With $f =$ const. the development of plastic deformations leads to harder test conditions as under a constant load the true stress in the specimen increases. As far as test conditions were concerned the maximum load in the cycle was the same both for low-modulus and for high-modulus rubbers. The greatest deformation with $f =$ const. coincides for high-modulus rubbers practically with the greatest deformation with $\varepsilon =$ const. and for low-modulus rubbers it exceeds it considerably; the test method $f =$ const. is therefore harder for low-modulus than for high-modulus rubbers.

The durability of rubbers with a different starting molecular weight, but with the same high equilibrium modulus (13 kg/cm^2) is in both methods practically identical; it decreases somewhat with the decrease of the starting molecular weight. The explanation for the latter is that the structure of elastomers with a low molecular weight is more heterogenous.

The increase of the degree of vulcanization (increase E_∞) leads, according to data by Rezinkovskii and others,[23] to a steady growth of strength and to a change of the curve which shows the maximum deformation at rupture under $\varepsilon =$ const. (the strength and deformations were defined by the amplitudes of stress and strain under alternating flexure with a predetermined "life" of 10^6 cycles for the specimen).

Fillers influence the fatigue properties of rubbers from various elastomers in a different way. For SKS-30 the fatigue strength grows with filling, for SKB it hardly changes, and for NK it even falls.[23] The fatigue strength of filled and unfilled natural rubbers, and also synthetic rubbers with different concentration of polar groups, has been studied by Gul' and others[25, 26] in connection with the influence of solvents and plasticizers. With an increase of the degree of swelling the resistance to fatigue grows, passes beyond a maximum and then decreases. This can be explained by the overlapping of two processes. The decrease of the inner friction and failure energy of the intermolecular bonds during swelling leads at first to an increase of the durability but then the usual effect of a decrease of strength with an increase of swelling makes itself felt.

In conclusion let us examine the influence of temperature on the fatigue of rubbers. Taking into consideration eqn. (8.9) for coefficient B', one can write eqn. (8.4) like this:

$$\tau' = C'\sigma^{-b} e^{\frac{U'}{kT}} \tag{8.14}$$

where τ' is the durability under cyclic loading; σ is the maximum true stress during the cycle; C' and b are constants; U' is the activation energy of the failure process; T is the absolute temperature.

If one examines the fatigue in the narrow temperature range $T - T_0 \ll T_0$, where $T_0 = 300$–400°K, one can write, approximately, restricting oneself to the second member of Taylor's series:

$$\frac{U'}{kT} \approx \frac{U'}{kT_0} - \frac{U'}{kT_0^2}(T - T_0).$$

As a result we obtain, instead of eqn. (8.14) with σ = const. the following approximate formula:

$$\log \tau' = C_1 - C_2 T.$$

This expresses the temperature dependence of durability under dynamics methods of testing which is usually observed in a limited temperature range.

Literature

1. J. Dillon, *Advances in Colloid Science*, vol. III, 1950; D. Dillon, *Ustalost' viysokopolimerov*, Goskhimizdat, 1957.
2. A. S. Kuz'minskii, N. N. Lezhnev, M. G. Maizel', *DAN SSSR*, **71,** 319 (1950).
3. A. S. Kuz'minskii, N. N. Lezhnev, *DAN SSSR*, **83,** 111 (1952).
4. A. A. Somervielle, *Ind. Eng. Chem*, **28,** 11 (1936).
5. H. Winn, I. Shelton, *Ind. Eng. Chem.* **37,** 1 (1945).
6. G. L. Slonimskii, V. A. Kargin, G. N. Buiko, Ye. V. Reztsova, M. L'yuis-Riera, *DAN SSSR*, **93,** 523 (1953); *v sb.* "*Stareniye i utomleniye kauchukov i rezin*", Goskhimizdat, 1955, p. 100.
7. G. M. Bartenev, F. A. Galil-Ogly, *DAN SSSR*, **100,** 477 (1955); *v sb.* "*Stareniye i Kauchukov i rezin*", Goskhimizdat, 1955, str. 119.

8. M. K. Khromov, L. S. Priss, M. M. Rezinkovskii, *v sb. "Fiziko-mekhanicheskiye ispiytaniya kauchuka i reziniy"*, Goskhimizdat, 1960, p. 5 (*Trudiy NIIShP*, No. 7); M. M. Rezinkovskii, K. N. Lazareva, *Kauchuk i rezina*, **22**, No. 3, 17, (1963).
9. G. M. Bartenev, B. I. Panshin, I. V. Razumovskaya, G. N. Finogenov, *Izv. AN SSSR, OTN*, No. 6, 176 (1960); B. I. Panshin, G. M. Bartenev, G. N. Finogenov, *Plast. massiy*, No. **11**, 47 (1960).
10. V. R. Regel', A. M. Leksovsky, *FTT*, **4**, 949 (1962).
11. A. V. de Forest, *J. Appl. Mech., Trans ASME*, **3** A, 23 (1936).
12. S. V. Serensen, V. P. Koganov, L. A. Kozlov, R. M. Shneiderovich, *Nesushchaya sposobnost' i raschet detalei mashin*, Mashgiz, 1954.
13. V. A. Kargin, G. L. Slonimskii, *DAN SSSR*, **105**, 751 (1955).
14. A. S. Kuz'minskii, L. I. Lyubchinskaya, *DAN SSSR*, **93**, 519 (1953).
15. M. M. Rezinkovskii, B. A. Dogadkin, *Khim. prom.* No. 4, 227 (1954); B. A. Dogadkin, M. M. Rezinkovskii, *Uspekhi khimii*, **24**, No. 7, 801 (1955).
16. G. M. Bartenev, Yu. V. Zelenev, *Kauchuk i rezina*, No. 8, 18 (1960); *Viysokomol. soyed.* **4**, 66 (1962).
17. M. M. Rezinkovskii, V. S. Yurovskaya, B. A. Dogadkin, *Kolloid. zh.* **14**, 444 (1952).
18. G. M. Bartenev, N. M. Novikova, *Kolloid. zh.* **21**, 3 (1959).
19. G. M. Bartenev, M. M. Rezinkovskii, M. K. Khromov, *Kolloid. zh.* **18**, 395 (1956).
20. M. M. Rezinkovskii, L. S. Priss, M. K. Khromov, *Kolloid. zh.* **21**, 458 (1959).
21. M. M. Rezinkovskii, Ye. G. Vostroknutov, M. K. Khromov, *v sb. "Stareniye i utomleniye kauchukov i rezin"*, Goskhimizdat, 1956, p. 68.
22. B. M. Gorelik, *Trudiy NIIRP*, No. 4, Goskhimizdat, 1957, p. 125.
23. M. M. Resinkovskii, L. S. Priss, M. K. Khromov, *Kolloid. zh.* **20**, 368 (1958).
24. G. M. Bartenev, A. S. Novikov, F. A. Galil-Ogly, *Kolloid. zh.* **18**, 7 (1956).
25. V. Ye. Gul', G. V. Dorokhina, B. A. Dogadkin, *Kolloid. zh.* **13**, 339 (1951).
26. V. Ye. Gul', D. L. Fedyukin, B. A. Dogadkin, *Kolloid. zh.* **15**, 11 (1953).

CHAPTER 9

THEORY OF THE TEARING OF RUBBERS

TEARING as well as breaking is one of the methods of testing the strength of rubbers. The difference between them is that in a tear a part of the greatest stress concentration is attained early by specially incising the specimen or by the choice of its shape.

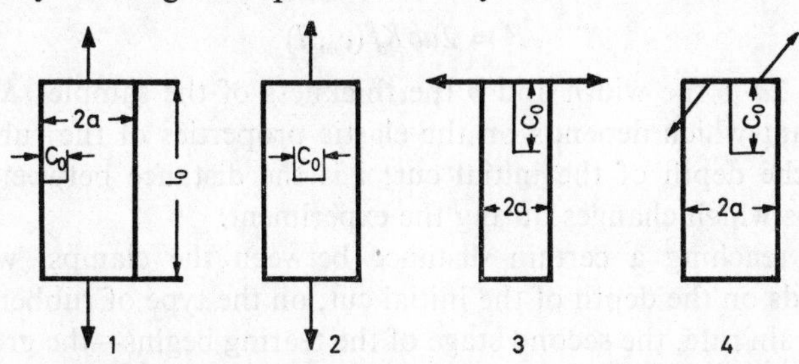

FIG. 131. Types of specimens with cut (l_0 is the initial length of the specimen). 1. Cross-cut from the edge, tested under tension. 2. Cross-cut in the middle, tested under tension. 3. Longitudinal cut, tearing by cleavage. 4. Longitudinal cut, tearing by shear.

In its simplest form the study of tears in rubber[1-8] is carried out with strips which have either longitudinal or transversal cuts of the length c_0 (Fig. 131) The place of the cut determines the place of the tear of the specimen during deformation.

9.1. The Two Stages of the Process of Tearing Rubbers

Experiments on tearing are usually carried out through pulling apart the clamps of the testing machine with a sufficiently high speed† and the moment when the tear begins is recorded. Not

† Usually a rapid visco-elastic rupture is observed in the rubber, without the formation of a rough zone on the fracture surface, i.e. virtually without a slow stage.

only a parting of the specimen during the experiment but also its deformation takes place. It is therefore difficult to separate the work of deformation of the specimen from the work of tearing. For some types of specimens one can, however, calculate the work of the deformation and then find the work which is expended only on the tearing of the specimen.[1-9]

One has to distinguish two stages of the tearing of the samples. In the first stage, practically, only a deformation of the specimen takes place with increasing distance between the grips, and there is no noticeable growth of the cut ($c = c_0 = \text{const.}$, $dc = 0$).

The force F which is applied to the clamps grows at this stage and it depends in general on the elastic properties of the rubber, on the size of the initial cut, the thickness and width of the specimen, and on the distance between the grips:

$$F = 2abKf(c_0, l) \tag{9.1}$$

where $2a$ is the width and b the thickness of the sample; K is a constant which depends on the elastic properties of the rubber; c_0 is the depth of the initial cut; l is the distance between the clamps which changes during the experiment.

On reaching a certain distance between the clamps, which depends on the depth of the initial cut, on the type of rubber and the strain rate, the second stage of the tearing begins—the growth of the cut. This stage proceeds in different ways for different types of specimens and methods of testing (Fig. 132). For specimens of the first two types† only tension arises ($l/l_0 = \lambda$ is the scale factor of tension). As the distance between the grips increases, the stress increases. When, at $l = l_K$ it reaches some critical value, a rapid growth of the cut across the whole specimen takes place with a speed near the speed of sound in rubber. In other words, one does not observe at first a growth of the cut during testing (the speed of growth in the first stage can be disregarded), but then an "instant" growth of the cut across the whole sample. The dependence of the stress on the distance of the grips is shown on Fig. 133 for samples of the first two types. The slow growth of the curve corresponds to the first stage of the test, and its rapid fall (dotted line) to the second stage, when a catastrophic growth of the cut can be observed. Natural rubbers will normally show such a type of tearing.

† Strips of rubber with $a \gg b$.

The tearing of specimens of the third and fourth type proceeds in a different way. In the first stage the sample will only deform. With increasing distance between the clamps the stress increases, and at a certain distance $l = l_K$, the cut begins to grow (second

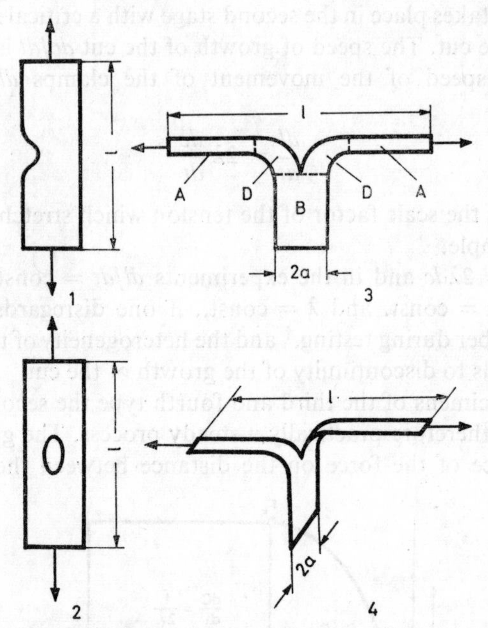

FIG. 132. Samples of the fourth type, indicated on Fig. 131, during tearing (l is the distance between the clamps of the testing machine for all samples). *A*, Region of tension. *B*, Non-deformed region. *D*, region of deformation by bending.

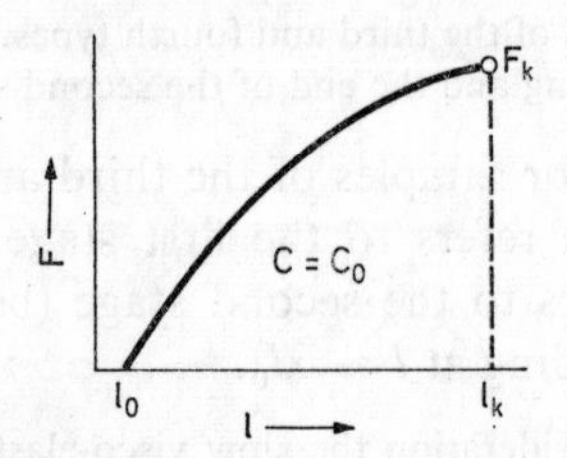

FIG. 133. Dependence of the stretching force F of samples of the first two types on the distance l between the clamps. The small circle marks the moment of transition to rapid rupture with the critical value of the force F_k.

stage of tearing). However, in contrast to the samples of the first two types the growth of the cut is not catastrophic but proceeds with a constant speed which depends on the speed of the movement of the clamps. In samples of this type more and more new parts of the specimen are stressed during the growth of the cut. The growth of the cut takes place in the second stage with a critical stress at the root of the cut. The speed of growth of the cut dc/dt is connected with the speed of the movement of the clamps dl/dt by the equation:

$$\frac{dl}{at} = 2\lambda \frac{dc}{dt} \tag{9.2}$$

where λ is the scale factor of the tension which stretches the ends of the sample.

As $dl = 2\lambda dc$ and in the experiments $dl/dt = \text{const.}$, virtually also $dc/dt = \text{const.}$ and $\lambda = \text{const.}$, if one disregards the creep of the rubber during testing,† and the heterogeneity of the material which leads to discontinuity of the growth of the cut.

For specimens of the third and fourth type the second stage of tearing is therefore practically a steady process. The graph of the dependence of the force on the distance between the clamps is

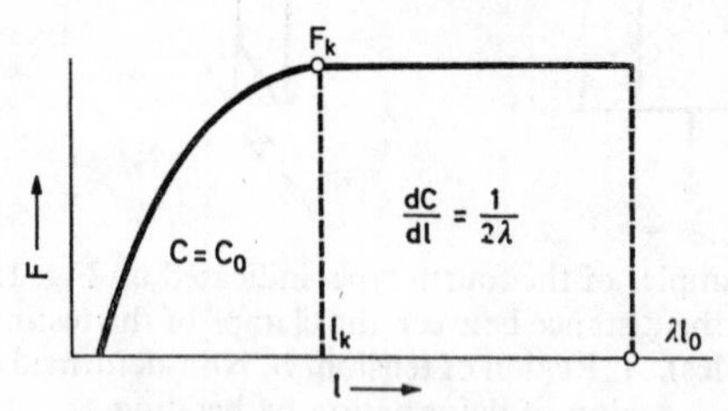

FIG. 134. Dependence of force F on the distance l between the clamps for samples of the third and fourth types. The small circles mark the beginning and the end of the second stage of tearing.

shown on Fig. 134 for samples of the third and fourth type. The growth of the curve refers to the first stage of tearing and its horizontal part refers to the second stage (beginning of tearing at $l = l_K$, and its ending at $l \approx \lambda l_0$.

† If one takes in consideration the slow visco-elastic deformation of the ends of the specimen between the root of the cut and the clamps, one has to consider the value $d\lambda/dt$ as constant at dl/dt; λ grows slowly in the second stage of the experiment and dc/dt and, consequently, the force F increase slowly.

9.2. Theory of the Tearing of Rubbers

Let us investigate the tearing of a sample of the first type (Fig. 131), stretched with a constant speed $dl/dt = \text{const}$. On reaching the critical distance between the clamps $l = l_K$, a transition to the second stage of tearing takes place, and the cut grows with nearly the speed of sound.

Two assumptions are introduced into the theory of tearing.

1. The specimen is strained so slowly that the inner friction losses in the mass of the specimen can be disregarded, but sufficiently quickly for the tearing to have a critical character in the second stage.

2. The visco-elastic after-effect (creep) can be disregarded.

The elastic energy of a sample W is in this case a function of two variables—its length l and the depth of the cut c:

$$dW = \left(\frac{\partial W}{\partial l}\right)_c dl + \left(\frac{\partial W}{\partial c}\right)_l dc \tag{9.3}$$

This equation means that the elastic energy of the specimen W increases due to the work of the external force F [first member $\partial \omega/\partial l)_c \, dl = F dl$], and decreases as a result of the formation of new surfaces with the growth of the cut (the second member is negative).

In the first stage of the tearing a deformation of the sample takes place, accompanied by an accumulation of elastic energy, whilst the cut does not grow ($dc = 0$). It follows, therefore, from eqn. (9.3) for the first stage:

$$dW = \left(\frac{\partial W}{\partial l}\right)_c dl = F\,dl \tag{9.4}$$

where F is the stretching force.

In the first stage we disregard the first member in eqn. (9.3) and consider that in a rapid break, whilst the cut grows with the speed of sound, the distance between the clamps does not change practically, i.e. $l = l_K = \text{const}$. Therefore

$$dW = \left(\frac{dW}{\partial c}\right)_{l_K} dc. \tag{9.5}$$

Here $(\partial W/\partial c)_{l_K} < 0$, as the material around the growing cut unloads, and a part of the elastic energy (which was stored in the specimen up to the moment of failure), is expended on the work of tearing, and is dissipated in the form of heat.

The transition from the first stage to the second proceeds at certain critical values of the length of the sample and of the applied force (l_K and F_K). The speed of the growth of the cut in the second stage is similar to the speed of sound in rubber and does not depend on the strain rate of the specimen.

According to Rivlin and Thomas $(\partial W/\partial c)_{l_K}$ is the work of the formation of two surfaces with the area $2b$ cm² (where b is the thickness of the sample) during the growth of the cut along the path of 1 cm. The work which is expended in the formation of a unit of the area of the new surface with a critical character of the tear, is

$$T_{\text{ch.e}} = \frac{(\partial W/\partial c)_{l_K}}{2b}$$

and is called the characteristic energy of tearing. It is a constant of the material and consists in the free surface energy and the energy dissipated† per 1 cm² of the surface of the tear. The dissipated energy depends on the speed of the process but at a constant (critical) speed of tearing it is a constant of the material.

For the calculation of the derivative $\partial W/\partial c$ in eqn. (9.5) Rivlin and Thomas[1, 2] suggested the following method. The elastic energy W, accumulated in the specimen up to the moment of the transition to the rapid stage of tearing when l reaches l_K, is found from the experimental curve of the dependence of force F on the length of the sample l in the first stage (see Fig. 133). In the first stage, in agreement with eqn. (9.4) $dW = F\,dl$, and consequently

$$W = W(c_0 l_K) = \int_{l_0}^{l_K} F\,dl.$$

The value W is determined by the area which lies below the curve $F = F(l)$ (see Fig. 133).

For different sizes of the initial cut c_0 the value W has a different size (Fig. 135), as the force $F = F(l_1 c_0)$, and c_0 is the parameter of the integration. With the increase of the length of the initial cut, the elastic energy, stored in the first stage, decreases, but not

† The unavoidability of the dissipation of energy during the formation of the new free surface was investigated in the theory of strength of solids.[10, 11]

linearly (see Fig. 135). Therefore the derivative $(\partial W/\partial c_0)_{l_K}$ is not a constant of the material. Rivlin and Thomas reckon, however, that $(\partial W/\partial c)_{l_K} = (\partial W/\partial c_0)_{l_K}$, and it is assumed that the tangent of the angle of the slope of the dotted line (see Fig. 135) equals approximately the mean value of the derivative $(\partial W/\partial c_0)_{l_K}$. They note that in their experiments the real value of the derivative for c_0 of different lengths changed within the limits of 10–15%.

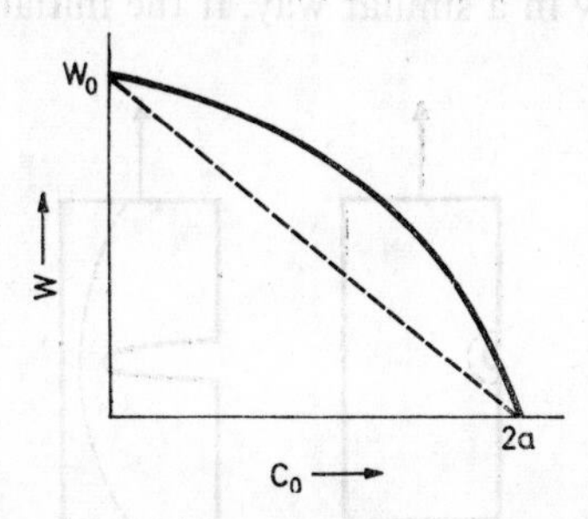

FIG. 135. Dependence of the elastic energy W, accumulated by the end of the first stage of tearing, on the initial length of the cut c_0 for samples with a cross-cut.

The substitution of $\partial W/\partial c$ for $\partial W/\partial c_0$ is not permissible. The derivative $\partial W/\partial c_0$ defines the difference in the accumulated energy during the transition from a sample with only the initial length of the cut c_0 to a sample with another length c_0 whilst both specimens are in a state of elastic equilibrium. Each point of the curve (see Fig. 135) corresponds to a critical value of the distance between the clamps l_K but this value is different for different lengths c_0 (the greater c_0, the smaller l_K). The rapid growth of the cut in the second stage of tearing proceeds with a practically constant value of l_K which is determined by the initial length of the cut. When the cut, in the second stage, increases to some value $c' > c_0$, the critical elongation of l_K is found to be greater than l_K of the other sample which has an initial length of the cut c'. Correspondingly, the elastic energy is found to be, in this moment of failure, different than it would follow from the graph of Fig. 135. The smaller c_0, the more will the dependence of the elastic energy on the length of the rapidly growing cut differ from the dependence of the stored elastic energy on the initial length of the cut.

There is one more objection to the method of determining the characteristic energy according to Rivlin and Thomas. The mean value of the derivative $\partial W/\partial c_0$ must depend essentially on the dimensions of the sample which they do not take into account. For instance, with the increase of the initial length of an uncut specimen l_0 ($c_0 = 0$) the stored elastic energy W_0 increases proportionally with l_0; consequently, the mean value of the derivative will also grow in a similar way. If the initial length of the cut

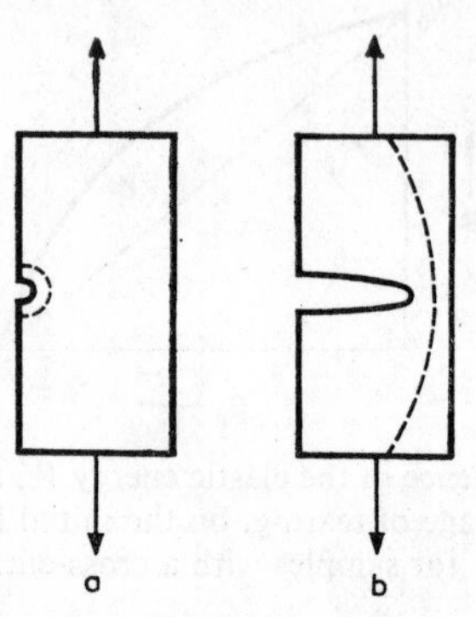

FIG. 136. Region of unloading around the cut in dependence on the initial length of the cut for samples of the first type (the region of unloading is marked by the dotted line): *a*, small initial length of the cut. *b*, great initial length of the cut.

c_0 is considerably smaller than the width and length of the sample ($c_0 \ll 2a$, $c_0 \ll l_0$), the region of unloading around the cut occupies a negligible part of the whole volume of the specimen (Fig. 136). The elastic energy of such a sample is, at the end of the first stage of tearing, almost the same as without a cut. The stored elastic energy of two specimens of the same width but of different initial length l_0' and $l_0'' = 2l_0'$ is, towards the end of the first stage, almost double, and at $c_0 = 0$ exactly double (see Fig. 137, curves 2 and 3). One can, therefore, assume that the accumulated elastic energy of the sample W with small cuts is straight proportional to its initial length l_0.

With increasing initial length of the cut the region of the unstressed material around it increases (see Fig. 136). The remaining stress spreads over a smaller and smaller volume of the sample, and the transition to the second stage of tearing takes place at smaller and smaller l_K. Owing to this the energy W decreases along a falling curve (see Fig. 135). When the length of cut c_0 is almost equal to the width $2a$ of the specimen, the latter is practically entirely unstressed, and $W \approx 0$ independent from the length of the specimen (see Fig. 137).

The constant for the whole length of the specimen is not the mean value $\partial W/\partial c_0$ but the initial value of that derivative at low values of c_0 (see Fig. 137). If the initial length of the cut c_0 in the

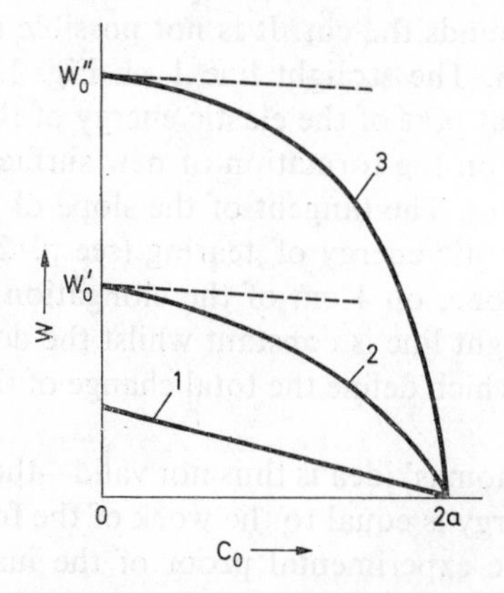

FIG. 137. Change of the elastic energy of the sample in dependence on the initial length of the cut. 1. Energy of the formation of new surfaces. 2. Elastic energy, accumulated by the end of the first stage for samples of the length l_0. 3. The same for samples of the length $2l_0$.

sample is considerably smaller than its linear dimensions, one can practically consider the cut as in an infinite lamina and any small change of the length of the cut c_0 causes an equal change of the elastic energy in samples of different lengths. This means that the derivative $(\partial W/\partial c_0)_{l_K}$ at small values of c_0 has the same value for all $l_0 \gg c_0$ (see dotted line in Fig. 137).†

† For instance, for solid brittle bodies at $c_0 = 0$ the derivative $\partial W/\partial c_0 = 0$. This follows from the work of Griffith[12] and Frenkel[13], according to which $W = W_0 - Kc_0^2$ (where K is a constant), and, consequently, $\partial w/\partial c_0 = 2Kc_0$. For elastomers, according to Bueche's and Berry's paper,[14] Griffith's formula is not valid, and for them the question of the calculation of the value of $(\partial W/\partial c_0)_{l_K}$ with $c_0 = 0$ remains open.

Finally, one more essential objection against Rivlin's and Thomas' method of calculation: if the derivative $(\partial W/\partial c)_{l_K}$ has been correctly established by some method or other from experimental data one should not, strictly speaking, connect it with the characteristic energy, as the derivative determines in fact the total decrease of elastic energy in the specimen during the growth of the cut. The elastic energy is expended not only on the formation of new surfaces and the dissipation of energy which is connected with it but also on the additional processes of unloading in the mass which surrounds the cut. It is not possible to allow for this in the calculation. The straight line 1 on Fig. 137 describes the dependence of that part of the elastic energy of the sample which is expended only on the formation of new surfaces on the initial length of the cut c_0. The tangent of the slope of the straight line 1 is the characteristic energy of tearing (see p. 240), calculated, as this is often done, on 1 cm of the elongation of the cut. The slope of this straight line is constant whilst the derivative $\partial W/\partial c_0$ (curves 2 and 3) which define the total change of the elastic energy changes with c_0.

Rivlin's and Thomas' idea is thus not valid—the decrease of the stored elastic energy is equal to the work of the formation of new free surfaces. The experimental proof of the inaccuracy of this suggestion is that $(\partial W/\partial c_0)_{l_K} \neq$ const.

According to Griffith's[12] theory a crack of the length c grows catastrophically if the decrease of the elastic energy dW with an infinitely small elongation dc of the crack is greater than the expenditure of energy on the formation of new surfaces. According to Griffith this expenditure of energy is equal to the free surface energy; however, as has been shown in the theory of strength, Griffith did not take into account the dissipation of energy (see Chapter 1). At critical failure

$$-dW = \delta A + \delta Q$$

where $dW < 0$ is the change of the elastic energy; $\delta A = 2\alpha_{surf}\, dS$ is the work of the formation of two new surface areas dS (where α_{surf} is the free surface energy); δQ is the dissipated energy.

In so far as the process of failure is a steady one in the second stage, i.e. the speed of failure is constant in the critical stage, the dissipated energy is in general the sum of two components:

$$\delta Q = 2q\, dS + \delta Q'$$

where q, the heat which is dissipated during the formation of 1 cm^2 of new surfaces, is connected with the mechanism of growth of the crack itself; $\delta Q'$ is the heat which, due to mechanical losses, is dissipated in the mass of the whole specimen.

The critical failure in this case is formulated as by Griffith, only, one has to put the sum $(\alpha_{surf}+q)$ instead of the free surface energy α_{surf}, i.e. the characteristic energy of tearing of the solid body (see § 1.5).

In a catastrophic failure the decrease of the elastic energy dW is obviously greater than the expenditure of energy on the formation of new surfaces $2dS(\alpha_{surf}+q)$. Consequently, the method suggested by Rivlin and Thomas cannot be considered as well founded.

Let us examine the tearing of a specimen of the third type (see Figs. 131 and 132). Stretching proceeds with a constant speed dl/dt where l is the distance between the clamps. The process of tearing, like that for the specimens of the first two types, can be divided into two stages (see Fig. 134).

1. The stretching of the ends of the sample without increase of the length of the cut, right up to a certain ultimate scale-factor of tension λ_K which corresponds to the critical value of the outside force F_k.

2. The growth of the cut with $F = F_k$ with a constant rate of tensioning whereby this speed is determined by the speed of the moving clamps (as $dl = 2\lambda_K dc$, $dl/dt = 2\lambda_K dc/dt$ if one uses λ_K, the constant of the second stage†).

In this case the characteristic energy $T_{ch.e}$ depends on the speed of failure of the rubber sample which is determined by the speed of the moving clamps. The greatest value of the characteristic energy corresponds to the critical speed of tearing which approaches the speed of sound in rubber, but in experiments with samples of the third type the characteristic energy will be considerably smaller, as the speed of the moving clamps is smaller than the speed of sound. The elastic energy W must in this case not be expressed as a function of the distance between the clamps 1, nor as a function of the length of the cut c, as for specimens of the third type in the second stage 1 and c are unambiguously connected with each other $(dl/2\lambda_K dc)$, and the derivative $(\partial W/\partial c)_l$ loses its meaning. We shall therefore regard the work A of the external

† We disregard the slow process of the visco-elastic after-effect of the rubber.

force F as a function of the scale factor of tension λ, and of the length of the cut c (these values do not depend on each other):

$$dA = \left(\frac{\partial A}{\partial \lambda}\right)_c d\lambda + \left(\frac{\partial A}{\partial l}\right)_\lambda dc \tag{9.6}$$

The temperature of the experiment and the strain rate are set.

In the first stage $c = c_0$ and $dc = 0$, and therefore, considering that $\lambda = l/l_0$, we obtain

$$dA = \left(\frac{\partial A}{\partial \lambda}\right)_{c_0} d\lambda = \left(\frac{\partial A}{\partial l}\right)_{c_0} dl. \tag{9.7}$$

But $(\partial A/\partial l)_{c_0} = F$, consequently

$$dA = F\,dl. \tag{9.8}$$

In the first stage the whole work of the external force F is expended on the increase of the elastic energy of the stressed ends of the sample and the energy which is accumulated towards the end of the first stage, is

$$W = \int_{2c_0}^{l_K} F\,dl \tag{9.9}$$

where $2c_0 \approx l_0$.

At a sufficiently great length of the cut c_0 the accumulated energy is proportional to c_0, if one can disregard the zone of bending D on Fig. 132. The dependence of the force F on the distance between the clamps l for both samples with an initial length of the cut c_0 and $4c_0$ (which corresponds to $l_0 = 2c_0$ and $l_0 = 8c_0$) is shown on Fig. 138. The elastic energy of the second

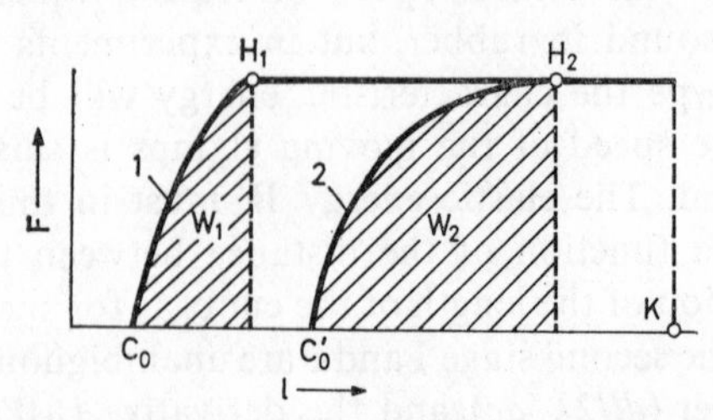

FIG. 138. Dependence of the force F and the elastic energy W, accumulated by the end of the first stage of tearing (shaded area), on the initial length of the cut c_0 for samples of the third type. 1. Sample $cl_0 = 2c_0$. 2. Sample $cl_0 = 2c_0' = 8c_0$.

specimen W_2 is almost four times greater than the elastic energy W_1 of the first specimen.

In the second stage $\lambda = \lambda_K$ and $d\lambda = 0$, therefore

$$dA = \left(\frac{\partial A}{\partial c}\right)_{\lambda_K} dc = F_K \, dl \qquad (9.10)$$

where $dl = 2\lambda_K \, dc$.

Therefore the external force F also performs work in the second stage of the tearing in contrast to the specimens of the first and second type. The work in the first stage (shaded area on Fig. 138) is expended only on the increasing of the elastic energy of the ends of the sample (like for specimens of the first type). The work in the second stage is expended both on the increase of the elastic energy of the ends of the sample (the ends of which increase due to the growth of the cut), and on the process of tearing itself which is accompanied by the formation of new surfaces. It is, therefore, suggested that one should determine the characteristic energy for specimens of the third type[1, 2] by using the equation of the balance of energy in the second stage:

$$dA = d\varepsilon + dW$$

where dA is the elementary work of the force F_k; $d\varepsilon$ is the work expended on tearing; dW is the increase of the elastic energy of the ends of the sample.

With the growth of the cut by the length dc the area of pure tension increases (see Fig. 132, area A) by the size of the volume $2ab \cdot dc$ and the unstressed area B decreases by the same volume. In agreement with eqn. (9.10) the work of the force F_k is

$$dA = F_K \, dl = 2F_K \lambda_K \, dc.$$

The increase of the elastic energy can be written

$$dW = 2ab \, dc W_1$$

where b is the thickness of the sample; W_1 is the specific work of the deformation of 1 cm³ of rubber at the extension from $\lambda = 1$ to λ_K.

As the energy of tearing is by definition $d\varepsilon = 2bT_{\text{ch.e.}} \, dc$, one can, in conclusion, write the balance of energy like this:

$$2F_K \lambda_K \, dc = 2T_{\text{ch.e.}} b \, dc + 2W_1 ab \, dc$$

or

$$F_K \lambda_K = T_{\text{ch.e.}} b + W_1 ab \qquad (9.11)$$

From eqn. (9.11) we find the characteristic energy $T_{\text{ch.e.}}$:

$$T_{\text{ch.e.}} = \frac{F_K \lambda_K}{b} - W_1 a \tag{9.12}$$

The characteristic energy can be calculated if we know the specific work of the deformation W_1. This can be calculated or determined by experiment. Several formulae for the deformation of rubber under tension have been suggested, and these formulae were compared with the experimental results by Bartenev and Vishnitskaya.[15] For instance, W_1 was calculated[9, 16] from the two-parameter equation of the deformation of rubber by Muni-Rivlin.

The following formula of the law of deformation, suggested by one of the authors,[15] is simple and exact at equilibrium and static stretching of unfilled rubbers by up to 100–200%:

$$f = E\frac{\lambda - 1}{\lambda} \tag{9.13}$$

where $f = F/ab$ is the apparent stress; E is the visco-elastic modulus of the rubber.†

In any existing case we have, for rubber,

$$W_1 = \int_1^{\lambda_K} f \, d\lambda \tag{9.14}$$

If we apply here eqn. (9.13) we get

$$W_1 = E(\lambda_K - 1 - \ln \lambda_K).$$

Substituting this result in eqn. (9.12) we find

$$T_{\text{ch.e.}} = \frac{F_K \lambda_K}{b} - aE(\lambda_K - 1 - \ln \lambda_K)$$

Whereas, according to (9.13) $F\lambda/b = aE(\lambda - 1)$

$$T_{\text{ch.e.}} = aE \ln \lambda_K \tag{9.15}$$

or

$$T_{\text{ch.e.}} = aE \ln \left(\frac{abE}{abE - F_K} \right) \tag{9.16}$$

† As the visco-elastic after-effect is disregarded from the very beginning, it is suggested with the same degree of certainty that the modulus E does not depend on the strain rate of the rubber, and that one can regard it simply as the equilibrium modulus of the rubber E_∞.

Taking the more precise law of deformation into consideration we obtain a more precise result for the characteristic energy.

Another method for finding W_1 is the following: at a predetermined strain rate we find the experimental relation between the apparent stress f and the scale factor of tension λ up to λ_K, and can calculate the area under the curve $f(\lambda)$. According to eqn. (9.14) this will also be W_1.

The characteristic energy of specimens of the fourth type is calculated by a similar method. One must, however, keep in mind that, as the failure of samples of the fourth type is caused by shear, the stress along the cross-section of the sample is very unevenly distributed, and the characteristic energy can differ considerably from the characteristic energy for samples of the third type.

9.3. Experimental Determination of the Characteristic Energy of Tearing

The characteristic energy $T_{ch.e.}$ is the sum of the free surface energy and the energy lost by dissipation, calculated on 1 cm^2 of the new surface. According to the data of a number of authors the characteristic energy has an order of magnitude of 10^6 erg/cm^2, whilst the free surface energy of elastomers is of an order of magnitude of 10^2 erg/cm^2. It is obvious that the characteristic energy consists practically only in the losses of the tearing process, calculated on 1 cm^2 of new surface.† From this point of view it is easy to explain the experimental facts which will be discussed below.

A series of papers by Rivlin, Thomas and Greensmith[1-8] has been dedicated to the determination of the characteristic energy for different types of specimens.

It follows from the preceding section that only specimens of the third and fourth type are suitable for the experimental determination of the characteristic energy of tearing. In this the actual process of failure of these specimens differs somewhat from the

† The dissipation of energy takes place not only in connection with the formation of new surfaces but also because of the losses in the mass of the elastomer during its deformation. The latter, obviously, make the main contribution to the value $T_{ch.\ e.}$; therefore one can tentatively relate it to the unit of surface.

projects which have been examined by us. For instance, due to the microheterogeneity of the elastomer the growth of the cut proceeds not strictly with the same speed so that F_k is the mean value of the force in the second stage of tearing. Besides, the visco-elastic after-effect of the rubber (creep) which has not been taken into account

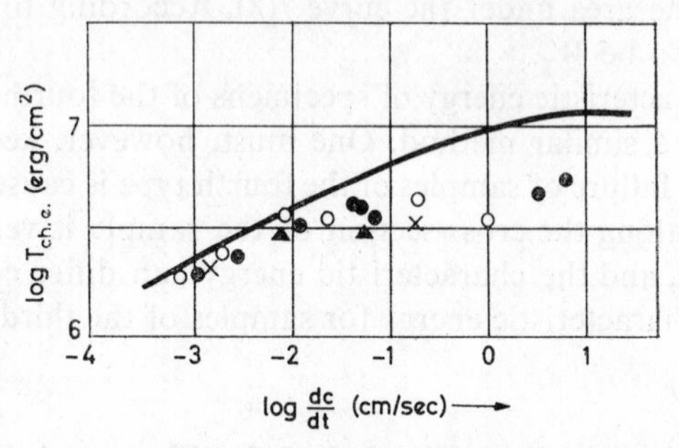

FIG. 139. Dependence of the characteristic energy on the speed of tearing (solid line: samples of the third type at constant speed of the clamps).[6] ● — samples of the fourth type under constant outside force; △ samples tested on shear, 5 cm long; ○ — samples 3 cm long; × — samples 1·5 cm long.

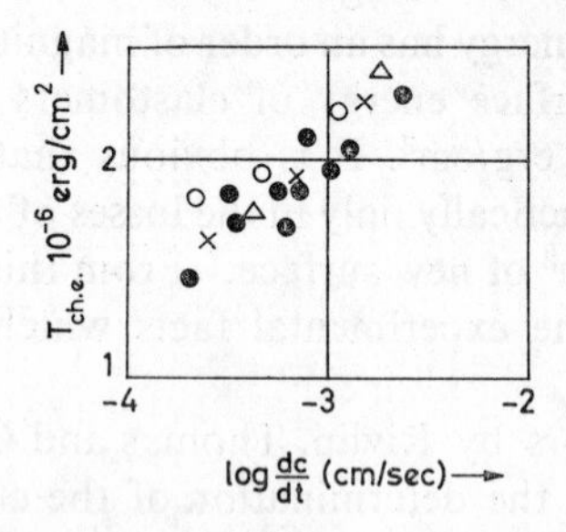

FIG. 140. Dependence of the characteristic energy on the speed of tearing in regions of low speeds.[6] ● — samples of the fourth type; ○ — samples, tested for shear, 5 cm long; △ — the same, 3 cm long; × — the same, 2 cm long.

in the preceding discussions, leads to an additional elongation of the ends of the specimen. As a result the growth of the cut slows down a little, as the moving clamps cause not only a growth of the cut but also an additional elongation of the ends of the specimen. The force F_k decreases, and instead of the horizontal part on Fig. 134 a slightly inclined straight line is obtained.

As the dissipation of the energy is the greater, the higher the speed of growth of the cut, one has to expect that with an increase of the speed of the clamps the characteristic energy increases as well. This principle has in fact been observed (Figs. 139 and 140).

The dependence of the characteristic energy on the speed of the clamps has a complicated character, as at low speeds a "fibrous" failure is observed with the formation of a rough surface of the tear, and at high speeds, like in the failure of solids, a smooth surface of the tear is formed (Fig. 141, facing p. 290).

The mechanical losses, as we have already noted, occur in the whole mass of the specimen. In particular, in experiments with samples of the third and fourth type losses occur which are connected with the transition of all new parts of the sample beyond the region of bending into the region of simple tension. The characteristic energy of tearing and the mechanical losses must be the greater, the harder the specimen is, i.e. it must grow with its thickness (Table 11).

TABLE 11

Characteristic Energy of Rubber[16]

Rubbers	Strain rate (mm/min)	$T_{ch.e.}$ (kg cm/cm^2) with the thickness of the sample (mm)		
		0·5	1	2
NK	50	22·5	33·3	38·8
	500	38	39·0	56·5
SKS-30	50	5·7	8·0	9·45
	500	7·6	10·0	10·45

The increase of the losses with increasing thickness of the sample is caused by the circumstance that to 1 cm^2 of new surface belongs a great mass of rubber with its great losses.

With rising temperature the characteristic energy decreases;[4] this is also in agreement with the known fact of the decrease of the mechanical losses in rubbers with rising temperature, starting at the glass transition point.

According to Rivlin and Thomas the characteristic energy does not depend on the dimensions and the type of the specimen at predetermined strain rate and temperature which is, in particular, confirmed by the figures shown on Fig. 139 and 140. According to the data on Table 11 the characteristic energy for samples of different thicknesses differs considerably. Such a contradiction can partly be explained by the fact that according to a number of reports the investigation of tearing was carried out without a proper analysis of the experiment. One of the mistakes in several papers[1-9, 16] is the calculation of the specific work of the deformation from the equation of the equilibrium deformation, instead of finding it by the curve of deformation with the same strain rate as during tests on tearing.

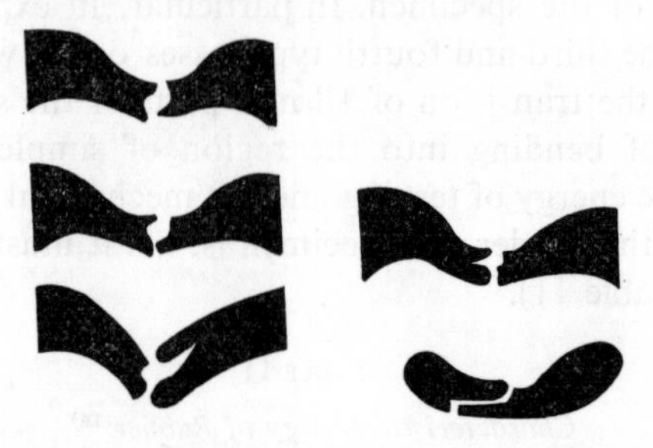

FIG. 142 "Knotty" tearing of filled rubbers on different specimens.

In some cases the tearing is complicated through being accompanied by a change in the structure of the rubber.[3, 4, 17, 18] For instance, during the stretching of natural rubber a crystallization takes place; filled rubbers attain an orientated structure under great tension which leads to an increase of the heterogeneity of the tear: to the so-called "jerky" and "knotty" tear (Fig. 142). These types of tearing appear with certain experimental combinations of temperatures and strain rates; this is evidently connected with the conditions of the formation of an orientated structure.

In regions of "knotty" and "jerky" tearing the usual principles of tearing are upset.

One has to draw the conclusion from all that has been said that the characteristic energy is an extremely complicated function of many factors. In contrast to the solid brittle bodies the elastomers possess clearly expressed relaxation properties and show great mechanical losses. The characteristic energy which has been determined experimentally, includes the following components: (1) free surface energy (approximately 10^2 erg/cm^2); (2) dissipated energy per 1 cm² of new surface (this is approximately of the same order as it is connected with the transition during the break of the chemical bonds across the barrier potential; the height of this is approximately one and a half times greater than the free surface energy, calculated on one bond); (3) the energy which is dissipated in the whole mass of the specimen, including that due to bending and straightening, and also that of the stretching of the ends of samples of the third and fourth types.† The latter component is by several orders of magnitude greater than all remaining ones.

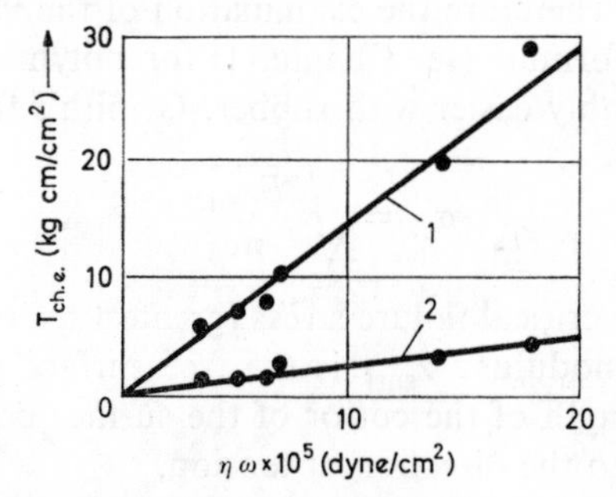

FIG. 143. Dependence of the characteristic energy of technical rubber (each point refers to a certain rubber) on the factor $\eta\omega$ (η is the dynamic viscosity, ω is the cyclic frequency), which defines the mechanical losses at dynamic experiments.[18] 1. At 25°C. 2. At 90°C.

The data shown on Fig. 143 point to a definite connection between the characteristic energy which was found experimentally and the value which characterizes the inner friction of the rubber. The greater the inner friction, the greater the experimentally found value of the characteristic energy of tearing. The extrapolation of the linear dependency to zero leads to the conclusion that the experimentally determined characteristic energy in a material which does not show inner friction, is practically zero. In fact it has such a small value in relation to the surface energy that one cannot represent on the scale of Fig. 143.

† The losses during the stretching of the ends of the samples are automatically taken into account if one does not calculate the specific work of the stretching of the rubber from the equation of the equilibrium deformation but determines it from a strain curve which is obtained at the same rate as prevailed during the tearing.

The characteristic energy which has been experimentally determined has therefore no fully defined physical meaning and is only of technological interest and characterizes the total mechanical losses during the tearing of rubber.

9.4. Examination of the Validity of Griffith's Formula for Elastomers

It is easier to make a cut of definite length and shape in rubber than in solids. Therefore the examination of the validity of Griffith's famous formula (see Chapter 1) for polymeric materials is made considerably easier with rubber. Griffith's formula is

$$\sigma_K = \sqrt{\frac{2E\alpha_{surf}}{\pi c_0}} \tag{9.17}$$

where σ_K is the critical failure stress (greatest technical strength); E is Young's modulus; α_{surf} is the free surface energy; and c_0 is the initial length of the cut or of the surface crack, which lies perpendicular to the direction of tension.

The examination of Griffith's formula was carried out on silicone rubbers not long ago by A. Bueche and Berry[14] who, using the experimental value of Young's modulus E and the maximum technical strength σ_K, determined the free surface energy of different rubbers according to eqn. (9.17). They assumed that rubbers have surface microcracks with the dimensions $c_0 \approx 10^{-3}$ cm. This assumption was confirmed by the experimental data of the dependence of strength on the depth of the cut. Rubbers without cuts show the same values of strength which are obtained by extrapolation of the experimental dependence up to a length of the cut of approximately 10^{-3} cm.

A. Bueche and Berry obtained for filled rubbers $\alpha_{surf} = 10^5$ erg/cm², and for unfilled ones from 1×10^3 to 6×10^3 erg/cm²† corresponding to the decrease of density of the spatial network. In fact the free surface energy of rubber is of an order of 10^2 erg/cm², and should not depend to any extent on the num-

† The calculation was carried out assuming that the modulus E and the breaking stress refer to the initial, and not the true, section of the specimen during rupture. Properly, the calculation ought to be carried out with reference to the true section, when the value α_{surf} increases.

ber of cross-links, as the number of the broken chains of the main valences per area unit of the cross-section does not depend on the number of cross-links. The order of magnitude of the energy which is indispensible for the formation of new surfaces of the rubber was found by the authors in the following way. Dimethylsiloxane chains have an area of cross-section of 64 $(\text{Å})^2$. 1 cm^2 of the cross-section of the sample intersects $1 \cdot 56 \times 10^{14}$ chains of macromolecules.† The dissociation energy for the Si–C bond is 106 kcal/mol, consequently, for the break of $1 \cdot 56 \times 10^{14}$ bonds it is necessary to expend 1154 erg/cm^2. As two surfaces are formed, each measuring 1 cm^2, this makes 580 erg on 1 cm^2. During contraction after the failure f both surfaces increase λ_f times, i.e. instead of 2 cm^2 measure $2\lambda_f$ cm^2 λ_f is the scale factor of tension during rupture.

In this calculation are two inaccuracies.

1. After the contraction of the sample there is a great number of chain segments on the free surface, besides the ends of the broken chains; the surface energy is, therefore, determined basically not by chemical forces, as the authors assume but by Van der Waals forces.

2. Even if the break takes place in the fully orientated state, when the energy of the break is determined only by chemical bonds, one has to keep the following in mind: (a) the energy of idssociation, as shown in the theory of strength,[10, 11] is not equal to the energy of rupture of the bonds between two newly formed surfaces, (b) after the break the radicals react, as a result of which oxygen-containing groups are formed on the broken ends of the chains. It is precisely the end groups which determine the size of the surface energy.

Considerations of these corrections cannot lessen the great difference between the actual size of the surface energy and the one calculated by eqn. (9.17). Nevertheless, the high values of α_{surf} obtained do not yet by themselves disprove the validity of Griffith's formula for rubbers, as the authors think. The surface energy per 1 cm^2 should enter into Griffith's formula instead of

† One has to distinguish two cases: (1) the break in the fully orientated state (in point of fact, this case is also examined by A. Bueche and Berry) when all chains are orientated, and transverse the cross-section of the sample only once, and (2) the break in the non-orientated state, when the chain can transverse the section many times.

α_{surf}, as the former is expended both on the formation of new surfaces and on the mechanical losses by which the rapid growth of the crack or cut is accompanied (for the detailed theory of this process see Bartenev and Razumovskaya[11]). The characteristic energy $T_{ch.e.}$, which is about 10^6 erg/cm², should probably be used for rubber instead of α_{surf}. It may depend on the number of cross-links as the dissipated energy which basically determines the size of the characteristic energy depends on the extent of cross-linking. For filled rubbers A. Bueche and Berry obtained a "surface energy" greater than for unfilled ones. This agrees with the fact that in filled rubbers the mechanical losses are greater than in unfilled ones.

The non-validity of Griffith's formula to elastomers is proved by other experiments by the same authors. A. Bueche and Berry investigated the dependence of the strength of silicone rubbers at break on the initial depth of the cut c_0 and Young's modulus E (they varied the modulus, changing the composition and the quantity of active filler). According to Griffith the strength σ_K is proportional to $1/\sqrt{C_0}$ and $\sqrt{E}$. A. Bueche and Berry found that $\sigma_K \sim c_0$ and $\sigma_K = A - BE$, where A, B are constants.

The analysis of the experimental data leads to the conclusion that at rupture the energy dissipates not only near the newly formed surfaces, but in the whole mass. The dissipating energy is then not proportional to the area of the new surfaces, and therefore one must not apply Griffith's formula to rubbers, even when replacing the surface stress by the characteristic energy of tearing. Besides, Griffith's theory which was developed by him for brittle materials does not take into account the influence of molecular orientation and the changes of the structure of rubber under tension.

In conclusion we note that in recent times papers appeared in which the influence of temperature, the frequency of cyclic loadings and other factors in the process of tearing of rubber was studied. For instance Patrikeev and co-workers[19] studied the influence of the cut on the resistance of rubber to tearing in the wide temperature range from -70 to $+160$°C. It was found that the strength properties of rubber in a certain range of low temperatures (but[20] above T_g) are very responsive to the cut. Recently, Andrews investigated the propagation of the cut under cyclic loading. Stripes of equal width δ show very clearly on the surface of the break of rubbers which crystallize under tension. By the width of the stripe δ one can judge the advance of the cut during one cycle. The value δ does practically not depend on the frequency of deformation but grows sharply with the increase of the maximum deformation during the cycle. The observed value of δ is 1 μ or more.

Literature

1. R. S. Rivlin, A. G. Thomas, *J. Polymer Sci.* **10,** 29 (1953).
2. A. G. Thomas, *J. Polymer Sci.* **18,** 177 (1955).
3. H. W. Greensmith, A. G. Thomas, *J. Polymer Sci.* **18,** 189 (1955).
4. H. W. Greensmith, *J. Polymer Sci.* **21,** 175 (1956).
5. A. G. Thomas, *J. Polymer Sci.* **31,** 467 (1958).
6. A. G. Thomas, *J. Appl. Polymer Sci.* **3,** 168 (1960).
7. H. W. Greensmith, *J. Appl. Polymer Sci.* **3,** 173, 181 (1960).
8. H. W. Greensmith, *The Rheology of Elastomers*, edited by P. Mason and N. Wookey, London, 1958, p. 113.
9. R. S. Rivlin, *Phil. Trans. Roy. Soc.* A **241,** 379 (1948).
10. G. M. Bartenev, *Izv. AN SSSR, OTN*, No. 9, 53 (1955).
11. G. M. Bartenev, I. V. Razumovskaya, *DAN SSSR*, **133,** 341 (1960).
12. A. Griffith, *Phil. Trans. Roy. Soc.* A **221,** 163 (1921).
13. Ya. I. Frenkel', *ZhTF*, **22,** 1857 (1952).
14. A. M. Bueche, J. P. Berry, *Fracture*, Cambridge, Mass. Technol. press, New York, 1959, p. 265.
15. G. M. Bartenev, L. A. Vishnitskaya, *Izv. AN SSSR, OTN, Mekhanika i mashinostroyeniye*, No. 4, 175 (1961); *Viysokomol. soyed.* **4,** 1325 (1962).
16. A. I. Lukomskaya, *v sb. "Fiziko-mekhanicheskiye ispiytaniya kauchuka i reziniy"*, Goskhimizdat, 1960, p. 121 (*Trudiy NIIShP*, sb. 7).
17. W. F. Busse, *Ind. Eng. Chem.* **26,** 1194 (1934); *Rubb. Chem. Technol.* **8,** 121 (1935).
18. H. W. Greensmith, L. Mullins, A. G. Thomas, *Trans. Soc. Theology*, **4,** 179 (1960).
19. G. A. Patrikeev, B. G. Gusarov, V. I. Konoplev, *Viysokomol. soyed.* **2,** 1438 (1960).
20. E. H. Andrews, *J. Appl. Phys.* **32,** 542 (1961).

CHAPTER 10

BASIC CONCEPTS OF FAILURE OF POLYMERS IN AGGRESSIVE ENVIRONMENTS

10.1. Types of Failure under the Influence of Aggressive Environments

In the preceding chapters the strength was examined as the resistance to the destructive action of external mechanical stresses. However, when studying the influence of physically and chemically active environments on strength, the failure of the material has to be examined in a more general way—as a result of the surmounting of the forces or interactions between atoms or molecules. From this point of view mechanical forces are only one of the means of surmounting of such interactions; a peculiarity of failure in this case is the development of the failure process in separate places of the specimen, as a result of which it parts in two or several pieces.

Apart from the effect of the mechanical forces the failure of materials (solids, elastomers) can be caused by the following factors:

1. Effect of heat (melting, evaporation, decomposition).
2. Dissolving.
3. Action of chemically aggressive environments.

Weaker effects are also possible which are usually not accompanied by a complete failure of the material (adsorption, swelling) but facilitate it.

Strictly speaking the chemical nature of materials cannot only change by dissolving or through the effect of heat (up to the decomposition temperature) but also by mechanical break of the intermolecular bonds (in its pure form this can be observed, for instance, in low-molecular paraffins). The active centres which form during mechanical break enter into a chemical interaction with the elements of the environment and the chemical composition of the body changes at the place of failure.

The failure of the material asa result of the interaction with the chemically aggressive environment can be of two types.

1. During the chemical interaction there occurs a complete change of the composition of the material and of its physical properties; a considerable number of chemical bonds break and are rearranged.

2. The failure is accompanied by the change of the chemical composition of the material only in separate places with a retention of practically unchanged properties of the material as a whole (local failure).

To explain the influence of the environment on the strength properties of the original material the case of local failure is of primary interest.

The active influence of the external conditions on the material appears especially clearly during a simultaneous application of mechanical stress.† The contribution of each of these factors depends on the applied strain rate and also on the active environment.

The influence of surface-active substances on the strength (and also on the resistance and the deformability of the stressed material) was discovered and widely explored by Rebinder,[1–3] beginning in 1928 (see § 1.9). Rebinder's effect (i.e. the decrease of strength under the influence of the adsorption of surface-active substances) was discovered on polycrystalline solids (metals,[4–6] rocks,[7] on glasses,[8] graphite[9] and others[10]).

This effect found a wide, practical application to facilitate the processing of metals, the drilling of rocks, pulverizing, etc., with the use of suitable surface-active substances.

It is not adsorption of the environment which has the greater influence on the mechanical, in particular strength—properties of polymers but its influence on the whole mass which appears as swelling or during the introduction of a plasticizer in the process of manufacturing of polymers. The strength of swelled poly-

† One has to keep in mind that the influence of an active environment on the undeformed material with a subsequent application of stress is, owing to its mechanism, qualitatively different from the influence of the environment on deformed material.

mers decreases tenfold. This change of strength is reversible as was shown by Pisarenko and Rebinder,[11] and investigated in greater detail by Luttrop.[12] The latter showed that the strength of elastomers after swelling is smaller by an absolute amount than after recovery (with the same liquid content in the polymer). This shows that the defects which exist on the surface play a predominant role during failure.

Due to the irregular distribution of the plasticizer during swelling, the surface layer turns out to be weaker; but during recovery the inner layer of the specimen is the weaker one. The data on the influence of swelling on the strength of elastomers indicates that it is to a considerable degree determined by the magnitude of the intermolecular forces.

The influence of swelling on the durability of some fibres (viscose, Nylon) and plastics [poly(methylmethacrylate)] was investigated in the works of Zhurkov and co-workers[13, 14] (see § 4.4). The swelling of these materials lowers the durability, and creates a sharper time dependence of strength. The energy of the failure activation process, which consists basically in the breaking of the chemical bonds, does not change with the introduction of plasticizers.

When determining the durability of rubbers which contain plasticizers, a different picture is observed.[15] The durability of these rubbers which conform to the relation[16]

$$\tau = B\sigma^{-b}$$

decreases with the introduction of a plasticizer, as into plastics, and the time dependence becomes more pronounced (Fig. 144). At the same time the failure activation energy changes as well. In SKN-26 rubbers the apparent activation energy U is 27 kcal/mol; with the introduction of 8·5% w/w kerosene, U decreases to 18 kcal/mol, and with 15·6% w/w to 16 kcal/mol. In Nairit (chloroprene rubber), U is 93 kcal/mol; with the introduction of 15·2% w/w dibutylphthalate, U grows to 130 kcal/mol. In the investigated cases, as in unplasticized rubbers, U dues not depend on the stress (with the exception of Nairit in the temperature range of 30–40°C).

These facts can be explained by the circumstance that the failure of elastomers consists basically in the surmounting of intermolecular forces, as a consequence of which the plasticizer which

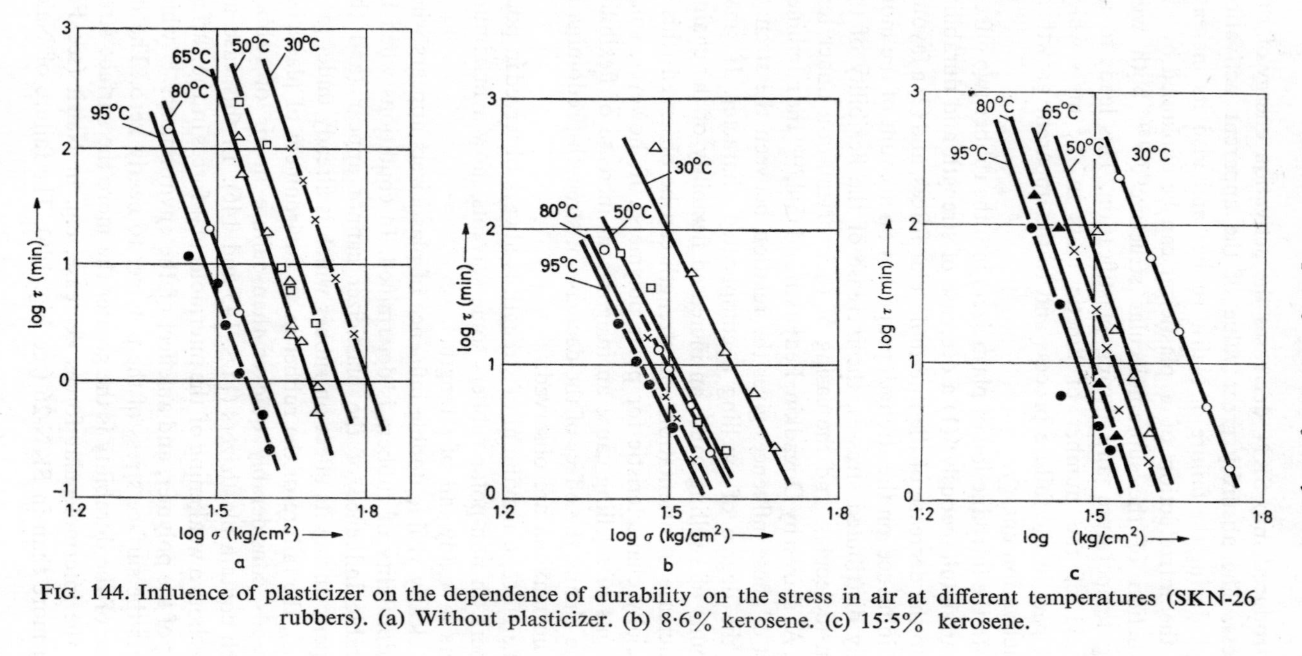

FIG. 144. Influence of plasticizer on the dependence of durability on the stress in air at different temperatures (SKN-26 rubbers). (a) Without plasticizer. (b) 8·6% kerosene. (c) 15·5% kerosene.

is introduced into SKN decreases the activation energy of this process. The absurdly great value of the apparent activation energy during the failure of Nairit rubber and also its increase with the introduction of a plasticizer can be explained by the destruction of the supramolecular structure[17, 18] both with rising temperature and during plastification. This leads to an increase of the number of kinetic units in the mass which takes part in the failure process and to the apparent growth of the activation energy.

With the introduction of plasticizers into the rubber two effects appear simultaneously: (1) a decrease of strength and durability due to the decrease of the intermolecular forces, and (2) a favourable influence on the strength properties on account of the more evenly distributed stresses, the increase of the flexibility of the chain molecules, and the easing of their orientation under tension. As shown by Dogadkin, Fedyukin and Gul,'[19] the combined effect of these influences gives the relation between the strength and the degree of swelling a complicated character. If small amounts of swelling cause an increased flexibility of the chains, an increase of orientation and strength can be observed (this is especially characteristic for polar polymers). If, however, small amounts of swelling cause an insignificant increase of flexibility of the chains, the effect of the decrease of strength predominates and no maxima are observed.

Therefore, a maximum of strength is always observed in polar rubbers, in non-polar rubbers some solvents show a maximum, others a steady fall of strength.

A clearly visible double influence of plasticizers appears when the durability of rubbers is determined. In conditions when favourable influence of the plasticizer cannot appear (and this happens during its effect on rubber which is already under great stress), the failure of the rubber in an environment of plasticizer proceeds considerably quicker than failure in air of rubbers which contain plasticizers (Figs. 145 and 146). In this case appears both a weakening of the intermolecular forces in the surface layer of the polymer, and an effect of the environment[1–3] which lowers the surface stress of the polymer to nearly zero. The decrease of the durability is the greater the more the surface stress and the intermolecular forces decrease, e.g. in Nairit (see Fig. 145) more than in SKN-26 (see Fig. 146). The failure of Nairit

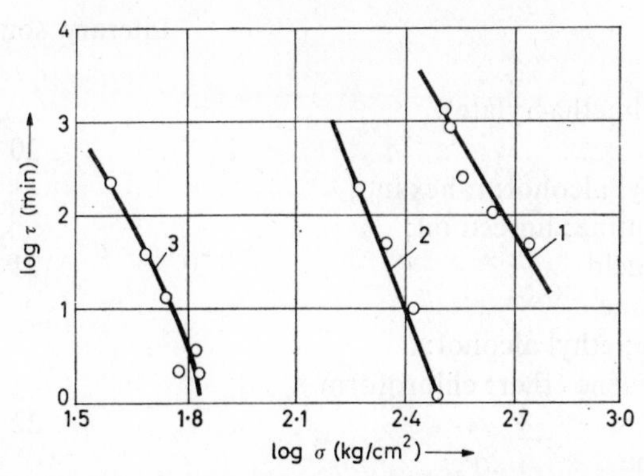

FIG. 145. Influence of the plasticizer on the durability of Nairit at 40°C. 1. Without plasticizer, test in air. 2. Plasticized (15·1% DBP) test in air. 3. Pre-loaded, test in DBP.

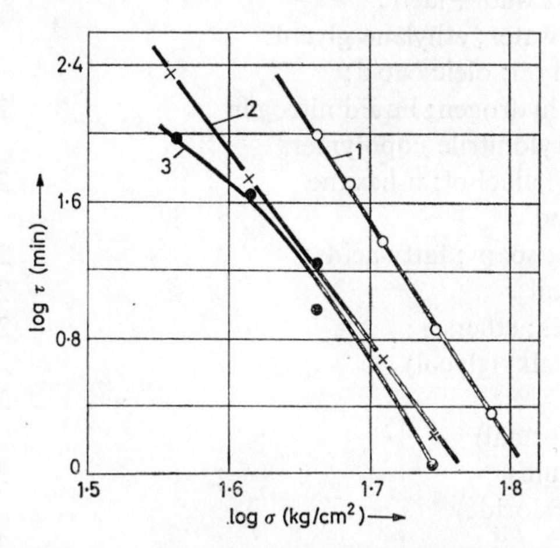

FIG. 146. Influence of the plasticizer on the durability of SKN-26 rubbers at 30°C. 1. Without plasticizer, test in air. 2. Plasticized (8·6% kerosene) in air. 3. Pre-loaded, test in kerosene.

is accompanied by the appearance of separate cracks on the surface.

Cracking under the influence of physically active (in particular, surface-active) substances appears to a considerable extent in stressed plastics.

Herewith a list of environments causing stress cracking of plastics:

	Literary source
Ply(methylmethacrylate)	
Water	20
Isopropyl alcohol; n-hexane; turpentine; linseed oil; oleic acid	21
Polyurethane	
Acetone; ethyl alcohol; diethylene ether; chloroform benzole	22
Polystyrene	
Heptane + petroleum ether; heptane + isopropyl alcohol	21, 23
Methyl alcohol + lactic acid + water; ethylene glycol; linseed oil; oleic oacid; liquid hydrogen; liquid nitrogen	24
Styrene acrylonitrile copolymers	
Isopropyl alcohol; n-hexane	21
Polyethylene	
Alcohols; soaps; fatty acids	25, 26
Silicone oil	27
Aldehydes; ethers	28
Poly(arylalkylglycol)	29
Oils	30
Poly(vinylformal)	
Polar liquids	31
Poly(vinylchloride)	
Solvents	32
Methanol; isopropyl alcohol; n-hexane	21
Carbon dioxide; liquid hydrogen; liquid nitrogen	24
Rubbers in glassy state	
Carbon dioxide; liquid hydrogen; liquid nitrogen	24
Polycarbonate	
Methanol; isopropyl alcohol; n-hexane turpentine	21

Chemically active environments have an even greater influence on the strength properties of the materials than purely physically active ones. The effect is sometimes so important that the failure of stressed materials under simultaneous effect of chemically active environments appears often as a phenomenon which is not connected with the strength properties of the material—but as a qualitatively different process. For instance the speed of the failure process under the action of ozone on stretched rubber may at a certain concentration of O_3 increase hundred thousand times compared with the speed of failure without ozone. The idea of the similarity of the processes of corrosion failure and static fatigue which has more than once been expressed by one of the authors and is also mentioned in this book, has recently been more and more widely accepted. For instance, the opinion has been expressed that there exists an analogy between ozone cracking of rubbers and the cracking of plastics under the influence of mechanical stresses.[33] In one of the Japanese papers[34] the process of the development of ozone cracks in stretched rubber is described with the help of the same method and is similar to the investigation into the development of cracks during the brittle failure of solids.[35]

According to the different types of effects of aggressive environment on stressed polymers, one can distinguish three basic types of failure.

1. Local action of the environment on the surface of the material, accompanied by the formation of microcracks. This effect leads to more and more unevenly distributed stresses, to their concentration in the roots of the cracks, and to catastrophic failure.

2. "Uniform" non-localized failure, not accompanied by visible cracking. The clearest manifestation of such a type of failure is the chemical stress relaxation.

3. Failure, accompanied by cracking in the whole mass of the material, including the surface (a phenomenon observed with fatigue of rubbers). This type of failure will not be especially investigated in view of the existence of sufficiently detailed reviews.[36, 37]

The following different factors are accepted as characteristics of "uniform" failure, in particular the speed of creep during constant stress, the durability and also the stress relaxation during constant deformation. The methods of study of local failures will be described in detail further on.

10.2. Chemical Relaxation

Tobolsky and co-workers[38] showed that on stressed unsaturated rubbers, under the influence of atmospheric oxygen at a temperature above 100°C an increasingly falling stress (chemical relaxation) and an increase of the speed of creep can be observed (these phenomena received the name chemorheology[39]). A similar phenomenon was also observed in vulcanized rubbers of other types: in polysulphide,[40] silicone,[40, 41] carboxyl,[42] fluorine,[43] and urethane rubbers.[40] Tobolsky's papers were discussed in a number of reviews,[44, 45] and especially fully in his monograph.[40]

At present the following basic conclusions can be formulated, whilst taking into account the latest data in this field. The decrease of the stress in deformed rubbers at high temperatures is the result of the failure processes and the rearrangement of the spatial network of the polymer which proceeds with the destruction of the chemical bonds and their subsequent appearance in new places. A direct evidence of such a rearrangement is the irreversible flow of spatial polymers with a small increase of the residual deformation. This proves also to a certain degree the great importance of the energy of the activation process of the chemical relaxation. Thus, according to Tobolsky's data,[38] the activation energy of this process is 30 ± 2 kcal/mol for sulphur containing rubbers based on natural, butadiene-styrene, polychloroprene and butyl rubbers. In a recent work[46] it was shown that after a preliminary exposure to atmospheric nitrogen at a high temperature nine different vulcanized sulphur containing natural rubbers the subsequent chemical relaxation in the presence of oxygen proceeded with an activation energy of 29 kcal per mol. Similar figures were also obtained by Berry and Watson.[47] The activation energy of the chemical relaxation of Thiocol rubbers lies within the range of 24–36·5 kcal/mol,[40] and of fluorine rubbers[43] (at a temperature of above 150°C) within the range of 30–40 kcal/mol.

In general the failure of the chemical bonds which causes the chemical stress relaxation in rubbers may originate under the influence of heat (at high temperatures), under the influence of heat and oxygen and also under the influence of catalysts. As Dogadkin and Tarasova[48] showed, the break of the polysulfide cross-links in vulcanized NK (natural rubber) and BSK occurs already at a temperature of 70°C in the environment of nitrogen. At higher temperatures even the C–S–C and C–C bonds break with a noticeable speed. At 130°C the speed of chemical relaxation of vulcanized natural rubbers which contain mainly polysulfide cross-links† is approximately 10–30 greater than that of vulcanized rubbers which contain mainly C–S–C and also C–C cross-links. For vulcanized butadiene-styrene rubbers the difference in the speed of chemical relaxation fluctuates within the range of 6–250.

Under the simultaneous effect of oxygen and high temperatures there proceeds in vulcanized unsaturated rubber an oxidizing destruction of the rubber chains. This appears most clearly in the presence of strong cross-links (C–C, C–S–S) which are formed during radiation or peroxide vulcanization, and during vulcanization with thiuram. The destruction of the macromolecules of rubber is also clearly proved by Tobolsky's data[40] which show that the number of breaks of the chain changes equally in unvulcanized natural rubber (i.e. which does not contain cross-links) and in vulcanized natural rubber which does contain a different number of C–C links. Dunn[50] showed that vulcanized natural rubber which contains strong cross links (peroxide and thiuram vulcanized rubber), extracted with acetone, relaxes in air in the same manner. The determining role played by the oxidizing degradation of the chain is confirmed by the decelerated chemical relaxation in air when inhibitors are introduced,[50, 51] and also to a certain degree by the data on the influence of the relaxation properties of rubber in oxygen[51] on the speed of this relaxation process. The speed of the chemical stress relaxation of thiuram vulcanized rubbers at 100°C is eight times greater for natural rubber than for SKB[51] and the speed of oxidizing these poly-

† As shown in the work of Studebaker and Nabors[49] in vulcanized NK the polysulfide bonds constitute 100% of the total quantity of cross-links when using sulphur and DFG as vulcanizing agents, but only about 3% when using Captax and sulphur, and about 6% when using thiuram without sulphur. C–S–C bonds are absent in the first case, in the second they constitute 97% and in the third 94% of all bonds.

mers differs in corresponding conditions by a factor of 7.[45] In experiments in vacuum (10^{-1} mm Hg) the speed of the chemical relaxation of natural rubber decreases five times. The comparatively small decrease of speed in contrast to Tobolsky's figures is connected with the fact that the rubbers contained antioxidants which retarded their oxidizing chemical relaxation.

The corroborating calculations of Berry and Watson,[47] based on the shape of the relaxation curves which maintained that in peroxide vulcanized natural rubber no destructive degradation of the chain but a break of the cross-links takes place during the chemical relaxation, provoked a number of categorical objections on the part of Tobolsky. The latter considers that: (1) the shape of the curve does not permit the drawing of any clear conclusions, (2) the cross-links in the investigated polymers are stronger than links by double bonds which are easily oxidized. Berry's and Watson's point of view is also disproved by direct experimental data which were obtained by Tobolsky[40] and Dunn.[50]

A more complicated case is the chemical relaxation of sulphur vulcanized unsaturated rubbers which contain a complicated collection of different types of cross-links. Tobolsky considers that in these vulcanized rubbers mainly oxidizing breaks of the polymer chain occur, and only in a few cases a break of the cross-links.[40] He justifies this by the fact that in butyl rubber and BSK the speeds of the chemical relaxation of extracted sulphur-free and sulphur vulcanized rubbers are practically equal, and for natural rubber they only differ by a factor of 4. At the same time he assumes for vulcanized modified divinyl rubber (polybutadiene in which 97% of the double bonds are substituted by SCH_3 groups) he assumes a considerable amount of breaking of the cross-links, as the relaxation speed for sulphur and sulphur-free vulcanized rubber of this type differs by a factor of 20 and more. One has to take into account that the results which were obtained by Tobolsky refer to vulcanized rubbers which contained mainly strong C–S–C and C–C cross-links and not unstable polysulfide links.

Based on Dogadkin's and Tarasova's data one can expect that at high temperatures a thermal degradation of polysulfide and disulfide links will take place. This is confirmed in the work of Delgoplosk and co-workers.[54] The inhibiting effect of the double bonds of polysulfide sulphur[45] on the process of oxidization of rubber must lead to an increase of the share of breaks under the influence of heat. As Dunn[50] remarks, extracted natural rubbers which contain mainly polysulphide cross-links differ from rubbers with strong cross-links by a small difference in the speeds of chemical relaxation in air and in vacuum, and also by a lesser effectiveness of antioxidants. On the other hand, Hilmer's[46] data that the activation energy of the chemical relaxation in air† is the same for different natural rubbers (including those obtained with thiuram and with sulphur only) and also for SKB and SKN rubbers, for butyl rubbers and polychloroprene, and amounts to 30 ± 2 kcal/mol, as determined by Tobolsky, are a confirmation of Tobolsky's point of view.

This question needs further clarification.

The chemical relaxation can also take place due to the failure and renewal of bonds which proceed easily under the influence of catalysts at ordinary temperatures and which lie at the base of the phenomenon of cold flow. For instance, in Thiokols the failure of the sulphur bonds is catalysed by impurities of an ionic character (mercaptide,[40] Lewis acid), in polysiloxane rubbers the failure of the Si–O links is catalysed by steam, CO_2, alkalis and acids, and does not depend on the presence of oxygen.[40,41] The failure of salt cross-links in carboxyl containing rubbers[40] which are vulcanized with metal oxides can also be accelerated under the influence of these catalysts. The formation of a small number of strong cross-links in these rubbers through vulcanization with thiuram or γ radiation leads to a pronounced deceleration of stress decrease,[42] similar to the effect of strong links in vulcanized rubbers which contain unstable polysulphide links.

Conducting experiments on the failure of cross-linked polymers under the condition $\varepsilon = \text{const.}$ permits to show the destructive processes which proceed under the influence of heat, of chemical agents and mechanical forces. Tobolsky reckons that the new cross-links which form during failure are not stressed and therefore they do not fail. The failure process and the process of formation of new cross-links can be quantitatively assessed by comparing the data on chemical relaxation and the periodical changes of the static modulus of the specimens which have been kept in a stressed state at constant temperature.

† After a preliminary and prolonged exposure of stretched rubbers in an atmosphere of N_2 at the temperature of the subsequent oxidization.

Which part does stress play in these processes? Under conditions of cyclic loading the stress can accelerate the processes of oxidization, failure, etc., through the lowering of the activation energy[52] or due to a rise of the probability of a collision between reactive groups.[53] It may assist failure due to the decrease of the probability of a recombination of free radicals, similar to the effect of solvents which weaken the so-called cell effect.[45, 54] Static deformation (within wide ranges: tensile strain up to 250%,[38, 46, 55] compressive strain from 50%[56]) has no influence of the magnitude of the initial stress on the relative speed of the chemical relaxation which indicates that this process is not activated by stress. This is indirectly confirmed by the fact that the value of the activation energy of the chemical relaxation is greater than that of the oxidization of rubbers. No comparative data exist in literature for these rubbers. For other rubbers based on natural rubber the value of the activation energy of chemical relaxation varies from 22 kcal/mol[57, 58] to 28 kcal/mol,[58] 29 kcal permol[46] and 30 ± 2 kcal/mol,[40] whilst the activation energy of the oxidization of NK rubbers with different vulcanizing groups has values from 19·5 to 23 kcal/mol.[56, 59]

Under great deformations (400–700%) a greater drop of stress occurs than under deformations up to 100%.[38] This may be connected with the circumstance that, apart from oxidizing degradation and mechanical failure of the comparatively weak bonds, there proceeds in the polymer also the failure of the stronger bonds under the effect of great mechanical stress.

10.3. Mechanical–Chemical Phenomena

It is known that under the influence of mechanical stresses a break of chemical bonds in the molecules of poly(isobutylene) and of cellulose takes place which is accompanied by a decrease of the molecular weight.[60] A mechanical breakdown of the molecules also takes place during roller-mixing of P.V.C., SKB and natural rubbers,[61] during pulverization of poly(methylmethacrylate) and polystyrene[62] in a mill, during milling of poly(methylmethacrylate), polystyrene, poly(tetrafluorthylene), poly(isobutylene), polyethylene, NK,[63] and starch with cryolite[64] at low temperatures (77°K).

The EPR method[63, 65–67] can be successfully employed for the determination of any free radicals which arise.

It has to be noted that radicals produced in vacuum are of a different type than those produced in air.[66]

As known, the strength of polymers, especially elastic ones, is to a large extent determined by the intermolecular forces.

However, during mechanical failure of polymers there occurs, without doubt, a breaking of chemical bonds[14, 68] and, consequently,[69] chemical changes of the material take place in the moment of the breaking of bonds. These changes were directly observed during the failure of some transparent plastics. It was then established that the material of the surface is regenerated by the silver cracks which form during the failure.[70, 71] This happens, obviously, due to the interaction of the free radicals which form during the break with the surrounding environment. The effect of mechanical stress on polymers which causes activation of chemical reactions either through deformation of bonds without formation of free radicals, or with formation of free radicals (during the failure of covalent bonds) or of ionic compounds (during the failure of ionic bonds[72]), involves the development of various processes. For instance, free macroradicals interact with atmospheric oxygen,[63, 73] with fillers (for instance, with carbon-black a carbon-black rubber gel is obtained[73]); or under simultaneous mechanical failure of several polymers block copolymers are formed (for instance, block copolymers of nitrile rubber with epoxy or phenol-aldehyde resins,[74] poly(vinylchloride) with phenol-aldehyde resin,[75] polystyrene with poly(methylmethacrylate),[62] NK with phenol resins[76]).

The complex of the phenomena which are connected with the change of the chemical nature of the material under mechanical stresses, and also with the formation and further transformations of reactive compounds which are obtained during mechanical failure (grinding, cutting, milling, ultrasonic effects) received the name of mechanical–chemical phenomena. The spread of these phenomena and the wide possibilities for these different chemical reactions caused the appearance of a great number of papers in this field which are reflected in a number of reviews, articles,[62, 73, 77, 78] and in monographs.[79]

The mechanical–chemical phenomena are characteristic only for high polymers (in the widest sense of the word) as only on long molecules with decelerated relaxation capacity can the mechanical stresses be concentrated sufficiently to break the chemical bonds. The rate of the failure of the chemical bonds in the polymer depends on the magnitude of the applied force, the temperature and the activity of the environment. This process proceeds to a different degree under static fatigue, chemical relaxation and also under such strong mechanical effects as grinding, cutting, milling and ultrasonics. As a result of the failure of the spatial network, processes can take place in cross-linked polymers which are outwardly similar to the processes which are only characteristic for linear polymers (flow). For instance, under thermo-mechanical and thermo-chemical–mechanical effects a chemical relaxation was observed, and under the effect of great stresses a chemical flow of spatial polymers.[61, 80, 81] The chemical bonds broke then basically just due to mechanical effects.

Under the effect of mechanical forces the chemical bonds will break the easier, the smaller the capacity of the polymer for flow and for stress relaxation. With a lowering of temperature and with an increase of the frequency of the effect of the force this process is intensified, and the greatest part of the expended energy is used on the breaking of chemical bonds.

10.4. Corrosion Cracking of Non-elastic Materials

Cracking of stressed materials under simultaneous effect of the outer environment is one of the most widespread types of failure. In the case of metals the term "corrosion cracking" is accepted for this phenomenon. On account of the purely external appearance this term can also be used for other materials.

Corrosion cracking is, because of the very rapid development of the process, the most critical type of failure. It takes place under small deformations and stresses, for instance for Al–Mg alloys already under about 2% of the ultimate strength.[82] Corrosion cracking was observed, and studied in detail, on the following metals and alloys in a number of environments.[83–85]

Metals and Alloys	Environments
Aluminium alloys	Air, sea water
Magnesium alloys	Na_2SO_4, $CH_3 \cdot COOH + NaCl$
Copper alloys	Mercury, air, NH_3, SO_2, amines
Iron alloys	$NaNO_3$, H_2S, sea water
Alloys of gold with silver	$FeCl_2$
Alloys of platinum with silver	$FeCl_3$
Lead	Pb $(CH_3 \cdot COO)_2$
Titanium	HNO_3 (red), HCl (10%)
Nickel	NaOH, KOH (conc.)
Copper	Molten tin

Corrosion cracking has also been observed in elastomers. This process proceeds intensively under the chemical interaction of rubbers with sulphur, whilst plastics crack severely also in the presence of physically active substances.

As was earlier remarked, the fracture of solids proceeds in brittle failure by a gradual development of defects, microcracks, and during visco-elastic failure by a growth of the small tears which is accompanied by an additional orientation of the polymer at the root of the tears. The influence of the physically and chemically agressive environments leads first of all to an acceleration of the growth of those defects which are on the surface. Besides, under the chemical interaction with the environment new defects may also be formed. The cause of the simultaneous appearance and growth of a great number of cracks in polymers under small stresses was noted above (see Chapter 3). As corrosion failure is usually observed just under small deformations and stresses and, besides, the chemical action makes a simultaneous development of defects of different degrees of criticality more probable, the growth of a great number of cracks usually takes place under simultaneous effect of stress and aggressive environment.

10.5. Ozone Cracking of Rubbers

Rubber articles which are used in open air in a stressed state, are subjected to the so-called ozone cracking. Under the influence of atmospheric ozone cracks which are perpendicular to the direction of the tension are formed on the surface of such articles. In spite of the very small concentrations of ozone in air (of the order of millionth of per cents), cases are known when the rubber articles became unusable already after a few days or even during storage or during transport, i.e. before they were even put to use.

In recent years the investigation into ozone cracking, and especially the search for methods of combating it acquired a wider and wider scope. This has a number of reasons.

1. The progressive pollution of air by the exhaust gases of cars and by flue gases of industrial plants leads to a sharp increase of the concentration of ozone in the atmosphere of large towns and industrial centres (ozone forms through the photo-chemical reactions which take place in the presence of nitrogen oxides and the vapours of organic substances). For instance, in Los Angeles the concentration of ozone in the air reaches $6–9 \times 10^{-5}\%$, which is ten to fifty times greater than its average concentration in the atmosphere.(86)

2. The ceiling of jet aircraft, which are equipped with a great number of rubber articles, rose to 20–25 km, i.e. to the zone of the maximum concentration of ozone.

3. The study of the effect of ozone on rubber acquires an even greater importance in connection with the wide use of rubber articles in radiation-chemical investigations. Rubbers which are subjected to the effect of radiation are at the same time subjected to the effect of ozone which is produced under these circumstances in considerably greater quantities than exist in the atmosphere.

There can be no doubt at the present time that the basic reason for the cracking of stressed rubbers in the open air is the atmospheric ozone.(87)

Ozone, which is a very reactive substance, enters into a chemical interaction with many components of rubber, and especially actively with the double bonds of rubber molecules. According to Farmer's(88) data, it could be expected that C–S bonds also fail mainly through ozone, however, in Barnard's (89) work it is shown that the speed of the ozonization of organic sulphides is smaller by 2–3 orders of magnitude than the speed of the ozonization of olefines.

With double bonds ozone forms ozonides. Besides ozonization, reactions take place with the formation of free radicals into which ozone enters as oxidizer and catalyst of oxidization.(90)

If one assumes,[92, 93] in agreement with the data on the polar structure of ozone,[91] that the ozonization proceeds basically like an electrophilic non-chain reaction the process will take the following course.

Under the effect of the molecular ozone the π—bonds of olefines are polarized and the ozone joins by a double bond with the formation of primary ozonide (molozonide)

$$\begin{array}{ccccc} >\underset{\delta+\ \delta-}{C{=}C}< & \longrightarrow & >\underset{+}{C}{-}C< & \longrightarrow & >C{-\!-\!-}C< \\ {}^{-}O{-}O^{+}{=}O & & {}^{-}O{-}O{=}O & & O{-}O{-}O \end{array}$$

Molozonide (molecular ozonide) is the unstable form which either isomerizes into iso-ozonide, or polymerizes, or decomposes spontaneously. Isomerization can be regarded as a two-stage process: at first a bipolar ion and a carbonyl compound is formed; as a result of their subsequent reaction iso-ozonide is produced.[94]

$$\begin{array}{c} >C{-\!-\!-}C< \\ O{-}O{-}O \end{array} \longrightarrow >\overset{+}{C}{-}O^{-} \quad {}^{-}O{-}O{-}\underset{+}{C}< \;\rightleftarrows\; >C\begin{array}{c}O{-}O\\ O\end{array}C<$$

If the rubber is under stress, a degradation of the polymer molecule can occur during isomerization. Rising temperature and decrease of viscosity of the rubber also furthers the degradation.[95]

Ozone cracking of rubber has been reported in the literature quite separately from the cracking of other materials in the absence or presence of agressive environments. In a number of papers this phenomenon is even at present[38] considered unique and connected with the particularities of the chemical interaction of ozone with rubbers from unsaturated polymers, and a number of explanations are given for it. One of the explanations[96] is that on the surface of the rubber, as a result of the reaction of ozone with double bonds of rubber molecules, a layer of ozonide and of products of the oxidization of rubber is formed, which possesses a lower strength and elasticity than the original material. In individual places of the surface layer the stress can exceed the strength of the affected and altered material, and a break of the layer occurs.[97] The ozone reaches fresh surfaces of rubber through the embryo crack, and the failure process develops continuously right up to rupture. Although it is impossible to explain all cases of cracking from this point of view, some ozonides under normal conditions are really brittle,[45] and it has been shown experimentally[98] that such a layer does form in a number of cases. Newton[99] reckons that embryo cracks form during the failure of rubber molecules at the time of the regrouping of molozonide and iso-ozonide. With the growth of the deformation grows the probability of failure, and consequently, the number of nascent cracks.

In one of the papers an attempt is made[100] to answer the question why the number of nascent cracks is considerably smaller than the number of double bonds which exist on the surface. This arises from the circumstance that for the severing of the broken molecules and the prevention of their regrouping the presence of a tensile stress is needed. The ends of the broken molecules can only separate by some minimal distance if at the same time several neighbouring molecules break. If the molecules are not to break simultaneously, they must orientate themselves in the direction of the tensile stress in such a way that their double bonds are side by side and react simultaneously with the ozone. With the increase of stress the number of these simultaneous breaks grows, as owing to the increase of the degree of orientation the probability of the parallel arrangement of double bonds of adjoining chains increases. The number of cracks increases but it will always be smaller than the number of the existing double bonds on the surface.

As will be seen from a more detailed account further on, the part played by stress in corrosion cracking is complicated and leads not only to the prevention of the rearrangement of the broken molecules. Under the effect of stress the closed cracks open up eliminating thereby any hindrances: the brittle surface layer of the reaction product of the polymer with sulphur breaks and, finally, due to the stress concentration at the roots of the cracks they grow similar in manner to tears during mechanical failure.[101, 102]

It has been shown recently that the phenomenon of ozone cracking in elastomers is not unique. A failure of the same type (Fig. 147) can be observed under the effect of concentrated nitric acid on vulcanized butyl rubber and Nairit, of gaseous HCl and HF on vulcanized dimethylpolysiloxane (SKT) rubber, and of the dilute HCl, NaOH, H_2S on Thiokol rubbers.[103] An especially suitable object for investigation are vulcanized carboxilate rubbers (styrene-butadiene rubbers containing 2–5% methacrylic acid), as one can observe on them the cracking under the effect of ozone which causes the destruction of the molecular chains, as well as cracking under the influence of acids which destroy the surface bonds in vulcanized rubbers which contain metal oxides, and also failure, without cracking under the effect of acids on sulphur vulcanized rubbers.[102]

10.6. Characteristics of the Process of Corrosion Cracking

Many different methods are employed for the definition of the local failure of materials. It is proposed to define the cracking of plastics by the magnitude of the stress which causes the appearance of visible cracks in a specified time.[32] The cracking of metals is assessed in several ways: by the time before the appearance of cracks (if all samples crack); by the fraction of the samples which have cracked during a specified time;[83] by the number of cracks which have appeared during that time; by comparing the decrease of strength of the specimens which have been exposed to an aggressive environment both stressed and unstressed, and also after exposing the stressed specimens both to an aggressive environment and air.[82]

The following factors are useful in the investigation of ozone cracking in rubber.

1. *Extent of cracking*. For this an arbitrary scale of specimens of different degrees of cracking is prepared[99, 104–107] which is graded by a four-, six- or ten-point system, where *O* is given to the original specimen, and the highest number of points to the specimen with most cracks. For the rating of the extent of cracking of the specimen under investigation it is compared with this scale. The method is subjective and suitable only for the roughest estimate of the cracking of rubbers. In spite of this it is widely used abroad[108] and is there, perhaps, the basic method, reflected in two ASTM standards.[109, 110] To this method belong a number of others which provide a combined and more complicated estimate of the degree of cracking.[111]

2. *The time before the appearance of visible cracks.*[112] The results obtained depend on the observer, and on whether the

observation is carried out with the naked eye or through a magnifying glass. Therefore the method is somewhat arbitrary. However, this factor enters both the standards of the USSR[113] and of the USA.[110]

3. *Number of forming cracks.*[99] The method is extremely laborious, and with great stress, ozone concentrations and elevated temperatures it is inaccurate, as very many cracks are formed which fuse rapidly together as they grow.

4. *The time before the break of the specimen under constant load.*[114, 115] The method is objective, but does not permit to determine the kinetics of the ozone cracking. Besides, under constant load during the experiment the stress in fact changes, and its magnitude determines the speed of the process.

5. *Depth of the forming cracks.* It is suggested to determine the depth of the cracks with the help of a measuring microscope.[116] The method is objective, and gives the possibility of following the kinetics of the growth of the cracks during ozonization. However, in connection with the purpose for which the method is proposed by the authors the determination of the depth of the cracks is connected with the failure of the sample. Consequently, the kinetics can be followed only on different samples which greatly increases the clumsiness of the method and lowers its accuracy. The depth of the cracks can also be determined by a radiometric method.[117] With the help of radioactive powders (γ and α emitters) the volume of the cracks and the areas of their surfaces can be determined. The relation of these values is proportional to the depth of the cracks. The method is very labour-consuming, connected with frequent interruptions of the ozonization, and with the removal of the specimens from the testing chamber.

6. In recent times it has been suggested[118] to characterize ozone cracking by the *determination of the total volume of the cracks and by the change of the thickness of a specimen which is stretched to a constant deformation.* As the cracks develop the thickness increases due to the decrease of deformation of the parts between the cracks. The thickness is usually determined under a strain of the samples by 100%, and with their periodical removal from the ozone chamber. In this way each test is connected with a number of lengthy operations on the specimen (removal, stretching) which can influence even the course of the experiment.

7. *Growth of electric resistance.* As a result of the decrease of the solid part of the cross-section of the sample due to cracking, the electric resistance grows in proportion to the ozonization which can be measured.[119] This method is applicable only to conducting rubbers. It must be noted that the results are badly reproducible.

8. *Elongation under the effect of periodically interrupted loads.*[119] The increase of the elongation in proportion to the ozonization is measured. The method is objective, and permits the determination of the kinetics of the growth of the cracks. However, the specimen has to be withdrawn from the ozone chamber to measure the elongation under the effect of the specified load which disturbs the test conditions and makes the job much more laborious. Besides, the interrupted loading causes by itself a propagation of the cracks, especially towards the end of the process which gives a false idea of the kinetics of the effect of ozone.

9. *Rate of creep in ozonized air.* The test is carried out under the effect of constant load[120–122] or, in a more perfect modification, under the effect of constant stress.[123] These methods are objective. However, one has to keep in mind that in proportion to the cracking the cross-section of the solid part of the sample decreases due to cracks formation, and the stress in fact grows continuously, to a lesser degree it is true than under the effect of constant load.

10. *The speed of the growth of cracks at constant deformation.*[124] The speed of the growth is assessed by the drop of the tension in the strained rubber samples. The method is objective and has to be acknowledged as the most successful one. By it the basic quantitative principles of the process of ozone cracking have been discovered. In recent times it has begun to be employed also in Western countries for the assessment of the effectiveness of substances[125–128] which protect from ozone, and the investigation of the mechanism of cracking.[129]

Under this method a rubber specimen is kept under constant deformation (without an aggressive environment) up to the end of the rapid relaxation processes and until a practically constant value of the stretching force P_0 is established. Then the specimen is brought into contact with the aggressive environment, cracks form on it, and the force P_t which is necessary to maintain the deformation, decreases as the cracks grow.† From the decrease of the stress one can calculate for each moment the mean effective size of the cracked portion of the cross-section of the specimen S_t. According to these data the curve of the growth kinetics of the cracked layer can be drawn which shows the mean speed of crack propagation.

The value S_t, which is proportional to the mean effective depth of the cracks, is calculated from

$$S_t = \frac{P_0 - P_t}{\sigma} \text{ mm}^2$$

where P_0 is the initial (before the appearance of the cracks) stress in grams; P_t the same after the appearance of the cracks at the moment of time t; σ is the true stress in the sample (g/mm²).

Instead of S_t one can employ the relation S_t/S_0, where S_0 is the original cross-section of the sample. This factor is more convenient for the comparison of the kinetics of the process in samples of different thicknesses.

This method can be considered as basic under the following conditions:

1. The apparent equilibrium modulus of the sound part of the rubber specimen does not change under the effect of ozone.
2. The cracked part of the specimen is virtually completely destroyed.

It has been shown experimentally that both these conditions are fulfilled. The noticeable change of the chemical structure of the material which is due to oxidizing processes, and the change of its equilibrium modulus is only observed with a prolonged (10 hr) effect of ozone in comparatively great concentrations (0·5% and more) on thin (less than 100 μ) films of rubber or elastomers.[130] This was also confirmed by further tests in which SKN-18 and SKS-30 rubbers, stretched 20% and covered by unstressed layers of SKS-30, 60 μ thick, were ozonized for 10 hr (ozone concentration $2{\cdot}5 \times 10^{-3}$%). After the removal of the layer no cracks were observed on the rubber specimens, and the equilibrium modulus remained unchanged. Therefore, under ordinary conditions of testing the kinetics of ozone cracking (dura-

† The decrease of the stress in the sample, due to the slow relaxation process, can be disregarded.

tion of test less than 10 hr, sample 0·3 mm thick and more) the change of the equilibrium modulus can be disregarded.

A direct examination of the dependence of the size of the stress on the number and depth of the cracks in stretched rubber was also carried out. The cracks were simulated by cuts of a specified depth which were made on the edges of the samples which measured 60×10×0·5 mm. Below we show the drop of the stress in the rubber samples, with the cuts on opposite edges of the specimen having a depth of about 4·5 mm (when the stress in the specimen should decrease by 45%):

Number of cuts	Decrease of stress %	Number of cuts	Decrease of stress %
0	0	20	28·7
2	2·7	24	32·3
4	9·0	28	33·5
6	14·2	36	36·2
10	19·9	62	41
16	25	98	45
18	27·5		

The experiment shows that the stress falls gradually and reaches the calculated value with the application of about 100 cuts. A sharp deceleration of the drop of the stress can be observed after approximately 20 cuts. Such a behaviour of the specimens was observed with different rates of deformation. As ozone cracks do not appear simultaneously, their dimensions are not identical. The influence of the number of cracks of different depths on the size of the stress was, therefore, examined; at first deep cuts were made, then shallower ones, and yet shallower ones. The change of the stress with the greatest depth of cuts was calculated as 40%. The results of the experiments show (Table 12) that with a sufficiently great number of cuts (about 20) the appearance of new cuts does not change the stress.

Under the actual conditions of ozonization the fact that the cracks do not appear at the same time will have an even smaller influence on the size of the stress as the number of cracks is very great (at 15% deformation up to 9000 cracks appear on the sur-

TABLE 12

Dependence of the Magnitude of the Stress on the Number of Cuts of Different Depths

Depth of cut (mm)	Number of cuts of specified depth	Number of cuts	Change of stress (measured) (%)
2	2	2	7
2	2	4	14
2	2	6	23
2	2	8	28
1·5	4	12	40·5
1·5	4	16	43
1·5	4	20	43
1·5	4	24	44
0·5	6	30	43
0·5	6	36	44

face area of 600 mm^2).[99] The results obtained prove that with a small number of cuts the cracked part of the sample has a positive, increasing influence on the stress, acting like an additional sound section. Under actual conditions when the number of cracks is greater by several orders of magnitude this influence ceases. The method of investigating ozone cracking of stressed rubbers by a change of stress permits therefore to determine exactly the mean effective size of the cross-section of the cracked part of the specimen.

It is essential to emphasize the difference between the proposed method and the known method of investigating the kinetics of the change of structure of rubbers at elevated temperatures by the decrease of their stress[40, 48] (chemical relaxation). The observed effects appear to be identical, i.e. the external force which is necessary to keep the sample at a constant deformation, decreases gradually in both cases. The following is the essential difference of the proceeding processes.

1. During chemical relaxation the structural change takes place in the whole mass of the specimen reasonably uniformly,† and

† If there are no inhibitions to the diffusion of the chemically active agents into the specimen.

the stress decreases steadily. Under the effect of ozone on rubber the structural change starts with the surface whilst the properties of the inner layers remain practically unchanged. The mean nominal stress which in fact acts only on these layers, remains constant.

2. The structural change during chemical relaxation can be connected in a certain way with the change of the density of the spatial network. The change of stress (or tensioning force which in this case is the same, as the cross-section of the specimen remains unchanged) during the chemical relaxation makes it possible to judge the speed of the chemical processes in the rubber within certain limits. During ozone cracking of rubber macroscopic

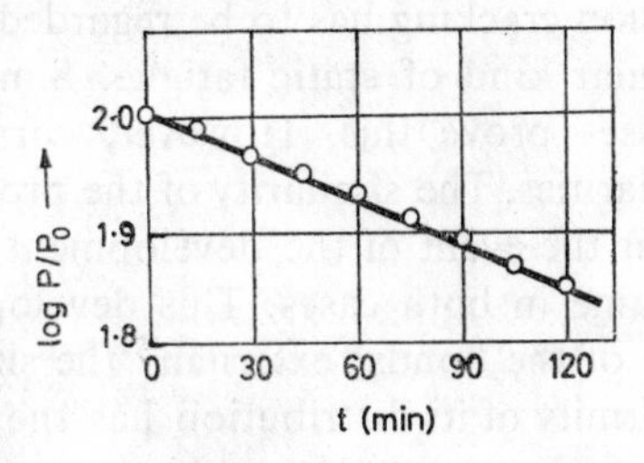

FIG. 148. Kinetics of the drop of force with ozone cracking in NK rubber under conditions ε = const.[119]

changes, together with the molecular changes, take place in a limited zone of the sample (on the surface), i.e. growth of cracks. The decrease of the stress in the specimen is connected with this process, and not with the change of the structure of the whole specimen. This adds certain difficulties to the establishing of the connections between the speed of the reduction of the stress, and the speed of the chemical reaction of the ozone with the rubber.

3. The stress decreases in these two cases at a different rate. During chemical relaxation the process, after the initial transitional part, is usually described by a straight line in logarithmic coordinates $\log \sigma/\sigma_0 - \log t$ (where σ_0 is the initial stress), but with ozone cracking by a straight line in linear coordinates $P/P_0 - t$[101, 124, 131] (see p. 323) or in semi-logarithmic coordinates $\log P/P_0 - t$ (Fig. 148).[125]

In principle any device which is suitable for the measuring of the drop of stress in specimens at ε = const. can be used for the determination of the kinetic growth of ozone cracks. However, the usual equipment does not permit the simultaneous combination of hermetic sealing and the possibility of measuring the stress without disturbing the experimental conditions whilst enabling large numbers of tests. These conditions are fulfilled by a specially constructed apparatus.[124] The measuring of the stress in the specimens can be carried out with the help of a spring dynamometer.[124] Recently, a device was used with automatic recording of the stress by transducers which were placed inside the testing chamber on steel rings, and joined to the rubber samples.[131]

The analysis of the obtained results in the light of the principles of the failure of rubbers in aggressive and neutral media shows that corrosion cracking has to be regarded as a phenomenon of a particular kind of static fatigue. Similar features of these two processes prove this. However, corrosion cracking has its own peculiarities. The similarity of the processes is founded on the fact that the event of the development of microcracks is basically the same in both cases. This development proceeds due to the failure of the bonds; externally the size of the stress, and also the uniformity of its distribution has the same quantitative and qualitative influence on the cracking process and the time to rupture. This similarity permits the successful use of such properties of static fatigue as durability for the objective assessment of corrosion cracking.

The difference between the processes is connected with the fact that different types of bonds fail, and that the speed of corrosion cracking is considerably greater than the speed of failure under static fatigue. With this is also connected the selectivity of corrosion cracking, the existence of a dependence of the speed of failure on the concentration of the aggressive environment and the possibility to investigate its principles in a wide range of deformations, down to very small ones.

In Chapter 11 we shall examine the principles which are common to static fatigue and to corrosion failure, in Chapter 12 and 13 the peculiarities of corrosion failure of visco-elastic materials.

Literature

1. P. A. Rebinder, *Izd. AN SSSR* (1958).
2. P. A. Rebinder, *Z. Phys.* **72,** 191 (1931); *Izv. AN SSSR, OTN*, Ser. khim., No. 25, 639 (1936); *Sbornik k XXX – letiyu Oktiabr'skoi revolyutsii*, Izd. AN SSSR, 1947, p. 533; *Fiziko-khimicheskaya mekhanika*, "Znaniye", 1958.
3. P. A. Rebinder, N. A. Kalinovskaya, *ZhFT*, **2,** 726 (1932).
4. B. I. Lichtman, Ye. D. Shchukin, P. A. Rebinder, *Fiziko-khimicheskaya mekhanika metallov*, Izd. AN SSSR, 1962.
5. Ye. D. Shchukin, P. A. Rebinder, *Kolloid. zh.* **20,** 645 (1958).
6. B. I. Lichtman, L. S. Bryukhanova, I. A. Andreyeva, *DAN SSSR*, **139,** 359 (1961); *FTT*, 3, 2774 (1961).
7. P. A. Rebinder, L. A. Shreiner, K. F. Zhigach, *Poniziteli tverdosti v burenii*, Izd. AN SSSR, 1944.
8. M. S. Aslanova, *DAN SSSR*, **95,** 1215 (1954); M. S. Aslanova, P. A. Rebinder, *DAN SSSR*, **96,** 299 (1954).
9. Ye. B. Matskevich, L. Yu. Butiagin, *Kolloid. zh.* **20,** 665 (1958).
10. G. I. Fuchs, *Kolloid. zh.* **22,** 256 (1960).
11. A. P. Pisarenko, P. A. Rebinder, *DAN SSSR*, **73,** 129 (1950).
12. H. Luttrop, *Kautschuk und Gummi*, **12,** WT 147 (1959).
13. S. N. Zhurkov, *Z. physik. Chem.* **213,** 183 (1960).
14. S. N. Zhurkov, S. A. Abasov, *Viysokomol. soyed.* **3,** 450 (1961).
15. Yu. S. Zuyev, S. I. Pravednikova, L. S. Zherebkova, V. D. Zaitseva, *Viysokomol. soyed.* **5,** 1201 (1963).
16. G. M. Bartenev, L. S. Bryukhanova, *ZhTF*, **28,** 287 (1958); G. M. Bartenev, *Plast. massiy*, No. 9, 48 (1960).
17. V. A. Kargin, A. I. Kitaigorodskii, G. L. Slonimskii, *Kolloid. zh.* **19,** 131 (1957).
18. V. A. Kargin, N. F. Bakeyev, Kh. Vergin, *DAN* **122,** 97 (1958).
19. B. A. Dogadkin, D. L. Fedyukin, V. E. Gul', *Kolloid. zh.* **19,** 287 (1957).
20. N. Suiny, P. Brauer, P. Scukower, *J. Dental Res.* **34,** No. 3 (1955).
21. R. L. Berger, *SPE Journal*, **18,** No. 6, 667 (1962).
22. H. Reiser, F. Glander, *Farbe und Lack*, **62,** No. 8, 361 (1956).
23. E. Ziegler, *Materials and Methods*, **39,** No. 6, 93 (1954).
24. Yu. S. Lazurkin, *Doktorskaya dissertatsiya, Institut fizicheskikh problem AN SSSR, M.* (1954).
25. De Coste, *et al.*, *Ind. Eng. Chem.* **43,** 117 (1951).
26. A. Rudin, A. Birks, ASTM Bull. No. 245, 60 (1960).
27. A. O'Connor, S. Turner, *Brit. Plast.* **35,** No. 9, 452 (1962).
28. F. K. Connors, *Austr. Plast.* **10,** No. 113, 23 (1960).
29. P. Kittmair, R. Villman, *J. Appl. Polymer Sci.* **6,** 1 (1962).
30. H. Bennett, *Corrosion*, **16,** No. 4, 99 (1960).
31. Masudzava, *Kobinsu Kadaki (Chem. High Polymers)*, **13,** No. 133, 211 (1956).
32. V. R. Regel', Yu. N. Nedoshivin, *ZhTF*, **23,** 1333 (1953).
33. G. Salomon, F. van Bloois, *Proceedings of the Fourth Rubber Tech-

nology Conference, London, May, 1962, Preprint 58; *J. Appl. Polymer Sci.* 7, 1117 (1963).
34. T. Siga, Yu. Inagaki, T. Arai, *J. Soc. Rubb. Ind., Japan*, **36**, 613 (1963).
35. G. M. Bartenev, *Izv. AN SSSR, OTN*, No. 9, 53 (1955).
36. *Ustalost' viysokopolimerov*, Sb. Goskhimizdat, 1957.
37. M. M. Resinkovskii, *Khim. nauka i prom.* **4**, No. 1, 73 (1959).
38. A. Tobolsky, J. Prettyman, J. Dillon, *J. Appl. Phys.* **15**, 380 (1944).
39. M. Mochulskii, A. V. Tobolsky, *Ind. Eng. Chem.* **40**, 2155 (1948).
40. A. Tobolsky, *Svoistva i struktura polimerov*, "Khimiya", 1964.
41. R. Osthoff, A. Bueche, M. Grubb, *J. Am. Chem. Soc.* **76**, 4659 (1954).
42. Z. N. Tarasova, B. A. Dogadkin, M. I. Arkhangel'skaya, S. B. Petrova, *Kolloid. zh.* **22**, 253 (1960).
43. A. S. Novikov, V. A. Kargin, F. A. Galil-Ogly, *Kauchuk i rezina*, No. 1, 33 (1959).
44. T. Alfrei, *Mekhanicheskiye svoistva Vlysokopolimerov*, Izdatinlit, 1952.
45. A. S. Kuz'minskii, N. N. Lezhnev, Yu. S. Zuyev, *Okisleniye kauchukov i rezin*, Goskhimizdat, 1957.
46. K. Hilmer, W. Scheele, *Kautschuk und Gummi*, **11**, No. 8, 210 (1958); **9**, No. 4, 82 (1956).
47. J. Berry, W. Watson, *J. Polymer Sci.* **12**, 201 (1955).
48. B. A. Dogadkin, Z. N. Tarasova, *Kolloid. zh.* **15**, 347 (1953).
49. M. L. Studebaker, L. G. Nabors, *Rubb. Chem. Technol.* **32**, 941 (1959).
50. J. Dunn, *Rev. Gén. Caout.* **37**, No. 3, 361 (1960).
51. L. I. Lyubchanskaya, A. S. Kuz'minskii, *DAN SSSR*, **135**, 1436 (1960).
52. A. S. Kuzminskii, M. G. Maizel's, N. N. Lezhnev, *DAN SSSR*, **71**, 319 (1950).
53. G. L. Slonimskii, V. A. Kargin, L. I. Golubenkova, *DAN SSSR*, **93**, 311 (1953).
54. B. A. Delgoplosk, B. L. Yerusalimskii, Ye. B. Milovich, G. P. Belonovskaya, *DAN SSSR*, **120**, 783 (1958).
55. G. Beatty, A. Yuve, *India Rubb. World*, **121**, No. 5, 537 (1950).
56. A. S. Kuz'minskii, L. I. Lyubchanskaya, *Kauchuk i rezina*, No. 6, 3 (1958).
57. O. Svein, *Acta Chem. Scand.* **9**, No. 6, 1024 (1955).
58. L. I. Lyubchanskaya, *Kandidatskaya dissertatsiya*, *MITKhT im. M. V. Lomonosova* (1962).
59. L. G. Angert, *Kandidatskaya dissertatsiya*, *MITKhT im. M. V. Lomonosova* (1959).
60. Löbering, *Koll. Z.* **98**, No. 2, 156 (1942); Hess, *Koll. Z.* **98**, No. 2, 182 (1942).
61. V. A. Kargin, T. I. Sogolova, G. L. Slonimskii, Ye. V. Reztsova, *ZhFKh*, **30**, 1903 (1956).
62. N. K. Baramboim, *ZhFKh*, **32**, 806 (1958).
63. S. Ye. Bresler, S. N. Zhurkov, E. H. Kazbekov, Ye. M. Saminskii, E. Ye. Tomashevskii, *ZhTF*, **29**, 358 (1959).

64. A. A. Berlin, E. A. Penskaya, G. I. Bolkova, *v sb. "Medzhunarodniyi simpozium po makromolekuliarnoi Khimii v Moskve"*, M., 1960, p. 334.
65. M. Pike, W. Watson, *J. Polymer Sci.* **9**, 229 (1952).
66. S. N. Zhurkov, E. Ye. ToMashevskii, V. A. Zakrevskii, *FTT*, **3**, 2841 (1961).
67. P. Yu. Butiagin, *DAN SSSR*, **140**, 145 (1961).
68. N. V. Mikhailov, V. A. Kargin, *Trudiy 4-i Konferentsii po viysokomolekuliarniym*, Izd. AN SSSR, 1948, p. 138.
69. G. L. Slonimskii, V. A. Berestnev, *Khim. Nauka i prom.* **4**, No. 4, 543 (1959).
70. M. I. Bessonov, Ye. V. Kubshinskii, *FTT*, **1**, 1441 (1959); *Viysokomol. soyed.* **1**, 1561 (1959); *Viysokomol. soyed.* **1**, 485 (1959); *FTT*, **3**, 1314 (1961).
71. G. A. Lebedev, Ye. V. Kuvshinskii, *FTT*, **3**, 2972 (1961).
72. A. A. Berlin, *Uspekhi Khimii*, **27**, 94 (1958).
73. W. Watson, *Kautschuk und Gummi*, **13**, WT 160 (1960).
74. V. A. Kargin, B. M. Kovarskaya, L. N. Golubenkova, M. S. Akutin, G. L. Slonimskii, *Khim. nauka i prom.* **2**, No. 2, 13 (1957).
75. A. A. Berlin, V. S. Petrov, V. F. Prosvirkina, *Khim. nauka i prom*, **2**, No. 2, 522 (1957).
76. R. Ceresa, *Trans. Proc. Inst. Rubb. Ind.* **36**, No. 5, 211 (1960).
77. N. K. Baramboim, *Uspekhi Khimii*, **28**, 877 (1959).
78. G. L. Slonimskii, *Khim. nauka i prom.* **4**, No. 1, 73 (1959).
79. N. K. Baramboim, *Mekhanokhimiya polimerov*, Rostekhizdat, 1961.
80. V. A. Kargin, T. I. Sogolova, *DAN SSSR*, **108**, 662 (1956); *ZhT*. **31**, 1328 (1957).
81. G. L. Slonimskii, Ye. V. Reztsova, *ZhFKh*, **33**, 480 (1959).
82. J. Leoyd, *J. Appl. Chem.* **4**, 1 (1954).
83. V. V. Romanov, *Korrozionoye rastreskivaniye metallov*, Mashgiz, 1960.
84. A. V. Bobiylev, *Korrozionnoye rastreskivaniye latuni*, Metallurgizdat, 1956.
85. A. Bailey, R. King, *J. Inst. Met.* **82**, No. 8, 105 (1954).
86. B. Biggs, *Rubb. Chem. Technol.* **31**, 1015 (1958).
87. I. A. Prokof'yeva, *Atmosferniyi*, Izd. AN SSSR, 1951.
88. E. H. Farmer, *IRI Trans.* **21**, 122 (1945).
89. Barnard, *J. Chem. Soc.* (1957), 4547.
90. Yu. S. Zuyev, V. F. Malafeyevskaya, *v sb. "Stareniye i zashchita rezin"*, Goskhimizdat, 1960, p. 3 (*Trudiy NIIRP*, No. 6).
91. Ya. K. Siyrkin, M. E. Diatkina, *Khimicheskaya sviaz' i stroyeniye molekul*, Goskhimizdat, 1946.
92. J. Wibaut, *Ind. Chim. belge*, **20**, No. 1 (1955); *Rec. trav. chim.* **74**, 431 (1954).
93. Y. E. Lefler, *Chem. Rev.* **45**, 399 (1949).
94. N. Milas, P. Davis, J. Nolan, *J. Am. Chem. Soc.* **77**, 2536 (1955); R. Criegee, *Rec. Chem. Progr.* **18**, No. 2, 110 (1957); R. Bailey, *Chem. a. Ind.* No. 34, 1148 (1957).
95. T. Newby, *Chemistry in Canada*, **11**, No. 11, 52 (1959).

96. E. F. Powell, V. E. Gough, *Rubb. Chem. Technol.* **19**, 406 (1946).
97. H. Tucker, *Rubb. Chem. Technol.* **32**, 269 (1959).
98. Yu. S. Zuyev, A. S. Kuz'minskii, *DAN SSSR*, **89**, 325 (1953).
99. R. Newton, *Rubb. Chem. Technol.* **18**, 504 (1945).
100. D. Smith, V. E. Gough, *IRI Trans.* **29**, 219 (1953).
101. Yu. S. Zuyev, S. I. Pravednikova, *DAN SSSR*, **116**, 813 (1957).
102. Yu. S. Zuyev, A. Z. Borshchevskaya, S. I. Pravednikova, U. Yue-tsin', *Viysokomol. soyed.* **3**, 164 (1961).
103. Yu. S. Zuyev, A. Z. Borshchevskaya, *DAN SSSR*, **124**, 613 (1959).
104. E. W. Ford, L. W. Cooper, *India Rubb. World*, **125**, 55 (1951).
105. J. M. Ball, R. A. Joumans, A. F. Russell, *Rubb. Age (New York)*, **55**, 481 (1944).
106. L. A. Smirnova, *v sb.* "*Stareniye kauchukov i rezin*", Goskhimizdat, 1952, p. 28.
107. Bayer, *Alterungsschutzmittel für die Gummi – Industrie*, **15**, 6 (1961), *Prospekt.*
108. A. Buswell, Y. Watts, *IRI Trans.* **37**, 175 (1961).
109. *ASTM Standards on Rubb. Prod. D1171 – 57* (1958), p. 627.
110. *ASTM Standards 1149 – 55T.*
111. D. C. Edwards, E. B. Storey, *Rubb. Chem. Technol.* **28**, 1096 (1955).
112. A. S. Kuz'minskii, Yu. S. Zuyev, *Khim. prom.* No. 9, 272 (1950).
113. *GOST 6949-63. Rezina, metodiy ispiytanii*, Standartgiz, 1964.
114. F. Norton, *Rubb. Age (New York)*, **47**, 87 (1940).
115. Yu. S. Zuyev, *DAN SSSR*, **93**, 483 (1953).
116. J. S. Rugg, *Anal. Chem.* **24**, 818 (1952).
117. J. L. Kalinsky, T. A. Werkenthin, *Rubb. Age (New York)*, **75**, 375 (1954).
118. C. Schaef, *Rubb. World*, **145**, 79 (1962).
119. R. F. Shaw, S. R. Adams, *Anal. Chem.* **23**, 1649 (1951).
120. H. Leeper, C. Gable, *Rubb. World*, **133**, 79 (1955).
121. N. N. Znamenskii, *ZhFKh*, **30**, 1092 (1956).
122. A. Veith, *Rubb. Chem. Technol.* **32**, 346 (1959).
123. D. Buckley, B. Robinson, *Rubb. Chem. Technol.* **32**, 257 (1959).
124. Yu. S. Zuyev, S. I. Pravednikova, *Trudiy NIIRP*, No. 3, Goskhimizdat, 1956, p. 114.
125. H. Vodden, M. A. A. Wilson, *IRI Trans.* **35**, 82 (1959).
126. J. Payne, *Chemistry in Canada*, **13**, No. 3, 29 (1961).
127. S. Eller, A. Stein, *Rubb. Age*, **89**, 972 (1961).
128. G. E. Decker, R. W. Wise, *Rubb. World*, **146**, 66 (1962).
129. J. H. Gilbert, *Proceedings of the Fourth Rubber Technology Conference*, London, May, 1962. Preprint 56.
130. Yu. S. Zuyev, *DAN SSSR*, **74**, 967 (1950).
131. Yu. S. Zuyev, V. D. Zaitseva, *Kauchuk i rezina*, No. 2, 22 (1936).

CHAPTER 11

CORROSION FAILURE AND STATIC FATIGUE

THE cracking of a solid, irrespective of whether it occurs in an inactive or in a physically or a chemically active environment, is characterized by many common features. There is in the first place the statistical character of the phenomenon, the development of cracks in the direction perpendicular to the direction of the acting force, the increase in the number of cracks, and the decrease of their average dimensions with growing stress. Besides this the cracking process is also characterized by other features which resemble each other with which we shall deal in greater detail.

11.1. Influence of the Type of Deformation (Strain)

It has been established that the cracks develop only under tensile stress. In the case of predominant compression no cracks will grow in the transversal direction.

The existing reports that ozone cracking can also be observed under compression have not been confirmed.[1] It was found that under compression only those parts of the specimen crack which are subjected to tensile stress. If radial compression predominates, no cracks at all will appear.

The creation of surface compressive stresses by hammer hardening of metal or chilling of glass and plastics is used as a method of increasing their resistance to the formation of cracks and the increase of their strength. The resistance of rubbers to ozone cracking can be raised in a similar way—by the creation of compressive stresses through the swelling of the surface layer of the rubber, or by purely mechanical methods. In all cases of deformation by stretching the cracks develop perpendicular to the direction of the acting force. On Fig. 149 the external appearance of rubber samples is shown which have been subjected to deformation by torsion in the presence of ozone which corroborates this situation. Corrosion cracks have a similar appearance; they are formed on metal rods which are subjected to torsion.[2] Under biaxial deformation the cracks have no distinct orientation.[3]

11.2. Part Played by Destructive Processes

The failure processes of the chemical bonds must predominate under the effect of stress in the cracking and breaking of polymeric materials in the absence of chemical action. In the presence of a chemically active environment this is a more complicated matter. Apart from the processes which are accompanied by the breaking of chemical bonds, there proceed reactions of substitution which do not cause the destruction of the polymer molecule. The chemically active environment has not, therefore, in all cases a simultaneous effect, and it is the stress which causes the cracking of rubbers perpendicular to the direction of the tension which is characteristic for the phenomenon of static fatigue. For instance, if a very intensive reaction of the polymer with the environmental medium takes place, accompanied by the complete chemical degradation of the material (for instance, the effect of concentrated nitric acid on NK), a completely destroyed (powdery or sticky) layer is formed on its surface.† Under the effect of bromine, iodine, hydrochloric and sulphuric acids and hydrogen fluoride[4] on stressed NK and SKB rubbers which is accompanied, as known, by substitution and cyclization, a hard layer without cracks forms on the surface, which shows wrinkles after the removal of the stress from the specimen (Fig. 150).

The type of cracking which interests us most can be observed only in those cases when the failure of the polymer is the predominating process.[5] As has been pointed out, such a cracking occurs under the effect of ozone on polymers with double bonds, under the effect of acids on carboxilate rubbers, acids and alkalis

† The formation of a a brittle surface layer is sometimes accompanied by cracking, however, here appears an irregular network of cracks. This also takes place, for instance, in the last stages of the process of photo-oxidation of rubber.

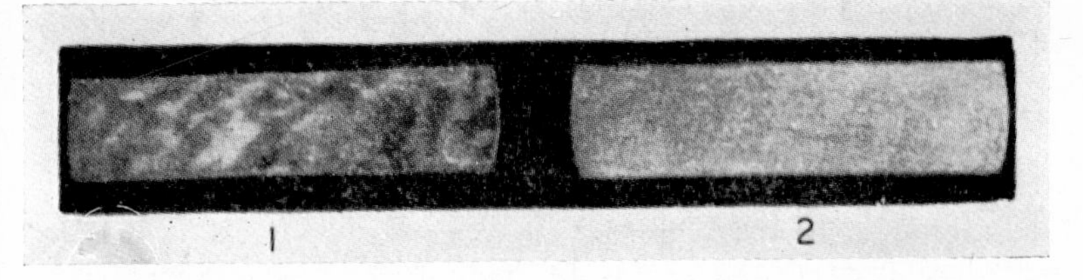

FIG. 141. Surface of tears in rubber at low (1) and high (2) speeds of tearing.[6]

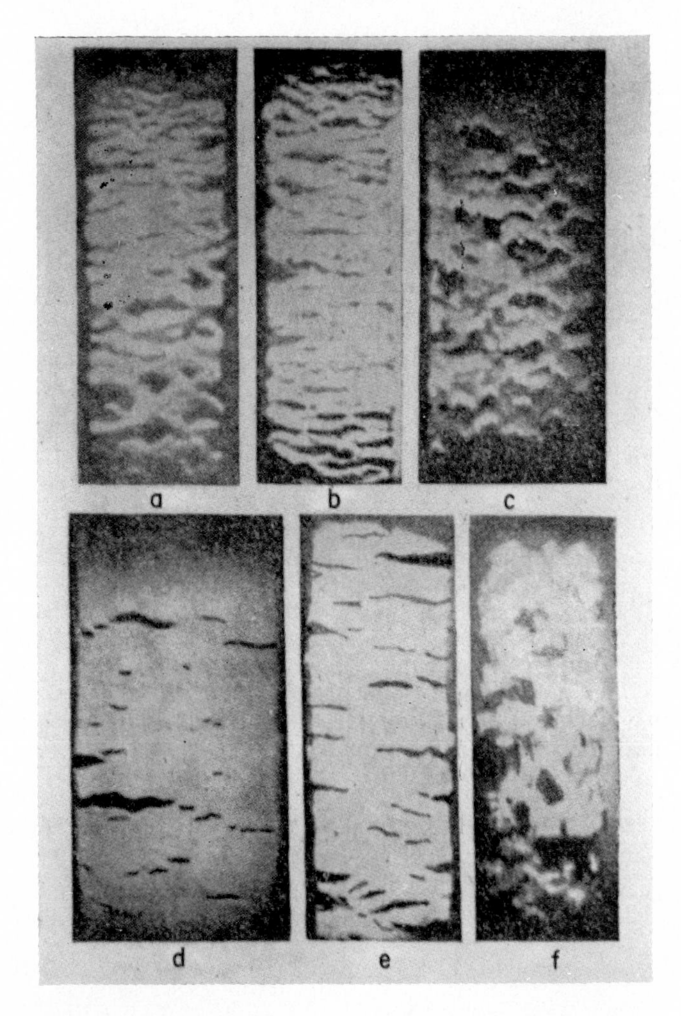

FIG. 147. Outside aspect of various rubbers after corrosion cracking in different environments. (a) Nairit, nitrogen oxide. (b) SKT, hydrochloric acid. (c) Butyl rubber, nitric acid. (d) Thiokol, NaOH solution. (e) SKS-30, ozone. (f) SKS-50-K, acetic acid.

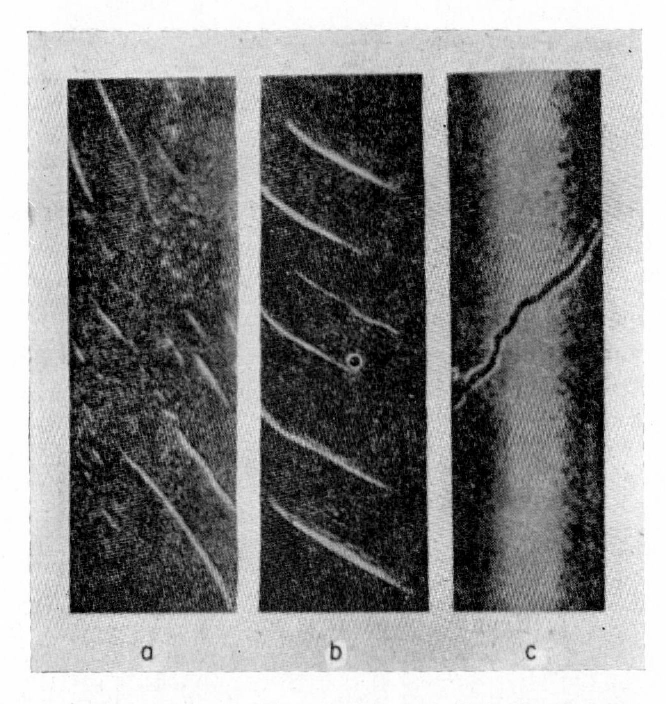

FIG. 149. Corrosion cracking of rubber and of Al–Mg alloy under torsion. (a) Rubber in ozone, 90° torsion. (b) The same, 30° torsion. (c) Al–Mg alloy in NaCl–H_2O_2 solution, 45° torsion.

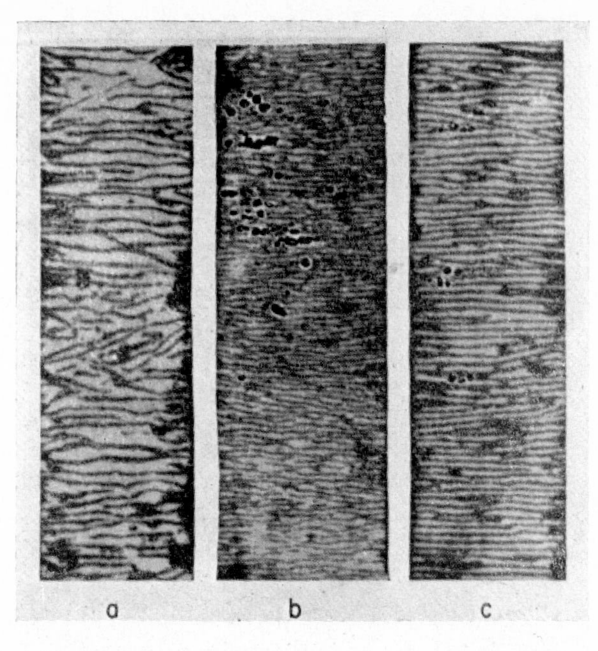

FIG. 150. External view of rubber samples after exposure (stretched) to aggressive agents. (a) NK rubbers, after exposure to bromine. (b) SKB rubbers, after exposure to bromine. (c) SKB rubbers, after exposure to iodine.

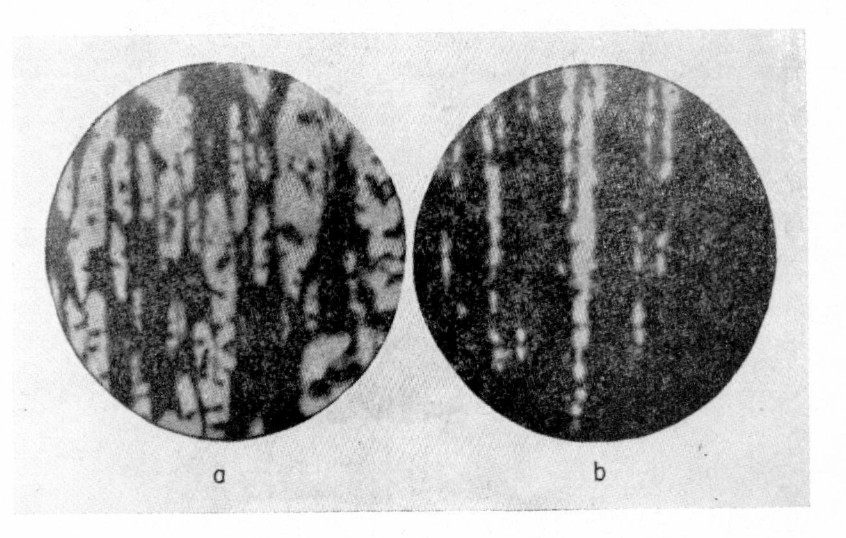

FIG. 156. Longitudinal cracks of highly stretched rubbers due to ozone cracking. (a) NK. (b) Nairit.

on Thiokol, etc. In these cases this is precisely the failure process which predominates.

The destructive action of ozone appears very clearly on natural rubber and vulcanized rubbers based on it. For instance, this can be seen from experiments on ozonization (1 hr at a concentration of ozone of 0·30–0·35 vol%) of unstretched samples, with the subsequent determination of their creep with σ = const. The deformation was determined after 30 min loading:

	Deformation, arbitrary unit	
	Experiment 1	Experiment 2
Natural rubber (0·1 mm thick)		
Before ozonization	17	25
After ozonization	208	274
Vulcanized NK (0·05 mm thick)		
Before ozonization	8	10
After ozonization	33	91

The great increase of the deformation after ozonization proves the considerable degradation of the material.

Carboxilate rubbers, vulcanized with metal oxides[(6–8)] which contain O–Me–O links, fail there under the effect of acids. Those vulcanized by liquid Thiokol of the structure

$$\ldots -CH_2CH_2OCH_2OCH_2CH_2S- \\ -SCH_2CH_2OCH_2OCH_2CH_2S-S-\ldots$$

fail due to hydrolysis under the effect of acids on the acetal bonds.[(9)]

$$\ldots -C_2H_4OCH_2OC_2H_4S-\ldots \xrightarrow{H_2O} \ldots -C_2H_4OH+\ldots - \\ +\ldots -C_2H_4SH+HC\begin{matrix}\nearrow O\\ \searrow OH\end{matrix}$$

and under the effect of alkalis or of hot Na_2S these rubbers may fail through the breakage of the S–S bonds:

$$\ldots -CH_2S-SCH_2-\ldots \xrightarrow{Na_2S,\ H_2O} \ldots -CH_2SH+HSCH_2\ldots$$

During the action of hydrogen fluoride or chloride on SKT rubber a layer of water droplets can be observed on its surfaces[10] which proves the failure on the silicon–oxygen bonds:

$$\ldots-\mathrm{O}-\underset{\mathrm{R}\ \ \mathrm{R}}{\mathrm{Si}}-\mathrm{O}-\underset{\mathrm{R}\ \ \mathrm{R}}{\mathrm{Si}}-\mathrm{O}-\ldots \xrightarrow{\mathrm{HF}}$$

$$\longrightarrow \ldots-\mathrm{O}-\underset{\mathrm{R}\ \ \mathrm{R}}{\mathrm{Si}}-\mathrm{F}+\mathrm{F}-\underset{\mathrm{R}\ \ \mathrm{R}}{\mathrm{Si}}-\mathrm{O}-\ldots+\mathrm{H_2O}$$

The predominance of the degradation processes over the other, simultaneously proceeding processes is, however, not sufficient for local failure. Conditions must also be created for stress concentration. Stress concentrations will promote an adequate mobility of the broken macromolecules. If their transfer (local plastic flow) is impeded, cracking is either reduced or does not take place at all. For instance, carboxilate rubber vulcanized with MgO (SKS-30-1, containing 1·5% methacrylic acid) which has a fairly sparse network, cracks intensively during tensioning in the presence of acid. With the formation of an additional number of stronger cross-links which are resistant to the action of acids (this is achieved by simultaneous vulcanization of the rubber with thiuram and MgO), the local displacement is impeded and the cracking in the acids ceases.

The introduction of polar groups into rubber has a similar influence. Carboxilate SKN-26-1 rubber, vulcanized with magnesium oxide, cracks only little or not at all under the action of acids, in spite of the severe degradation which is proved by the sharp increase of the speed of creep of loaded specimens.

If structural elements (for instance C = C bonds) contained in each unit of the polymer molecule are subjected to the action of an aggressive medium (for instance O_3), the decrease of the mobility of the chains due to the stronger intermolecular forces cannot stop the cracking but can retard it. And precisely with this can be explained the retarding of ozone cracking when changing from natural rubber to polychloroprene rubber.

11.3. Influence of Swelling on Corrosion Failure of Rubber

Under slight swelling the distribution of stresses becomes more uniform which leads to an impediment to cracking and to the increase of the durability of the rubber (this corresponds to the increase of strength with small amounts of swelling).[11] It is because of this that aggressive gases react usually considerably stronger than these same agents as liquids or as solutions[5] (Table 13). At the same time the presence of a solvent hampers the adsorption of the aggressive material and this also slows down the degradation process.

The slowing down of ozone cracking in water compared with that in air was observed already in 1953.[12] It was shown that during ozonization in water of stretched rubbers based on SKN-26, SKB and SKS-30, no cracks appeared during 2 hr, whereas in air under the same conditions cracking could already be observed after a few seconds. The results obtained were qualitatively confirmed also by foreign investigators.[13, 14] Recently, this question has been examined in greater detail.[15]

Rubber strips 0·5 mm thick were tested in an atmosphere of ozonized air under frequent deformations. The time to failure

TABLE 13

Durability of Stressed Rubber under the Effect of Aggressive Agents in the Form of Aqueous Solution and Vapours

Rubbers	Aggressive agent	Concentration[a] (mmol/mol)	Time to failure (min)	
			in solution	in vapour phase
SKS-301 (MgO)	HCl	1·9	>2 days	10
	CH_3COOH	1·9	120	28
	C_3H_7COOH	1·9	34	26
	HCl	19	228	–
	HCl + wetting agent (Acrosol OT)	19	97	–
SKT	HCl	2·4	>2 days	20

[a] The concentration of the aggressive agent is expressed in millimoles per "mol" of air, if the aggressive agent is in a gaseous state, or on a "mol" of water if it is contained in an aqueous solution.

of the rubber which was produced with standard formulations from different rubbers, was compared in ozonized air which had been dried with $CaCl_2$, and in air of 100% R.H. at room temperature. The experimental results with different ozone concentrations showed that for SKB, SKS-30, SKS-30-1, SKI and SKN-40 rubbers the resistance to ozone cracking is the same in dry and wet air (Fig. 151). Rubbers based on natural rubber and Nairit

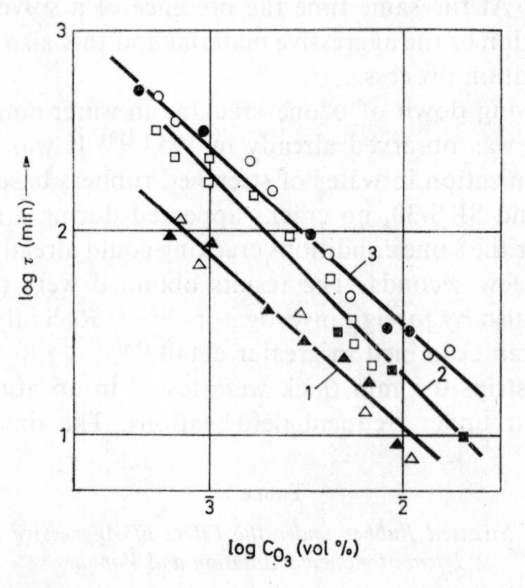

FIG. 151. Dependence of durability of various rubbers in dry and humid air on ozone concentration under frequent deformations. 1. SKS-30. 2. SKB. 3. SKI. Black points: humid air; white: dry.

are more ozone-resistant in wet air than in dry (Fig. 152). These data were confirmed in 1962 on NK, PCHP (benzene polychlorides) and BSK by the determination of the speed of drop of stress in dry and wet ozone.[19]

It is important to establish for the explanation of the obtained data whether the moisture enters into a chemical reaction with the rubber during the ozone cracking or whether it has a purely physical effect.

It has been established that the speed of the absorption of the ozone by the rubber grows considerably in the presence of moisture.[16, 17] However, the nature of this process is unclear:

1. If during the absorption of ozone in the presence of moisture the same chemical reactions took place as in the absence of moisture, the acceleration of the absorption of ozone ought to lead to an acceleration of cracking, and not to its deceleration.

2. In the presence of moisture the process of degradation of ozonides is usually accelerated which also ought to promote the acceleration of cracking, and not its deceleration.

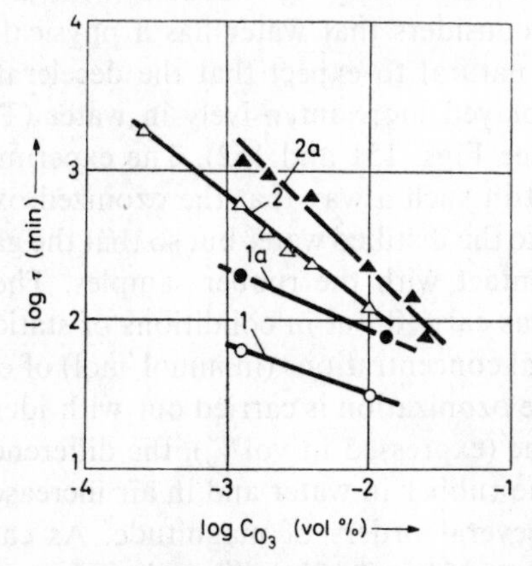

FIG. 152. Dependence of durability of natural rubbers and Nairit in dry and humid air on ozone concentration under repeated deformations. 1. Natural rubber, dry air. 1a. Natural rubber, humid air. 2. Nairit, dry air. 2a. Nairit, humid air.

3. The hydrogen peroxide which is formed during the interaction of ozone and water does not cause the appearance of cracks. As experiments showed,[15] not one of the rubbers which were tested in a strained state—natural and SKB rubber and Nairit (deformation 80%)—cracked, neither in concentrated solutions of hydrogen peroxide (30–80%), nor in its vapour.

4. The influence of moisture on the chemical reaction during ozone cracking ought to show up equally on rubbers of a similar chemical structure, but this is not the case, as can be seen on the example of NK and SKI rubbers.

Therefore, even if moisture has an influence on the chemical side of the process of ozone cracking, the mechanism of this influence is not fully understood. On the other hand, as the decelerating effect of moisture appears on more hydrophilic rubbers, it is logical to assume that moisture does have a physical effect. It can lead to the surface layer of the rubber slightly swelling in water; as a consequence the tensile stress decreases and becomes more uniformly distributed.

To verify further the correctness of the stated theory concerning the character of the influence of moisture on ozone cracking, experiments on ozonization in water were carried out on different rubbers, including on rubbers with a hydrophobic surface. If one considers that water has a physical effect on the process, it was natural to expect that the decelerating action of moisture is displayed more intensively in water (Table 14) than in its vapour (see Figs. 151 and 152). The experiments in water were carried out in such a way that the ozonized oxygen was fed continuously into the distilled water but so that the gas bubbles did not come in contact with the rubber samples. The ozonization of the rubber was carried out in conditions of static deformation with nearly equal concentrations (in mmol/mol) of ozone in water and in air. If the ozonization is carried out with identical concentrations of ozone (expressed in vol %), the difference in the time to rupture of the rubber in water and in air increases under these conditions by several orders of magnitude. As can be seen on Table 14 and Figs. 151 and 152, rubbers based on natural rubber show a considerably greater slowing down of the cracking in water and in steam compared with air than SKI rubbers, due to the latter being less hydrophilic. Ozone cracking in water is also slowed down in SKB rubbers (see Table 14), on which atmospheric moisture has practically no influence (see Fig. 151).

TABLE 14

Results of Testing NK, SKI and SKB Rubbers for Ozone Cracking in Water and in Air

Rubbers	Ozone concentration (mmol/mol)	Durability τ (min)		$\delta = \frac{\tau_{water}}{\tau_{air}}$
		in water	in air	
SKB	~0·001	520	225	2·3
NK	~0·004	2410	180	13
SKI	~0·002	1005	420	2·4

If the rubber surface becomes less hydrophilic the protective action of moisture ceases. Experiments in dry and humid ozonized air on SKI rubbers (see Fig 151) (which are less hydrophilic than natural rubber, and also than rubbers based on NK and Nairit, see Fig. 153) the surface of which was made water-repellent by the introduction of paraffin wax, showed that in these cases the inhibiting effect of moisture on the cracking of rubber disappears.

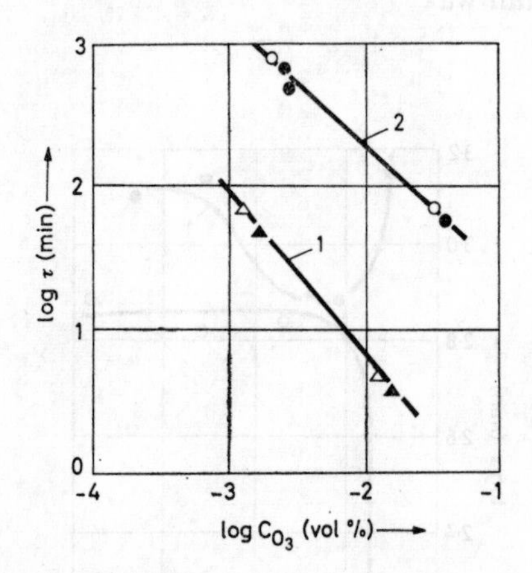

FIG. 153. Dependence of durability of paraffin wax containing rubbers in dry and humid air on ozone concentration under frequent deformations. 1. Natural rubber. 2. Nairit. Black points: humid air; white: dry.

This is also confirmed by the results of experiments under static deformation and ozonization in water. On Fig. 154 are shown the influence of ozone in water on NK with paraffin wax (5 g paraffin wax on 100 g rubber) and without paraffin wax. Whilst with small deformations paraffin wax in air has a strong protective effect,[18] the durability of the rubber which contains paraffin wax decreases sharply irrespective of the size of the deformation. A similar result was obtained with SKI rubber.

These results are correlated to the following data on the amount of swelling of rubbers in air of 100% R.H. (48 hr at room temperature):

	Swelling (%)
NK	
without paraffin wax	2·1
with paraffin wax	1·6
Nairit	
without paraffin wax	5·5
with paraffin wax	2·4
SKI	1·0
SKB	0·9

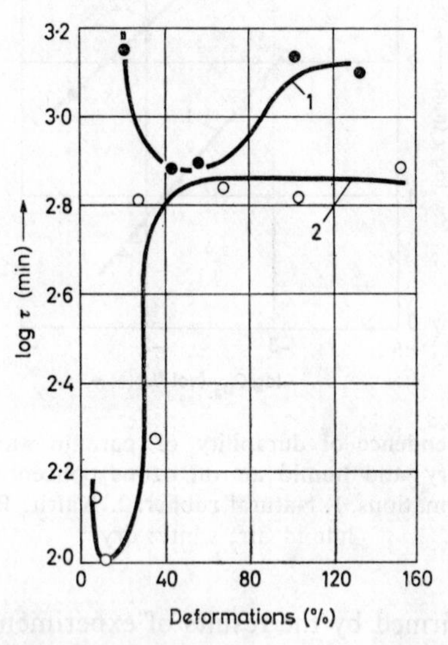

FIG. 154. Influence of paraffin wax on the durability of natural rubbers under static deformation under ozonization in water (ozone concentration 0·001 mmol/mol). 1. Without paraffin wax 2. With paraffin wax.

The beneficial role played by the swelling in water appears particularly clearly when the failure of the polymer takes place due to the action of water itself. This can be observed on SKS-30-1 rubber, vulcanized with MgO. As can be seen on Fig. 155 the speed of creep of this vulcanized rubber at comparatively high humidity is considerably greater than in air. Although in this case the process is not accompanied by the appearance of visible cracks, obviously the failure takes place due to the development of microcracks, and the retarding action of the water is connected with the swelling of the surface layer of the vulcanized rubber and with the increasingly uniform stress distribution.

The decrease of the non-uniformity of the stress distribution can be attained, not only by the swelling of the surface layer, but also by the decrease of the number of defects on the surface or by a complete removal by dissolving of the cracking surface layer. It was shown, for instance, that rubber articles which were produced in moulds which had a smoother surface resisted ozone cracking better. Similarly, metals with an improved surface finish show an increased resistance to corrosion cracking.[20]

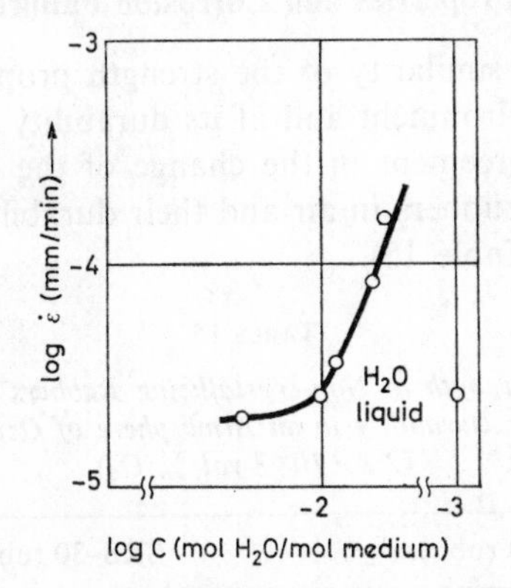

FIG. 155. Dependence of speed of creep $\dot{\varepsilon}$ of vulcanized SKS-30-1 on relative humidity. Temperature 20°C, σ = 225 kg/cm². The isolated point refers to water.

The increase of strength obtained by the dissolving of the faulty surface layer of materials was already observed in the well-known experiments of Joffé. The action of concentrated ammonia on stressed brass in contrast to the action of a diluted solution (which causes corrosion cracking), leads to a complicated failure (dissolving) of the brass surface layer and to the growth of its durability compared with the durability accompanied by cracking.[21, 22]

A similar phenomenon, made complicated, it is true, by the swelling of the rubber which then takes place, has been observed under the action of ozone on stretched carbon black filled NK, SKB, SKS-30 and SKN-26 rubbers, which had been immersed in glacial acetic acid and in carbon tetrachloride.[1] In spite of the comparatively great concentration of ozone (0·3%), no cracks appeared even after several hours of ozonization, whereas in air the specimens begin to crack within 10–15 min at ozone concentrations which are five hundred times smaller. This sharp increase of the resistance of rubber to ozone cracking is caused by the dissolution in CCl_4, and CH_3COOH of the surface of the layer which is formed under the action of ozone, and which contains microcracks; this was proved by the formation of suspensions of carbon black.

11.4. Strength Properties and Corrosion Failure of Rubbers

A proof of the similarity of the strength properties of rubber in an inactive environment and of its durability in an aggressive medium is the agreement in the change of the true strength of non-crystallizing rubbers in air and their durability in an atmosphere of ozone (Table 15).

TABLE 15

Mechanical Strength of Non-crystallizing Rubbers† in Air and their Durability in an Atmosphere of Ozone ($2{\cdot}2\times10^{-3}$ vol % O_3)

SKB rubbers		SKS–30 rubbers	
True strength (kg/cm²)	Durability (min) (at $\sigma = 2$ kg/cm²)	True strength (kg/cm²)	Durability (min) (at $\sigma = 2{\cdot}5$ kg/cm²)
8·6	25	346	15
247	54	401	17
516	86	430	19
758	105	757	23
830	116		

† Rubbers with different strength were obtained by a change of the vulcanization times.

This similarity is also confirmed by the formation of longitudinal cracks during exposure of natural rubber and Nairit, stretched by 500–600%[(23)] in ozone (Fig. 156, between pp. 290/1). The orientation and crystallization under tension leads, as is known, to a strengthening of the rubber, the growth of the cracks perpendicular to the direction of the orientation is hampered, and the formation of cracks through the growth of parallel slips is facilitated. A similar phenomenon is the formation of longitudinal silver cracks, observed at low-elastic deformation of a number of fibres and plastics during their crystallization and orientation of the molecular groups.[24]

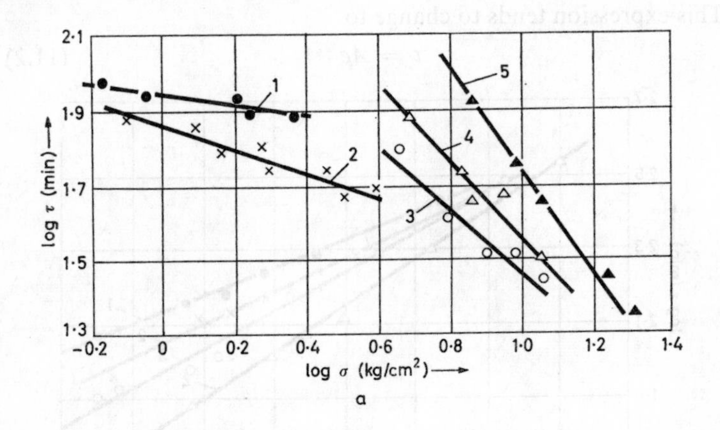

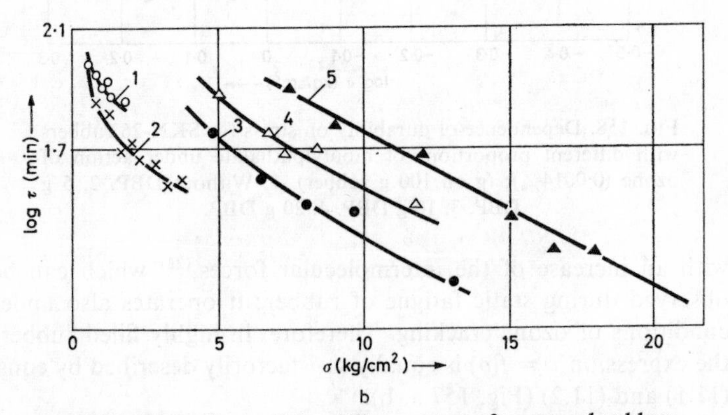

FIG. 157. Dependence of durability on stress for natural rubbers with different quantities of channel black (g on 100 g rubber) under the action of ozone (0·002%). 1. Without carbon black. 2. 30 g carbon black. 3. 60 g carbon black. 4. 90 g carbon black. 5. 120 g carbon black. (a) Coordinates: log τ and log σ. (b) Coordinates: log τ and σ.

Apart from the qualitative resemblance there is also a similarity in the quantitative dependence on stress of the durability between the processes of static fatigue and corrosion cracking.

In all investigated cases of corrosion cracking in the region of small deformations, well below the point of rupture, the expression

$$\tau = B\sigma^{-b} \tag{11.1}$$

which is characteristic for the static fatigue of rubber,[25] is valid. This expression tends to change to

$$\tau = Ae^{-\alpha\sigma} \tag{11.2}$$

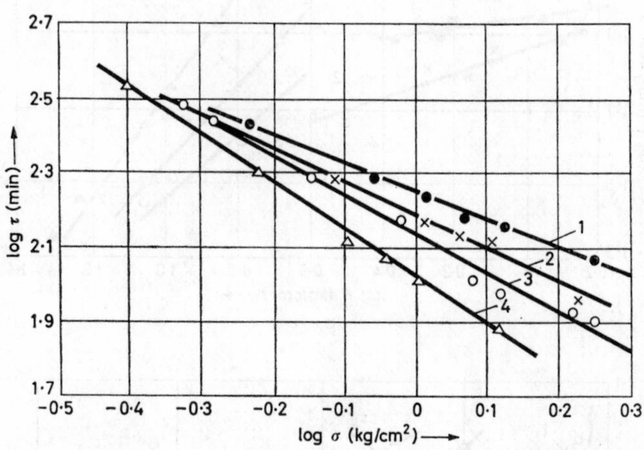

FIG. 158. Dependence of durability on stress for SKN-26 rubbers with different proportions of dibutylphthalate under action of ozone (0·0014%); (g on 100 g rubber). 1. Without DBP. 2. 5 g DBP. 3. 10 g DBP. 4. 20 g DBP.

with an increase of the intermolecular forces,[25] which can be observed during static fatigue of rubber; it operates also under conditions of ozone cracking. Therefore, in highly filled rubbers the expression $\tau = f(\sigma)$ is equally satisfactorily described by eqns. (11.1) and (11.2) (Fig. 157 a, b).

Most interesting are those results which are connected with the change of the incline of the straight lines $\log \tau - \log \sigma$ for different rubbers, i.e. with the change of the value of the constant b, with the introduction of fillers (see Fig. 157), plasticizers (Fig. 158), and with the transition from small to great deformations. In the range of ozone concentrations under consideration the value b does practically not depend on concentration:

	Concentration of ozone (%×10³)	b
SKN–40 rubbers	1·1	1·20
	5·2	1·10
	12	1·15
SKN–40 Rubbers with 30 g channel black	1·1	1·45
	1·8	1·40
	5·2	1·50
	12	1·50

The physical meaning of the constant b is not yet well understood, neither in connection with static fatigue nor with corrosion cracking. Taking into account the resemblance between these processes, and also the fact that the process of static fatigue is simpler than corrosion cracking, it was necessary to analyse more thoroughly the existing data of the constant b referring to static fatigue of materials by attempting to connect this constant with well-defined physical conceptions which could then be transferred to the cases which relate to corrosion cracking. The attempt was therefore made in one paper[26] to understand the nature of this factor somewhat better by comparing eqns. (11.1) and (11.2).

If it is accepted that the durability of rubber, like that of solids, conforms to eqn. (11.2) with the difference, however, that for rubber $\alpha = f(\sigma)$, then it can be shown that $b = \alpha\sigma \ln c/\sigma$, where c = const., i.e. b is proportional to α.

Indeed, from eqn. (11.2) follows:

$$-\alpha = \frac{d \log \tau}{d\sigma}$$

As $\log \tau = \log A - 1/2{\cdot}3\ \alpha\sigma$, therefore

$$\frac{d \log \tau}{d\sigma} = -\frac{1}{2{\cdot}3}\left(\frac{d\alpha}{d\sigma}\sigma + \alpha\right)$$

From eqn. (11.1)

$$b = \frac{d \log \tau}{d \log \sigma} = 2{\cdot}3 \frac{d \log \tau}{d \ln \sigma} = 2{\cdot}3 \frac{d \log \tau}{d\sigma} \sigma$$

from this follows

$$b = \sigma \left(\frac{d\alpha}{d\sigma} \sigma + \alpha \right)$$

Calling $\alpha\sigma = y$, we obtain

$$\frac{dy}{d\sigma} = \sigma \frac{d\alpha}{d\sigma} + \alpha)$$

$$\left(\text{i.e.} \quad b = \sigma \frac{dy}{d\sigma} \quad \text{or} \quad dy = b \frac{d\sigma}{\sigma} \right).$$

After integration we obtain

$$y = b \ln \left(\frac{\sigma}{c} \right)$$

or finally

$$\alpha = \frac{b}{\sigma} \ln \frac{\sigma}{c} \tag{11.3}$$

From the theoretical relation $\tau = f(\sigma)$ for solids[27] it follows that

$$\gamma = \beta\omega \tag{11.4}$$

where β is the coefficient of stress concentration; ω is the volume of the elementary kinetic particle. If one proceeds from the fact that eqn. (11.4) is valid also for the static fatigue of rubber, and, considering eqns. (11.3), (11.4) and the relation $\alpha = \gamma/RT$, one can thus establish a straight connection between the coefficient of stress concentration and the constants b and α. However, in so far as the failure mechanism of rubber is a different one from that of solids, such a connection is scarcely probable. Experiments confirm that there is a connection between b and α on the one hand, and that β, on the other hand, has a more complicated character.

If one summarizes the data which are connected with the change of b and α in the absence of aggressive media, one can note the following. The presence of a straight connection of β with α and b with an unchanged value of ω ought to lead to an increase of α and b with the growth of β, whereas in fact the reverse is usually observed; in those cases where one can *a priori* assume an increase of β, the values α and b decrease. This happens, for instance, when a rubber with a regular structure is replaced by a rubber with a less regular structure (Fig. 159), during a lowering of the test temperature of the rubber[26] (Fig. 160), with the introduction of a coarsely dispersed filler (see Fig. 159), or by making incisions (a similar picture can be observed when cuts are made on metals[28]).

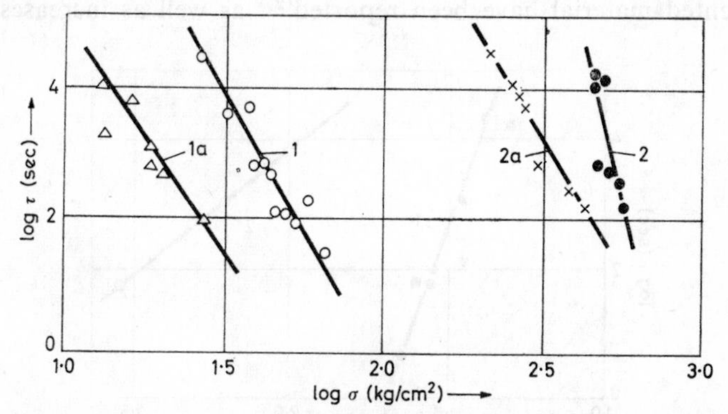

FIG. 159. Influence of regularity of the structure of rubber and of the addition of coarsely dispersed filler on the relation log τ – log σ in air at 55°C. 1. Unfilled SKS-30 rubber. 1a. The same + 50 g of powdered Al_2O_3 Nr.3. 2. Unfilled natural rubber. 2a. The same + 25 g of powdered Al_2O_3 Nr.3.

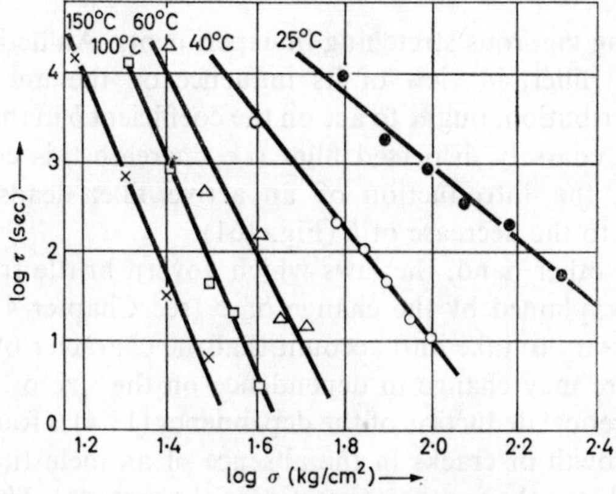

FIG. 160. Influence of temperature on the relation log τ – log σ for radiation vulcanized SKN-40 rubber (E_∞ = 24 kg/cm²) in air.

When a plasticizer is introduced, and if one takes into account only the decrease of β and does not consider the growth of ω, one could expect a decrease of b and α; however, in fibres an increase[29] of α is observed. In Nairit rubbers the value b decreases with the introduction of up to 4·3% of dibutylphthalate (DBP), but with a further increase of DBP it becomes greater.[30]

It would seem that with orientation the heterogeneity in the stress distribution should increase; however, decreases of α for oriented material have been reported[29] as well as increases[31]

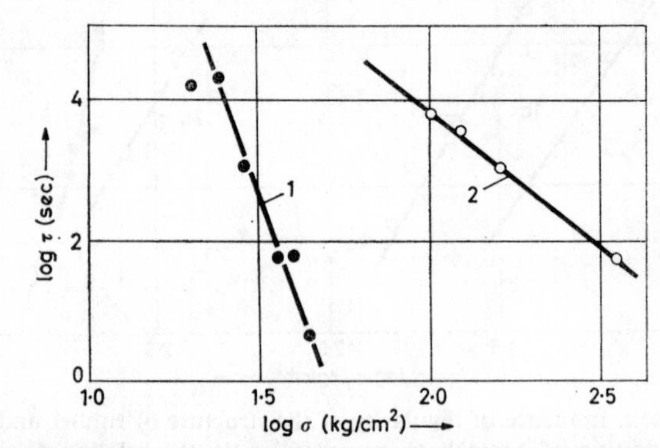

FIG. 161. Influence of active filler on the relation $\log \tau - \log \sigma$ for radiation vulcanized SKN-26 rubber in air (at 60°C). 1. Unfilled vulcanized rubber. 2. Vulcanized rubber with 30 g of channel black on 100 g rubber.

of α during vigorous stretching in experiments. An active (highly dispersed) filler, in view of its influence on the uniformity of stress distribution, ought to act on the coefficient b in the opposite way to a coarsely dispersed filler (i.e. increase this coefficient). However, the introduction of an active filler leads, on the contrary, to the decrease of b (Fig. 161).

On the other hand, the laws which govern brittle fracture can often be explained by the change of ω (see Chapter 4). Besides, it is necessary to take into account that the character of the polymer failure may change in dependence on the size of the stress. The theoretical deduction of the dependency (11.4) is founded only on the growth of cracks in the absence of an inelastic deformation which can change the properties of the material. The development of inelastic deformations is very clearly shown in visco-elastic materials in which, dependent on the conditions, different types of fracture[25] can be observed. In elastomers in the visco-elastic state the relation between the speed of creep (accompanied by the orientation and strengthening of the material without failure), and the speed of failure (development of cracks or tears)[32, 33] changes easily. This implies that the visco-elastic fracture can in fact be more or less a "brittle" one. The shift towards a more "brittle" failure is accompanied by a smaller deformation and weakening of the material which under small stresses ought to make itself felt to a greater extent than under high stresses, as the creep (and, consequently, the orientation) has under high stresses and short times of action, no time to develop sufficiently. Consequently the angle of slope (and with it b and α) of the straight line $\tau = f(\sigma)$ must decrease.

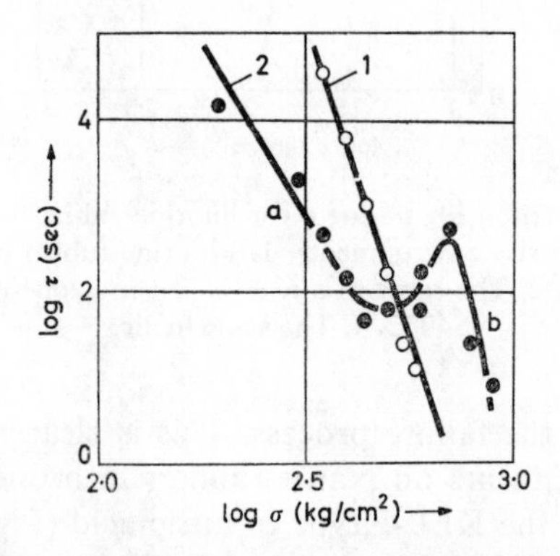

FIG. 162. Influence of perforation on the relation log τ − log σ for vulcanized natural rubber in air (at 60°C). 1. Samples without perforations. 2. Samples with three perforations.

Obviously, a shift towards a more "brittle" fracture must favour the decrease of the mobility of the macromolecules of the polymer with a lowering of temperature, with introduction of a filler and with orientation (including additional orientation at the tips of the cracks when incisions are made). This agrees with the existing data concerning the fracture of specimens with incisions having a more brittle character than that of specimens without incisions.[34] In all these cases b decreases (see, for instance, Fig. 162).

When investigating the failure of rubbers in the presence of aggressive media it has been successfully shown that the fracture of filled rubbers has a more "brittle" character than that of unfilled rubbers; this applied also to the effect of higher stresses. In experiments on rubbers in a chemically aggressive environment the size of b decreases,[35, 36] in comparison with the results in air, as the relative role played by creep decreases due to the sharp

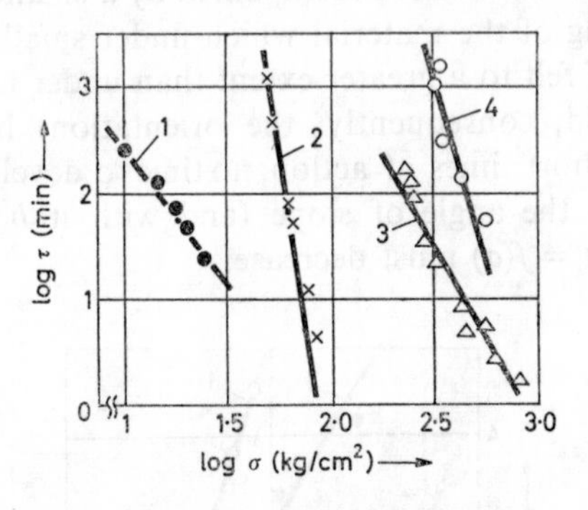

FIG. 163. Relation $\log \tau - \log \sigma$ for fluorine rubber and Nairit in air and aggressive environments. 1. Fluorine rubber in 54% nitric acid at 50°C. 2. The same in air. 3. Nairit in ozone (0·0006%) at 40°C. 4. The same in air.

acceleration of the failure process. This is clearly shown by the results of experiments on Nairit rubbers in ozone, and on fluorine rubbers of the KEL-F type in nitric acid (Fig. 163). Similar data have been obtained concerning the decrease of the value of α under the effect of water on untreated glass fibres and glass fibre reinforced plastics (in contradistinction to its effect on hydrophobic specimens[37]), and also under the effect of light on Nylon fibres, and without light.[33] In the latter case the light which destroys chemical bonds acts similar to a chemically aggressive medium.

It is also easy to understand the reason for the change of the value of the constant b of identical rubbers under different experimental conditions. The comparison of results which were obtained with the ozonization of identical rubbers in different experimental conditions, shows that under the condition $\sigma =$ const. (σ calculated on the full cross-section, without allowing for defects) the value of b is greater than under the condition ε = const. With ε = const. Nairit rubbers show a value of $b = 0{\cdot}5$ (deformation 550–700%), and with σ = const. in the same range of deformation $b = 3{\cdot}5$. A similar picture has been observed for natural rubbers. With ε = const. the value of $b = 0{\cdot}35$, and according to previously obtained results[12] with σ = const., $b = 1{\cdot}2$. This is connected with the fact that experiments with ε = const. are made mainly after the completion of the relaxation processes, due to which the uniformity of the stress distribution increases, whilst with σ = const. the relaxation processes can only be completed in cases of great durabilities, i.e. under small stresses. The greater σ, the smaller is the durability; the relaxation processes get completed to a lesser degree, and the non-uniformity of the stress distribution is greater. As a result the degree of overstress on the defects in regions of high stresses increases considerably with σ = const., and τ decreases more markedly than under the same stresses with ε = const. This leads to an increase of the coefficient b.

The strengthening of the intermolecular forces in polar rubbers (SKN-26, Nairit) in comparison with non-polar ones (NK, SKB) leads to a greater value of b, being $b = 0{\cdot}82$ and $0{\cdot}76$. For ozonized polar rubbers and for ozonized NK and SKB rubbers $b = 0{\cdot}35$ and $0{\cdot}33$. Such an increase of b occurs also with the change from SKN-26 to SKN-40 (0·80 and 1·20); the same can be observed during static fatigue of rubbers.[25] The same factors influence the value of the constant b under the effect of aggressive media on rubbers as under static fatigue. Apart from this, however, some rubbers show specific effects. For instance, in aggressive media the particular role played by the filler may manifest itself in two ways:

(1) The fillers change a particle of the chemically active centres of the system: fillers which are inert to the environment, reduce the particle of the chemically active centres; fillers which interact with the environment, and which also form weak chemical bonds with the polymer, enlarge this particle.

(2) Fillers have an influence on the speed of diffusion of the aggressive medium into the filled rubber: through the formation of continuous chains consisting of filler particles they can serve as lead-ins for the aggressive medium; the particles of an inert

filler reduce the diffusion of the aggressive medium in the absence of such structures.

In connection with corrosion cracking it is only important to consider the influence of the fillers on the particles of the chemically active centres, as the diffusion of the aggressive medium does not here determine the speed of the process. In the case of breakdown of the polymer chain brought about by the environ-

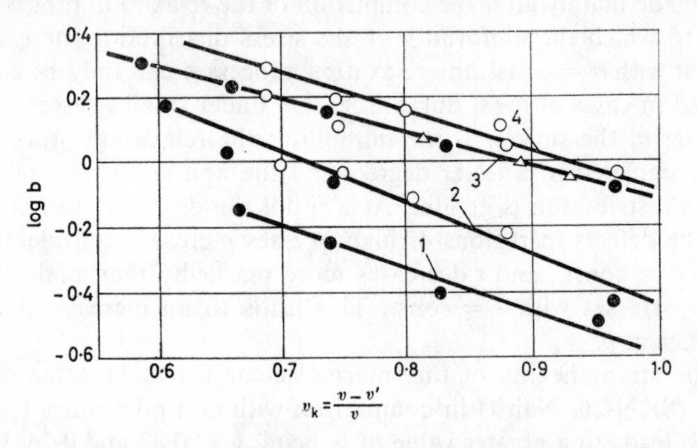

FIG. 164. Dependence of the value *b* on the volume content of raw rubber in rubbers with various fillers under ozone cracking. 1. SKB. 2. NK. 3. SKN-26. 4. SKS-30. ○ — calcium carbonate. ● — channel black. △ — dibutylphthalate.

ment, and the existence of corrosion cracking, the introduction of a filler which decreases the mobility of the molecules, must lead to a decrease of *b*; but, decreasing the particles of the chemically active centres must lead to a growth of *b*. A great number of active centres which exist in highly unsaturated rubbers which are cracking in ozone, leads to a predominance of the second influence, and the value *b* grows. This can be clearly seen on Figs. 157a and 164. With an increase of the quantity of channel black from 0 to 120 g on 100 g rubber, *b* grows in NK from 0·20 to 1·53, in SKN-26 from 0·82 to 2·10, in SKB from 0·33 to 0·70.[35]

With the introduction of a diluent *b* (independently from the nature of the diluent—active or inactive filler, plasticizer (see Fig. 158)—changes in the same way, proportional to the decrease of the rubber content in the volume. This can be seen clearly on Fig. 164, where for SKN-26 rubber the dots which refer to channel black, chalk and dibutylphthalate, lie on one curve. Such a dependence proves that other influences of the diluent on b which are not connected with the change of the particle of the chemically active centres of the polymer, are here not of material importance.

The dependence of b on the volume content of raw rubber V_K in the rubbers:

$$b = b_0 e^{-\varsigma V_k}$$

where b_0 and ς are constants; $V_K = V - V'/V$ (V is the total volume of the rubber, V' the volume of the non-rubber ingredients).

For different rubbers the value b_0 changes from 6 to 14, and the value ς from 2·5 to 4·5.

Data on the influence of fillers on the durability of rubbers under the action of ozone show that under small stress (1–5 kg/cm^2) the durability increases with the proportion of fillers, especially of active ones. Most pronounced is the increase of the durability of rubbers the hardness of which grows sharply with the introduction of a filler (SKB and SKN-26 with channel black). Under great stresses (25–50 kg/cm^2) the durability decreases with increasing filler.

The introduction of a filler which forms different bonds with the polymer, including bonds which fail under the influence of an aggressive medium, leads to a dependence of the corrosion resistance on the proportion of filler which is qualitatively different. For instance, as experiments show, under the effect of HNO_3 on Kel-F type fluorine rubber which contains "white carbon black" (like Aerosil), the presence of unstable bonds promotes a greater development of creep due to these bonds than in unfilled rubbers. With small proportions of carbon black this effect is overlaid by the decrease of the mobility of the polymer molecules (owing to the polymer filler bonds which are resistant to the aggressive medium), as a result of which the size of b does not change; but with large proportions of carbon black the size of b grows.

In experiments on rubbers in regions of great deformations[35] in contrast to experiments in regions of small deformations, a strong orientation and strengthening of the polymer takes place. A very considerable structural change and strengthening takes place in natural rubbers. Proof of this are the values of the constant B in the formula $\tau = B\sigma^{-b}$, which indicates the durability at $\sigma = 1$ kg/cm² and an ozone concentration $c = 1\%$:

	Deformation (%)	b	B
Rubbers based on NK	5– 15	0·35	72·5
	530–680	1·0	$6{\cdot}6 \times 10^5$
Rubbers based on Nairit	5– 55	9·75	417
	550–700	0·50	1660

In natural rubbers the value B changes approximately by a factor of 10,000 with the transition from small to great deformations, whereas in Nairit the factor is only 4 (i.e. it remains of the same order of magnitude).

The growth of orientation which leads to a decrease of the mobility ought to promote the decrease of the size of b; this is confirmed by experimental data. However, for NK rubbers the failure mechanism changes, obviously, at very great deformations and with it grows the size of b.

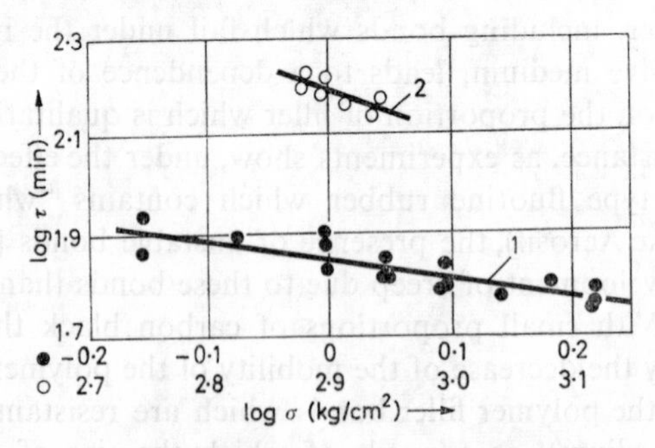

FIG. 165. Relation log τ – log σ for natural rubbers in the presence of ozone (0·002%). 1. Under small deformations. 2. Under great deformations.

Without wishing to discuss the reason for such a change of the value b, it is absolutely necessary to note the qualitative similarity between the behaviour of natural rubbers during ozone cracking and under conditions of static fatigue. As can be seen on Fig. 162

the dependence of the durability in air of perforated NK samples on the stress is described by a curve with two extremes, whereby in regions of small deformations (section *a*) the value b is less than in regions of great deformations (section *b*). Also a clearly expressed extreme can be observed on the curve (see Chapter 12) referring to ozone cracking; on the linear parts of which $b = 0{\cdot}35$ under small deformations, and under great ones $b = 1$, i.e. it grows (Fig. 165).

Literature

1. Yu. S. Zuyev, A. S. Kuz'minskii, *DAN SSSR*, **89**, 325 (1953).
2. V. V. Romanov, *Korrozionnoye rastreskivaniye metallov*, Mashgiz, 1960.
3. Z. F. Ossefort, W. Touhey, *India Rubb. World*, **132**, 62 (1955); *Rubb. Chem. Technol.* **28**, 1119 (1955).
4. D. H. E. Tom, *J. Polymer Sci.* **20**, 381 (1956).
5. Yu. S. Zuyev, A. Z. Borshchevskaya, *DAN SSSR*, **124**, 613 (1959).
6. H. P. Brown, C. F. Gibbs, *Ind. Eng. Chem.* **47**, 1006 (1955).
7. B. A. Dolgoplock, Ye. I. Tiniakova, *Kauchuk i rezina*, No. 3, 11 (1957); B. A. Dolgoplosk, V. N. Reikh, *Kauchuk i rezina*, No. 6, 1 (1957).
8. I. Poddubniyi, Ye. Erenburg, Ye. Starovoitov, *DAN SSSR*, **120**, 535 (1958).
9. N. P. Apukhtina, R. A. Shliakhter, F. B. Novoselov, *Kauchuk i rezina*, No. 6, 7 (1957).
10. K. A. Andrianov, *Kremniiorganicheskiye Soyedineniya*, Goskhimizdat, 1955.
11. B. A. Dogadkin, D. L. Fedyukin, V. Ye. Gul', *Kolloid. zh.* **19**, 287 (1957).
12. Yu. S. Zuyev, *DAN SSSR*, **93**, 483 (1953); *Khim. prom.* No. 9, 325 (1953).
13. I. M. Buist, *Rev. Gén. Caout.* **31**, No. 6, 479 (1954).
14. B. S. Biggs, *Rubb. Chem. Technol.* **31**, 1015 (1958).
15. Yu. S. Zuyev, V. F. Malafeyevskaya, *Kauchuk i rezina*, No. 6, 26 (1961).
16. J. H. Gilbert, *Proceedings of the Fourth Rubber Technology Conference*, London, May 1962, Preprint 56.
17. H. Tucker, *Rubb. Chem. Technol.* **32**, 269 (1959).
18. Yu. S. Zuyev, S. I. Pravednikova, G. V. Kotel'nikova, *Kauchuk i rezin*, No. 3, 21 (1962).
19. M. Braden, A. N. Gent, *J. Appl. Polymer Sci.* **3**, 90, 100 (1960); *Kautschuk und Gummi.* **14**, WT157 (1961); *Proc. Inst. Rubb. Ind.* **8**, No. 4, 88 (1961).
20. V. V. Romanov, *ZhPKh*, **34**, 1995 (1961).

21. A. V. Bobiylev, *Korrozionnoye rastreskivaniye latuni*, Metallurgizdat, 1956.
22. G. B. Klark, A. Ye. Gopius, A. Smirnova, *Trudiy Instituta fizicheskoi khimii AN SSSR*, pt. 8, Izd. AN SSSR, p. 110.
23. Yu. S. Zuyev, S. I. Pravednikova, *DAN SSSR*, **116**, 813 (1957).
24. A. I. Maklakov, G. G. Pimenov, R. Ya. Sagitov, *Viysokomol. soyed.* **3**, 1410 (1961).
25. G. I. Bartenev, L. S. Bryukanova, *ZhTF*, **28**, 287 (1958); G. M. Bartenev, *Plast. massiy*, No. 9, 48 (1960).
26. Yu. S. Zuyev, G. M. Bartenev, N. I. Kirshenshtein, *Viysokomol. soyed.* **6**, No. 9, 16 (1964).
27. G. M. Bartenev, *Izv. AN SSSR, OTN*, No. 9, 53 (1955).
28. M. I. Chayevskii, L. S. Bryukhanova, V. I. Lichtman, *DAN SSSR*, **143**, 92 (1962).
29. S. N. Zhurkov, S. A. Abasov, *Viysokomol. soyed.* **3**, 450 (1961); *FTT*, **4**, 2184 (1962).
30. Yu. S. Zuyev, S. I. Pravednikova, L. S. Zherebkova, V. D. Zaitseva, *Viysokomol. soyed.* **5**, 1201 (1963).
31. M. I. Bessonov, Ye. V. Kuvshinskii, *Viysokomol. soyed.* **2**, 397 (1960).
32. S. N. Zhurkov, T. P. Sanfirova, *ZhTF*, **28**, 1719 (1958).
33. V. R. Regel', N. N. Cherniyi, *Viysokomol. soyed.* **5**, 925 (1963).
34. G. A. Patrikeev, B. G. Gusarov, V. N. Konoplev, *Viysokomol soyed.* **2**, 1438 (1960).
35. Yu. S. Zuyev, S. I. Pravednikova, T. V. Lichtman, *Viysokomol, soyed.* **5**, 262 (1963).
36. Yu. S. Zuyev, A. Z. Borshchevskaya, *Kauchuk i rezina*, No. 10, 23 (1963).
37. V. A. Bershtein, L. A. Glikman, *FTT*, **5**, 2270, 2278 (1963).

CHAPTER 12

FAILURE KINETICS OF RUBBERS IN AGGRESSIVE ENVIRONMENTS AND CRITICAL DEFORMATIONS

12.1. Influence of the Reaction and Adsorption Properties of the Environment on the Speed of Failure of Rubber

IN CONTRAST to the effect of mechanical forces which may lead to the failure of any bonds in polymers, selectivity is characteristic for chemically aggressive media, the effect of which on the polymer is a chemical reaction. For instance, ozone strongly attacks only unsaturated rubbers, and does not cause the cracking of saturated rubbers.

The capacity of the polymer to react and the chemical activity of the environment has a great influence on the speed of corrosion failure. It has been shown on the example of SKS-30-1 rubbers[1] that with the increase of the dissociation constant of acids (with, in order to exclude the influence of diffusion, similar molecular weights) the durability of the polymer decreases (Table 16).

TABLE 16

Durability of SKS–30–1 (MgO) Rubbers in Acids with Different Dissociation Constants K

Acid	$K \cdot 10^4$	τ (min)	Concentration of acid (mmol/mol)
α-chloroacetic acid	14·0	12	7
Isovaleric acid	0·15	64	7
Pyroracemic acid	32·0	65	28
Butyric acid	0·15	115	28

The influence of the chemical activity of the medium on the failure speed of rubber is usually greatly complicated by the influence of its capacity to be adsorped on the rubber.

The decrease of strength and durability of rubbers which adsorb surface-active substances in the absence of corrosive action has been discussed in Chapter 10. In order to explain the role of adsorption of the environmental medium during corrosion failure the process of cracking was observed in SKS-30-1 (MgO) rubbers in alcoholic solutions of fatty acids of the same homolog series which had the same dissociation constant. It was found that pro-

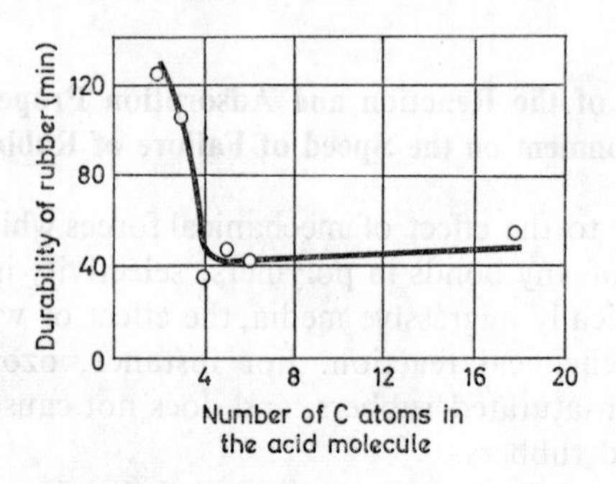

FIG. 166. Dependence of durability of SKS-30-1 (MgO) rubber in acid solutions on the number of carbon atoms in the acid molecule.

ceeding from acetic to propionic and butyric acids a sharp in crease of the speed of failure is observed (τ decreases), whilst τ does practically not change with the transition to the following member of the series (Fig. 166). The increase of the speed of the cracking process is connected both with the increase of adsorption of acid by the rubber and with the increase of the speed of swelling of the rubber in the series of acids which change inversely with the durability of the rubber.

The explanation of the relative role of the strength of acids and their sorptive capacity in the limitation of the durability of rubber was found with acids which greatly differ in chemical activity (acetic and butyric versus hydrochloric), in acidic gases and in their aqueous solutions. As can be seen on Table 13 (see Chapter 11), the speed of cracking of SKS-30-1 (MgO) rubbers in hydrochloric acid gas is greater than in the gases of acetic and butyric acids and in the two latter ones is practically identical: in aqueous solution and in the gas of butyric acid of equal concentration the process proceeds almost with equal speed, in solutions of acetic acid noticeably slower, and in solutions of hydrochloric acid much slower than in its gas.

The sharp difference in the corrosive action of the acids in solution and in gas form is connected with their different capacity to being adsorbed by the rubbers. With the solution of acids in water their adsorption is hampered. Therefore, the particularly soluble hydrochloric acid causes in solution, in spite of its high corrosive activity, a slower failure of the rubber than acetic acid.

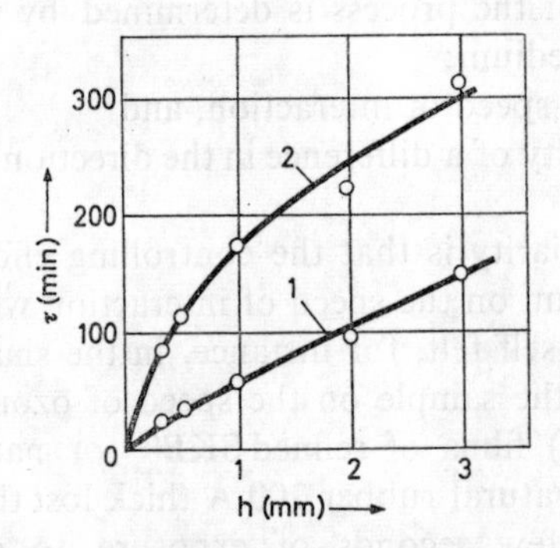

FIG. 167. Dependence of durability τ on the thickness h of the specimen (SKB, 50 g channel black per 100 g rubber). 1. Stress 4·0 kg/cm², ozone concentration 0·0042%. 2. Stress 3·0 kg/cm², ozone concentration 0·0028%.

The introduction of a wetting agent (Aerosol OT) into the solution accelerates the action of the hydrochloric acid on the rubber. In the gas phase where the factor of solubility is absent, hydrogen chloride is found to be more active than CH_3COOH gas.

One of the characteristic features of corrosion failure which is connected with the surface action of physically and chemically active substances is the fact that the growth of the cracks begins on the surface, whereas with static fatigue it begins both on the surface and in the mass. This is also the reason for the increase of durability of thicker specimens during corrosion failure[2] (Fig. 167), whilst with static fatigue the reverse picture pertains.

12.2. Kinetics of Failure in Unstressed Rubbers

The great speeds of failure of rubber in the presence of a chemically aggressive agent give the possibility to show the principles of this process in a wide range of stresses, beginning with unstressed rubber. The processes of interaction with the aggressive agent proceed in a different way in stressed and unstressed rubber, whereby one can distinguish three fundamental peculiarities of the interaction between the aggressive agents and undeformed as compared with deformed rubber:

(1) the speed of the process is determined by the diffusion of the aggressive medium;

(2) the smaller speed of interaction, and

(3) the possibility of a difference in the direction of the chemical processes.

The first peculiarity is that the controlling effect of the diffusion of the medium on the speed of interaction with undeformed material makes itself felt, for instance, in the sharp influence of the thickness of the sample on the speed of ozonization of thin (order of 100 μ) films of refined SKB[3] or natural rubber.[4] Films of refined natural rubber 200 Å thick lost their rubber-like properties in a few seconds of exposure to an atmosphere which contained 0·01% ozone; films 1 μ thick retained their elasticity under the same conditions for several hours. The oxidized layer of polymer which forms on the surface of the specimen hinders a further development of the reaction; if this layer is continuously removed, the intensive reaction with ozone will continue. This has been shown[5] on ozonized carbon black filled SKN, SKS and SKB rubbers which were immersed in chloroform (a good solvent for rubber ozonides). An intensive formation of a carbon black suspension takes place which proves the interaction of the polymer with ozone.

The direct comparison of the speed of interaction of the aggressive medium with stressed and unstressed polymer samples[6] was made by exposing vulcanized SKS-30-1 (MgO) to HCl. The speed of Mg^{2+}ions enrichment in the solution was measured; this describes the fracture speed of Mg surface bonds under the effect of hydrochloric acid. It can be seen on Fig. 168 that the process slows down on unstressed samples (dotted curves), when two sections can be noted on the kinetic curves: (a) the section on which the speed of interaction gradually decreases due to the penetration of H^+ ions into the polymer and, especially, due to the liberation of Mg^{2+} ions into the solution which limits the diffusion speed of these ions; and (b) the initial section on which the speed of the liberation of the Mg^{2+} ions is the same as under the attack by HCl on the deformed rubber (obviously, at first the reaction occurs on the surface of the vulcanized rubber). The

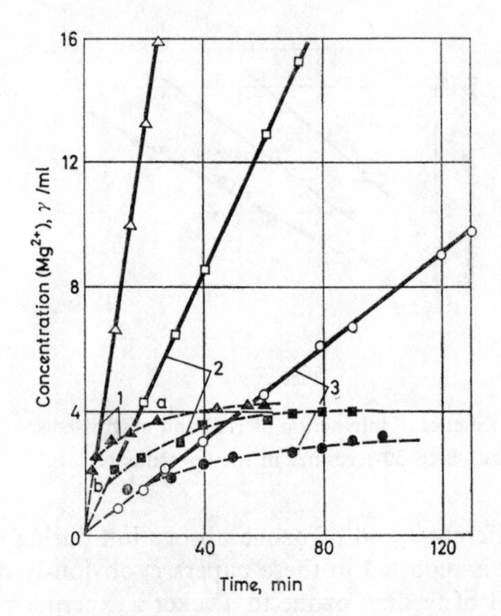

FIG. 168. Kinetics of interaction of HCl and stressed (solid lines, deformation 100%) and unstressed (dotted lines) vulcanized SKS- 30-1 rubber samples at different temperatures. 1. Al 45°C. 2. At 35°C. 3. At 25°C.

complicated shape of these curves, the straightening of section (a) in the coordinates $[Mg^{2+}]-\sqrt{t}$ (Fig. 169), and also the order of magnitude of the apparent activation energy (6·4 kcal/mol) bears witness to the diffusion[7] character of the failure process of unstressed vulcanized rubbers.

Two papers exist which report on the kinetics of ozone absorption by unstressed and strained vulcanized rubbers, both at the same temperature.[8, 9] Erickson and co-workers[8] exposed vulcanized technical BSK rubber which contained Neozone D to ozone concentrations of $2{\cdot}5$–25×10^{-5}% at a feeding rate of $2{\cdot}5\times10^{-8}$ mol/min. The speed of absorption of ozone by an unstressed specimen fell at first sharply, and then slower, to zero. Tucker[9] observed on ozonized expanded foam natural rubber (in order to increase the reacting surface) and polychloroprene samples initially a complete absorption of ozone, then a gradual decrease of the absorption speed to very small distinct values, but not to

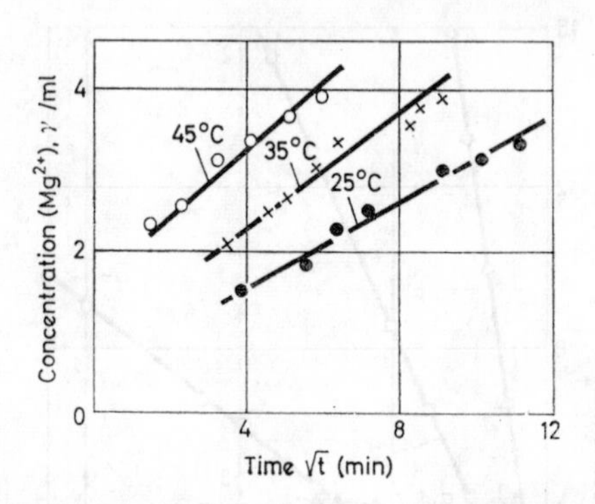

FIG. 169. Kinetics of interaction of HCl and unstressed samples of vulcanized SKS-30-1 rubber in the co-ordinates $Mg^{2+} - \sqrt{t}$.

zero. The different speed of ozone absorption during the initial stages which is reported in these papers is obviously due to the smaller speed of feeding ozone in Tucker's experiments than in Erickson's. The specimens, extracted with acetone, recovered after the termination of the experiment the capacity to absorb ozone, albeit to a somewhat lesser degree. The reported results agree with those of other authors and confirm the diffusion character of the interaction of the aggressive medium with unstressed polymers.

If macrocracks are formed under the attack of an aggressive medium on a stretched polymer, the failure process, due to the growth of these cracks, is not limited by the diffusion speed of the aggressive agent into the polymer. On the other hand, in the absence of corrosion cracking the failure speed may be limited by the diffusion speed of the aggressive medium into the polymer, as it occurs, for instance, in chemical relaxation of Nylon 6 under the influence of water vapour.[10]

The second peculiarity, namely the lower speed of attack of the aggressive medium on non-stressed than on deformed rubber is well known in the case of ozone, and is confirmed by electron-microscope observations,[11] and can also be found in the effect of hydrochloric acid on SKS-30-1 rubber. The difference in the failure speed can be seen clearly on Fig. 168, where the curves for the unstressed rubber lie below those for deformed rubber, in spite of the fact that the surface of the unstretched specimens was five times larger than that of the stretched ones.†

The third peculiarity of the kinetics of the failure of rubbers is the possibility that the stress may influence the direction of the chemical reactions. The laws which govern the growth of the equilibrium modulus of unstressed sheets of purified SKB under ozone prove the autocatalytic character of this process.[3] The introduction of phenyl-B-naphthylamine sharply slows down the cross-linking of SKB[3] sheets which is caused by ozone and also the effect of ozone on natural rubber sheets. Both these circumstances indicate that mainly a chain process of oxidation develops under the effect of ozone on unstressed polymers. In contrast, under the effect of ozone on strained polymers mainly an ozonization reaction takes place, evidently, at the same time as the oxidization processes. This relates to rubbers which contain inhibitors, in which after the initial transient period cracks develop with constant speed. The activation energy of the crack propagation is very low and corresponds to the activation energy of the interaction of ozone with olefines. There exist no data on the interaction of ozone with stressed pure vulcanized rubbers which contain no inhibitors.

12.3. Failure Kinetics of Stressed Rubbers

The investigation of corrosion cracking of rubber was carried out over a wide range of deformations under the effect of ozone and some acids. There exist, however, only few systematic data on the kinetics of this process.[2, 12–15]

Buckley and Robison,[14] when investigating the kinetics of ozone cracking of samples of vulcanized butyl rubber which

† The number of unstretched samples in the experiments was greater than that of stretched ones.

were under a constant stress, confined themselves to the initial stage of this process (before the appearance of visible cracks). The kinetics of ozone cracking of SKB rubber under the influence of a constant load[15] was studied in a wider time range (up to fracture). A special place is occupied by Braden and Gent's work,[16] in which they investigate the growth of a single incision on the rubber under the action of ozone.

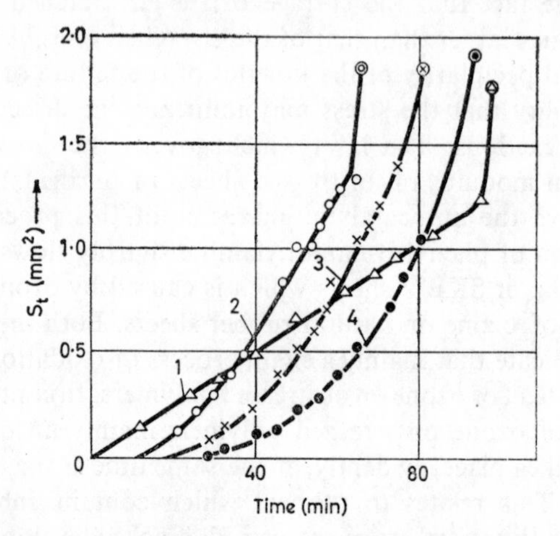

FIG. 170. Kinetics of crack growth under the attack of ozone (0·01%) on samples of vulcanized Nairit under different strains. 1. 100%. 2. 57%. 3. 50%. 4. 44%.

The system $\varepsilon =$ const. must be acknowledged as the most convenient method for the study of the kinetics of corrosion cracking of rubber, under which simultaneously also a constant nominal stress is maintained. The kinetics of the process of the drop of stress in the cracking specimen has been investigated under $\varepsilon =$ = const. with the help of an objective method which was developed by one of us.[2] It was shown that in the general case the kinetic curve for rubber $S_t = f(t)$ (see p. 280) consists of four plots,† the position and length of which is connected with the size of the deformation (Fig. 170).

† In the investigation of the kinetics of corrosion cracking of metals[17] the presence of three stages of the process is noted: the induction period, the period of approximately constant strain, and rapid fracture.

(1) The induction period, during which no visible reduction of stress is observed. The termination of this period (the incipient drop of the stress) coincides practically with the moment of the appearance of cracks which are visible to the naked eye. The induction period, being defined as the time before the appearance of cracks (τ_i) includes the time of the interaction of ozone with the rubber components which segregate on the specimen surface, and also the time during which the emergent cracks reach visible dimensions.

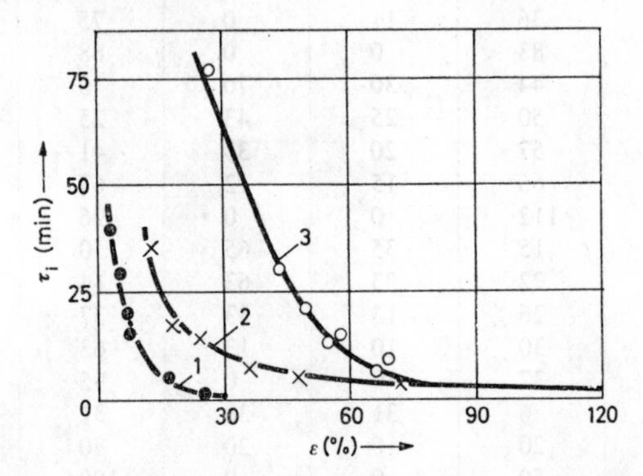

FIG. 171. Influence of the size of strain on the time to the appearance of cracks during ozone cracking of rubber. 1. Natural rubber (ozone concentr. 0·0033%). 2. SKS-30 (ozone concentr. 0·0027%). 3. Nairit (ozone concentr. 0·010%).

(2) The period when the speed of crack growth continuously increases (transient section). Evidently, the induction period is the hidden stage of this transient section (owing to the insufficient sensitivity of dynamometers).

(3) The period which is characterized by the constant growth of cracks (steady section).

(4) Rapid fracture of the specimen.

With increasing strain on the induction period and the transient section of the growth of cracks decreases gradually to zero and becomes the stage of rapid fracture, which did not exist during small deformations. The induction period (τ_i) decreases monotonously with the growth of strain (Fig. 171).

In the absence of substances in the rubber which protect from ozone the longest stages are those of transient and steady speed of the growth of cracks—shortest is the stage of rapid fracture (Table 17).

TABLE 17

Relative Duration (in % of Durability) of Individual Stages of Growth of Ozone Cracks

Rubber	Deformation (%)	Induction period	Transient stage	Steady stage	Rapid fracture
SKS–30	13	22	58	20	0
	18	18	26	50	6
	36	11	0	75	14
	83	0	0	88	12
Nairit	44	30	70	0	0
	50	25	43	25	7
	57	20	33	41	6
	66	15	12	67	6
	112	0	0	96	4
SKS–50	15	35	65	0	0
	22	23	63	14	0
	26	13	33	37	17
	30	10	13	63	14
	37	0	0	95	5
Natural rubber	6	31	38	31	0
	20	10	20	70	0
	50	0	0	100	0
	200	0	0	99	1

The stage of the steady growth of cracks is found with all rates of deformations (strain) and with all ozone concentrations. The presence of the transient stage, however, has to be related to the instability of the conditions of the process, primarily to the fact that the number of cracks changes. The growth of the number of cracks leads to a reduction[18] and more uniform stress distribution in the roots of the previously formed cracks which is accompanied by a deceleration of their growth. Besides, the cracks become deeper which must at first lead to an increase of the over-stress at their tips and to an acceleration of the cracking process. Therefore, according to Busse's[19] data which were obtained in the absence of ozone with an increase of the depth of the cut, the stress which is necessary to obtain a tear, at the beginning decreases sharply (i.e. the stress in the apex of the cut increases considerably).

Under ozone cracking the dependence of the stress at the roots on the depth of the cracks is complicated by the fact that not one but many "interacting" cracks are formed and their number changes. In order to verify whether the deepening of the cracks during ozone cracking leads to a growth of stress at their roots, experiments with samples of various thicknesses were made. These experiments showed that with a decrease of the thickness h of the specimen with the same absolute depth l of the cracks (the relative depth l/h increases) the speed of crack propagation grows. This is very clearly shown with the decrease of the thickness from 3 to 2 and 1 mm (Fig. 172), but less for thicknesses of 0·5 and 0·3 mm.

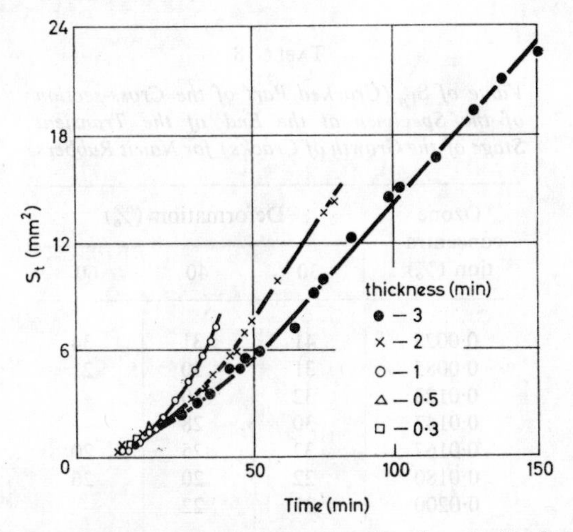

FIG. 172. Kinetics of crack growth under the attack of ozone (0·0042%) on rubber samples of different thickness (SKB, 50 g channel black per 100 g rubber) at a stress of 4·0 kg/cm².

The duration of the stage of transient speed of growth of cracks is determined by the time before the appearance of the maximum number of cracks, since in the virtually unstressed layer the subsequent decrease of the number of cracks (on account of their merging together) has obviously no influence on the size of stress at their tips. As known, an increase of strain shortens the time for the development of the maximum number of cracks. In connection with this the period of transient speed or rate of crack propagation decreases gradually, i.e. it ceases with an ever-decreasing value of S_t (see p. 280). The tendency to a decrease of S_{t_H} (i.e. the area of the cracking part of the cross-section towards the end of the transient section) can for the same reasons also be found with an increase of ozone concentration (Table 18).

In the transient stage the development of the process is described by the expressions

$$\frac{F_0-F_t}{F_0} = \left(\frac{t}{t_0}\right)^k \quad \text{or} \quad \log\left(\frac{F_0-F_t}{F_0}\right) = k \log\left(\frac{t}{t_0}\right)$$

where F_0 is the initial force; F_t is the force at the moment t; t_0 is a constant and k is the constant of speed.

TABLE 18

Value of S_{t_H} (Cracked Part of the Cross-section of the Specimen at the End of the Transient Stage of the Growth of Cracks) for Nairit Rubbers

Ozone concentration (%)	Deformation (%)		
	30	40	60
0·0023	41	31	34
0·0085	31	30	21
0·0123	32	30	–
0·0147	30	28	–
0·0167	32	25	20
0·0180	22	20	26
0·0200	22	22	–

From the data of Vodden and Wilson,[20] and also from those of Payne,[21] who obviously observed the drop of force only in the transient stage with small ozone concentrations (several parts of O_3 per million parts of air), the conclusion was drawn that $\log F$ decreases approximately linear with time.

Erickson,[8] who determined the kinetics of ozone absorption during the ozonization of stretched (20%) BSK, found that the speed of ozone absorption at first grows, passes a small maximum and thereafter remains constant which agrees in general with the existence of the transient and steady stages of crack propagation. Tucker[9] also observed the transient and steady regions of the curve of speed of ozone absorption; however, in contrast to

Erickson he found at its beginning not an increase but a decrease of the speed of absorption. The absence of a transient section which has been observed during the action of hydrochloric acid on vulcanized SKS-30-1[6] is connected with the fact that the experiment was carried out at great deformation (100%) under which the crack propagation begins normally at once with constant speed.

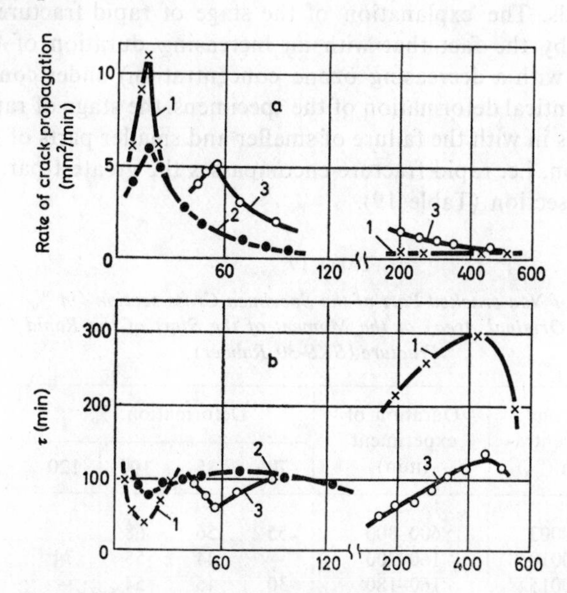

FIG. 173. Influence of strain on rate of crack propagation (a) and durability (b) of rubber during ozone cracking. 1. Natural rubber (ozone concentr. 0·0033%). 2. SKS-30 (ozone concentr. 0·002%). 3. Nairit (under small deformations, ozone concentr. 0·01%; under great deform. 0·02%).

In the steady section the speed of crack propagation changes rapidly, it passes beyond the maximum (Fig. 173) at a deformation which must be called a critical deformation (ε_{cr}). The speed of ozone absorption on the steady section has also a maximum value for vulcanized BSK[8] at a deformation of 20–30%. Similarly, the constant of the rate of crack growth changes also on the transient section.[2] The considerations which are related to the explanation of, and the detailed investigation into this phenomenon will be reported on p. 336.

The sharp increase of the rate of cracking which sets in after the steady stage can obviously be explained by the fact that during the time of the action of ozone a gradual accumulation and growth of internal defects, due to static fatigue, takes place in the sound part of the specimen. As a consequence, when the ozone crack becomes sufficiently deep, the rate of its growth rises sharply. An analogous phenomenon can be observed in the failure of solids. The explanation of the stage of rapid fracture is confirmed by the fact that with an increasing duration of the stress (and with a decreasing ozone concentration) under conditions of identical deformation of the specimens, the stage of rapid fracture sets in with the failure of smaller and smaller parts of the cross-section, i.e. rapid fracture encompasses the greatest part of the sound section (Table 19).

TABLE 19

Area of Non-cracked Part of the Specimen Cross-section (in % of the Original Area) at the Moment of the Start of the Rapid Fracture (SKS-30 Rubber)

Ozone concentration (%)	Duration of experiment (min)	Deformation, %			
		70	85	100	120
0·0003	600–800	55	56	68	–
0·0010	160–200	–	44	55	74
0·0015	160–180	30	46	54	–
0·0024	44–90	20	38	54	60
0·0046	37–43	15	41	48	55
0·0070	20–27	–	33	35	51

In experiments with $F =$ const. (or $\sigma =$ const.) the kinetics of cracking obey somewhat different laws than with $\varepsilon =$ const. An induction period has been observed in experiments on SKB rubber(15) and also on natural rubbers and BSK.(22) On vulcanized butyl rubbers(14) an induction period was not observed; the length of the specimen began to increase at once after contacting with ozone. In this case the lack of an induction period was confirmed by the great accuracy of the employed apparatus. Whilst in Znamensky's work(15) the increase of the length of the specimen was recorded with an accuracy of up to 10^{-2} cm, in Veith's[22] work with an accuracy up to 5×10^{-3} cm, in experiments on butyl rubber the increase of length was measured with an accuracy of up to 6×10^{-5} cm.

In view both of the transient conditions of experiments with the method σ = const., and especially at F = const., and even more of the continuously increasing strain and change of the geometrical dimensions of the specimen, the laws governing the kinetics of cracking of different rubbers differ from each other. For SKB rubbers, for instance, after the induction period the equation[15]

$$\varepsilon = \frac{e^{k_2 t}}{l_0}$$

is satisfied; for natural rubber Veith[22] proposes the equation

$$S = A_0 e^{-k_2 t}$$

and for BSK[22]

$$S = g - qt.$$

In these equations S is the area of the cross-section of the unbroken part of the sample; F is the load; l_0 is the initial length of the sample; t is the time of the application of the load; k_1, k_2, g, q and A_0 are constants.

As can be seen from these equations for the kinetics of cracking of rubber, the process develops with auto-acceleration in SKB and natural rubber, and in BSK with constant speed. The auto-acceleration of the process at F = const. agrees with the data obtained at ε = const. The region of the steady speed of crack growth which is observed at ε = const., degenerates at F = const., since the speed of cracking increases due to the continuous increase of stress. In view of the fact that at ε = const. the rate of crack propagation on the steady section or region passes beyond the maximum with the growth of deformation, and as experimenting at F = const. is accompanied by the growth of ε, one could expect at F = const. a more complicated dependence of rate on time. This appears in particular in the difference of the dependences for BSK rubber in comparison with SKB and natural rubbers and in the fact that in the case of BSK the process develops with constant speed.

In experiments on SKB rubbers[15] at $F =$ const., in contrast to experiments at $\varepsilon =$ const., the constant of speed of elongation (strain rate) grows continuously with the growth of the initial strain (deformation).† With an increase of deformation from 4 to 70% and from 3 to 84%[22] a decrease of the initial speed of cracking‡ was found for BSK and natural rubbers. In the case of natural rubbers this decrease tallies with the data[23] according to which the maximum speed of crack growth (at $\varepsilon =$ const.) is observed at deformations of 5–7%, and with an increase of strain it decreases (as will be shown later on, in testing at $F =$ const. the value ε_{cr} shifted[23] towards small deformations when compared with ε_{cr} at $\varepsilon =$ const.). Veith's[22] data confirm in fact also that the dependence of the speed of cracking on the size of the strain has maxima and minima. In so far as the initial speed increases with the decrease of strain, and at strains near zero the speed must approach zero, the curve $(dS/dt)_{t\to 0} = f(\varepsilon)$ must pass beyond the maximum in regions of small strains. On the other hand, at very large deformations the speed must grow, i.e. in that region the curve must pass beyond the minimum. In spite of this, Veith[22] denies the existence of any critical points on the curve.

In experiments to determine the kinetics of ozone cracking at constant stress[14] three sections appeared on the kinetic curve: at the beginning the process accelerated, then proceeded with a constant speed, and finally accelerated again. The authors[14] connect the first stage with the interaction of O_3 and the double bonds which exist on the surface, the second with the ozonization of the subsurface double bonds, and the subsequent acceleration with the development of the process in the depth of cracks which can be noticed with naked eye. These data can be

† The constant of speed of elongation of the specimen during its cracking is, obviously, not a characteristic factor of this process, as the obtained relation does not agree with the existence at $F =$ const. of a critical deformation of the rubber, during which the speed of the crack propagation passes beyond the maximum.

‡ Under the initial speed of cracking we understand the speed of the change of the initial cross-section of the specimen at dS/dt, calculated at $t \to 0$.

For natural rubber $dS/dt = -kA_0e^{-kt}$, at $t = 0$ correspondingly $dS/dt = -kA_0$. We have to bear in mind that the initial speed is an arbitrary characteristic, since its value depends on the geometrical dimensions of the specimen.

made to conform with the data which have been obtained at constant deformation,[2] if one assumes that the first two stages, which are observed at σ = const. are not recorded at ε = const. but enter the induction period. The last stage at σ = const. can then be identified with the beginning of the transient section at ε = const. The authors[14] bring no data on the dependence of the rate of ozonization on the magnitude of the applied stress.

Braden and Gent,[16] in order to simplify the cracking process investigated in their work the principles underlying the growth of the cut on the rubber sample at F = const. under the influence of ozone. Only the root of the cut was left accessible to ozone, the whole remaining surface of the sample was covered with silicone grease and in this way isolated from the influence of ozone. Due to the sharp localization of the process and the restriction of the surface and volume of failure the statistic character of the process is in this case not displayed to such an extent as with ozone cracking in normal conditions. The following conclusions[16] were drawn in the paper:

(1) A certain minimum load F_{cr} is necessary for the growth of the cut in an ozone atmosphere on a specimen which has been covered with silicone grease. The size of F_{cr} grows with decreasing depth of the cut, so that the stress at the tip of the cut σ_t which depends on the depth of the cut and on the effective radius, remains constant at the tip with 3·2 kg/cm². The strain corresponding to the smallest values of F_{cr}, measured on specimens with different surfaces, is 1%.†

The critical value of the stress σ_{cr} does not depend on the chemical structure of the polymer and the flexibility of its chains. The cause for this is that the formation of a new surface (during the growth of the cut) is connected with an increase of the free energy of the system and, consequently, as the authors assume, can take place only with a certain size of the specific elastic energy $W = E\varepsilon^2/2$, which was accumulated during the deformation and amounted, in the investigated cases, to $8{\cdot}2\times10^4$ erg/cm³. To these values of the specific elastic energy and critical load F_{cr} corre-

† The calculation of σ was carried out with a formula[24] derived from the theory of elasticity

$$\sigma_t = 2\sigma_r{}^{-1/2}l^{1/2} \qquad (12.1)$$

where σ is the applied stress, l is the depth of the cut, r is the effective radius at the tip of the crack and is smaller than l.

sponds a certain value of the characteristic energy $T_{ch.e.}$ which is absolutely necessary for the growth of the cut. Whilst in the absence of ozone the basic part of the energy $T_{ch.e.}$ is irreversibly dissipated, the whole energy passes in the presence of ozone into the energy of the two surfaces which are again being formed by the fracture of the ozonized molecules, and is,[11, 16] for BSK and natural rubbers, approximately 100 erg/cm^2.

The rate of growth of the cut depends neither on the size of the applied stress (if it is higher than the critical one), nor on the thickness of the specimen.

(3) If an incised specimen is exposed to ozone, and whilst a surface strip in front of the cut remains not covered with silicone grease, the rate of growth of the cut increases with the growth of the applied stress and with the decrease of the thickness of the specimen. These principles can be explained by the presence of a stressed region in front of the cut in which the stresses are sufficient for the development of ozone cracks. The greater the stress and the thinner the specimen, the more does this region spread over a larger surface, since not only the "aperture" of the cut is subjected to ozone attack but also the surface in front of it. Connected with this is also the influence of the stress and the thickness of the sample on the rate of growth of the cut on a surface which is exposed to the attack by ozone, and the absence of such an influence in the case of a protected surface. These data tally with those on the increase of the speed of crack propagation at ε = const. with a decrease of sample thickness.[2]

The statistic character of the flaw distribution on the rubber surface and its heterogeneity leads to a growth of the critical stress σ_{cr} with the reduction of the surface area exposed to ozone (experiments were carried out on samples without cuts, in the middle of which remained a circle of varying diameter which was not covered with silicone grease). σ_{cr} grows also with a change of the character of the surface (decrease of its imperfections). For instance, under the attack of ozone on the edge of the specimen $\sigma_{cr} = 0{\cdot}45$–$0{\cdot}9$ kg/cm^2, under ozonization of the flat surface $\sigma_{cr} = 1$ kg/cm^2 and under the attack of ozone on the flat specimen which was vulcanized in contact with a smooth glass surface, $\sigma_{cr} > 2{\cdot}5$ kg/cm^2, i.e. it approaches the maximum value of σ_{cr}.

The calculation of the depth of the defects on a flat sample surface, as indicated in eqn. (12.1) and starting from experimental values of σ_{cr}, leads to reasonable values of an order of $0{\cdot}5\text{–}2\times\times10^{-3}$ cm.

Some of the conclusions in these papers[16] are unclear and provoke objections.

To begin with, the assumption is incorrect that the growth of the crack or cut occurs only when the elastic energy which has been stored during the stretching, is sufficient for the formation of a new surface.† In fact the growth of the crack under the effect of ozone must be considered together with the dissipation of elastic energy and also the energy contribution of the chemical reaction of ozone with the double bonds of the rubber.

The calculation shows that during the reaction of a mole of the double bonds with ozone approximately 70 kcal/mol are liberated with the formation of ozonide. If the ozonization proceeds as shown by the equation

$$>C{=}C< + O_3 \rightarrow >\underset{|}{C}\!\!-\!\!-\!\!-\!\!\underset{|}{C}< \quad (\text{C–C bridged by } O{-}O{-}O)$$

38 kcal/mol energy are consumed on the transition of the C = C bond into C–C, and 45 kcal/mol on the breaking of the O–O bond in the ozone. The formation energy of two C–O bonds (15 kcal/mol) is liberated. The gain of energy is‡ approximately 70 kcal/mol. The greatest amount of liberated energy may be estimated by basing one's calculations on the number of double bonds per 1 cm² of rubber surface (approximately 10^{15} bonds).

Therefore, during the reaction of ozone with the double bonds on 1 cm² of surface with the formation of ozonides is liberated:

$$70{,}000\cdot\frac{10^{-23}}{6{\cdot}02}\cdot10^{15}\approx10^{4}\ \text{cal}=10^{-4}\cdot4\cdot10^{7}\ \text{erg}=4000\ \text{erg}.$$

The spontaneous decomposition of the ozonides which occurs continuously, is accompanied by an additional liberation of energy.

† This view of Braden and Gent[16] coincides with that of Buckley and Robison[14] who automatically transferred Rivlin's and Thomas's concepts about the phenomenon of tearing (see p. 240) to the phenomenon of corrosion cracking.

‡ If the liberated energy is, calculated whilst taking into account the formation of a stable ozonide $>C{-}O{-}C<$ (with the two C atoms also bonded through O–O), the energy gain is even greater —approximately 150 kcal/mol.

At the same time during the deformation, for instance by 5%, of a rubber with the modulus $E = 10$ kg/cm², energy is stored:

$$W = \frac{E}{2}\varepsilon^2 = 5\varepsilon^2 \approx 0{\cdot}01 \text{ cal cm/cm}^3$$

but on the surface† monomolecular layer (thickness approx. 10 Å) with an area of 1 cm²:

$$W = 0{\cdot}01\times10^{-7} \text{ kg cm} = 0{\cdot}981\times10^{-3} \text{ dyne cm} = 10^{-3} \text{ erg.}$$

This is also the greatest amount of elastic energy which is capable of turning into surface energy. As a matter of fact only the chemical energy which is liberated at the tips of the cracks, may be utilized for the crack propagation. As electron microscope data[11] show, microcracks develop under a strain of 5%, while on a surface of approximately 20 μ^2 approximately eighty cracks of a mean length of about 57 μ can be counted. Assuming[25] that three double bonds of polyisoprene exist on a length of 10 Å, we obtain that, on an area of 1 cm², $\dfrac{57\times3\times10^3}{2\times10^{-7}} \approx 8\times10^{11}$ double bonds will react with ozone with the formation of cracks, i.e. an energy of approx. 30 erg is liberated (greater by 4 orders of magnitude than the stored elastic energy)!

The amount of energy which is liberated during the reaction of ozone with rubber can also be estimated in a different way. We assume that the flat surface of the sample increases in proportion to the strain. Consequently, with a strain of 5% the growth of the surface per 1 cm² on one side of the specimen will be 0·05 cm². The cracks will propagate in the specimen until the tensile stresses on the surface layer disappear, i.e. until the projected area of cracks on the plane of the surface will be 0·05 cm². In this case the minimum size of the newly formed surface which must react with ozone with the formation of cracks, will be 0·05 cm². If one assumes that the free surface energy of the rubber is comparable to the surface energy of paraffin, which is approx. 50 erg/cm², it is easy to calculate that for the formation of a new surface $50\times0{\cdot}05 = 2{\cdot}5$ erg is needed.

The chemical reaction of ozone with the double bonds which are situated on the surface of 0·05 cm² (approx. 5×10^{13} bonds), gives $\dfrac{4\times10^3\cdot5\times10^{13}}{10^{15}} = 200$ erg, whilst the accumulated elastic energy on the surface layer is only 10^{-3} erg.

† The calculation must be carried out for the surface layer, as the interaction with the ozone takes place on the surface, and not in the volume.

As we can see, the amount of energy which is stored during the deformation of the specimen, even without taking into account its partial dissipation, is incomparably smaller than the energy which is liberated during the reaction of ozone with the polymer. In regions of small strains (20% and less) this quantity is quite insufficient for the compensation of the free energy of the new surface which is formed during the cracking, even allowing for the presence of local stress concentrations. The authors[16] ignore not only the quantitative consequences of the effect of ozone but interpret the observed phenomenon also qualitatively in such a way as if the ozone took no part whatsoever in the failure of the specimen. In fact, the authors consider if the value $T_{\text{ch.e.}}$ is approximately equal to twice the free surface energy of the rubber and does virtually not depend on the ozone concentration, on the temperature and the chemical nature of the polymer, that this means one can disregard the action of ozone. This agrees neither with the generally accepted explanation of ozone cracking at small strains, nor with the quantitative values of the activation energy (2–3 kcal/mol) of the growth of ozone cracks in SKS-30 and natural rubbers which agree in their order of magnitude with the activation energy of the reaction of ozone with olefines. The analysis of the numerical data shows that the values of the critical stress which were obtained by the authors[12] by two different methods (i.e. σ'_{cr} and σ''_{cr}), are sometimes qualitatively different for the same rubber at different temperatures. Thus σ'_{cr} increases by a factor of two for butyl rubber during the transition from 20°C to 50°C, whereas σ''_{cr} decreases by a factor of 1·5. This makes it necessary to approach the obtained data with caution.

It is also unclear why the rate of the growth of the cut is not a function of stress, as the time dependence of the strength of polymers must in this case unavoidably show up. Evidently, allowing for the great error in the determination of the rate of growth of the cut (±20%), it is more accurate to consider that the authors[16] failed to discover this dependence. Nevertheless, Braden's and Gent's papers are of definite interest. The method which they employ permits to establish in a simpler way the principles of the development of cracks, and besides, it has a practical value as it reproduces to a considerable degree the ozone cracking of rubbers which contain wax-like substances. The possibility of a methodical mistake is not excluded in the paper which may be due to the swelling of the surface layer of the rubber in the silicone grease, and the creation of a compressive stress which hinders the cracking up to the reaching of a certain tensile stress which exceeds the compressive one.

12.4. Critical Deformation (Strain)

The existence of the so-called "critical deformation" is most interesting and specific for corrosion cracking of visco-elastic materials. In investigations into the growth of ozone cracks it has been qualitatively established in a number of papers that for rubbers on the base of SKB, BSK, SKN and natural rubbers there exists such a region of deformation (10–50%) in which the extent of cracking is greater than under smaller or greater strains.[12, 26–32] This region of deformation has been called the critical deformation. However, in view of the absence of a quantitative criterion of cracking and also of the independent observations which indicate a dependence of the magnitude of the critical deformation on the composition of the rubber[29] and on the time during which the specimens are stressed,[29] the opinion has been expressed in the literature that the critical deformation is not a constant value,[12] that it is vague,[33] and that in rubbers which are not resistant to ozone cracking it does not exist at all.[15, 22, 27, 34, 35] For rubbers which are resistant to ozone cracking, such as Thiokol,[12] Neoprene,[12, 31, 34, 36] butyl rubber,[12, 34, 36] a continuous increase of the rate of cracking, and proportional to the growth of strain, has been noted in all papers, as well as an absence of a region of critical deformation. On the basis of this it has even been assumed that a different mechanism of ozone cracking is present in Neoprene than, for instance, in natural rubber.[34]

Before passing to the presentation of the quantitative results which have been obtained in the latest experiments in this field, it should be noted that there do exist some objective data on the existence of a critical deformation. For instance, when measuring (over equal lengths of time) the depth of the cracks which, under the effect of ozone, form with different deformations,[37] it was found that the smaller the strain the greater was the depth of the cracks. However, as there is no cracking at a deformation of zero, it is impossible to extrapolate the obtained curve to zero strain and, consequently, the curve of dependence of the depth of the cracks on the strain must have a maximum at some critical value of elongation.

Another quantitative confirmation of the existence of a region of critical deformation is the presence of a maximum on the curve describing the loss of strength of specimens which have been ozonized at various deformations. This maximum lies in the region of a 20% deformation; recently data have been obtained which show that at the same 20% strain the greatest decrease of the modulus is observed when ozonized natural rubber samples are flexed.[38]

The investigation of corrosion cracking of rubber employing objective characteristics (rate of crack propagation and durability) made it possible to formulate clearly the conception of "critical deformation" (ε_{cr}) itself, and to establish in which cases it arises and which laws govern its change under the influence of various factors.[6, 23, 39]

Under critical deformation we shall understand the region of deformation in which the maximum rate of crack propagation and the minimum durability τ is observed whenever the time to failure is determined by the duration of the growth of the visible cracks. The time to failure τ is the sum of the time τ_i before the appearance of the cracks and the time $\tau_{v.c.}$ of the development of visible cracks:

$$\tau = \tau_i + \tau_{r.c.}$$

If $\tau_i \gg \tau_{v.c.}$, i.e. $\tau \approx \tau_i$ and $\tau_i/\tau \approx 1$, the relation $\tau = f(\varepsilon)$ must be described by a monotonous curve, as well as the relation $\tau_i = f(\varepsilon)$. If, however, $\tau_i < \tau_{v.c.}$ i.e. $\tau \approx \tau_{v.c.}$ and $\tau_i/\tau < 1$, the relation $\tau = f(\varepsilon)$ will be described by a curve with a minimum in the region of critical deformation (see Fig. 173b).

Under small strains and before failure $\tau = B\sigma^{-b}$; the relation $\tau = f(\sigma)$ can, therefore, be described in a wide range of stresses by a curve with two extremes (Fig. 174).

For samples of the same thickness the relation τ_i/τ does practically not depend on the ozone concentration (Table 20), on the quantity of filler (Table 21), on the plasticizer (Table 22) and temperature; however, it increases sharply with the introduction of waxes and antioxidants† (Table 23). With increasing thickness of the specimen, τ in contrast to τ_i grows, and the relation τ_i/τ can change considerably with the change of this parameter. For instance, in experiments on wax containing rubber of thicknesses

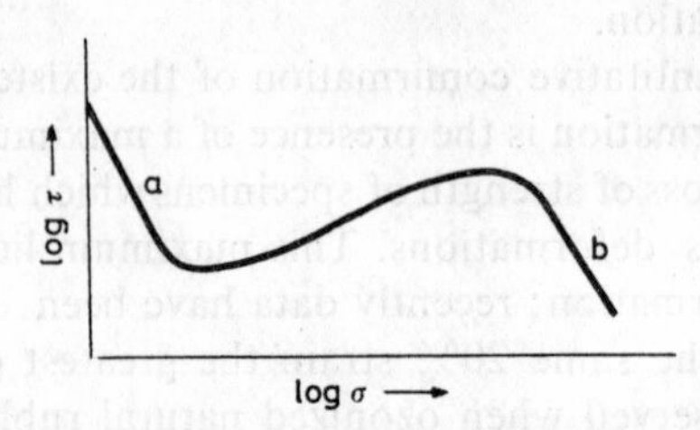

FIG. 174. Diagram of dependence of durability on stress in a wide range of stresses (a and b: straight line sections).

from 0·3 to 0·5 mm, a monotonous relation $\tau = f(\varepsilon)$ is obtained; however, an increase of thickness of the rubber sample to 2–3 mm leads to the appearance of a relation $\tau = f(\varepsilon)$ (Fig. 175) which contains a minimum.

TABLE 20

Relation τ_i/τ for Unfilled Vulcanized SKS-30 Rubbers at Different Ozone Concentrations

Concentration O_3 (%)	τ_i/τ	Concentration O_3 (%)	τ_i/τ
0·0012	0·14	0·0021	0·18
0·0014	0·10	0·0025	0·10
0·0016	0·09	0·0037	0·10
0·0017	0·10	0·0046	0·12
0·0019	0·15	0·0100	0·11

With a suitable choice of thickness of the specimen (in the investigated cases 0·3–0·5 mm) one can always create conditions when $\tau_i/\tau \ll 1$; in this case the change of τ with the deformation

† The introduction of Neozone D into Nairit rubbers which leads to a decrease of the relation τ_i/τ, is connected with a different protection mechanism of these rubbers from ozone as compared with the effect of other antioxidants.

TABLE 21

Relation τ_i/τ during Ozone Cracking for SKS-30 Rubbers Containing Different Amounts of Carbon Black

Deformation (%)	Channel black (g on 100 g rubber)			
	0	30	60	90
10	0·55	0·30	0·37	0·36
100	0·25	0·13	0·13	0·10

TABLE 22

Relation τ_i/τ during Ozone Cracking for Nairit Rubbers Containing Different Proportions of DBP

Deformation (%)	DBP (g on 100 g rubber)					
	0	5	10	20	30	40
40	0·30	0·30	0·28	0·27	0·36	0·34
75	0·21	0·20	0·22	0·21	0·20	0·25

TABLE 23

Value of τ_i/τ for Rubbers Containing Protective Substances

Rubber	Deformation (%)		
	10	20	100
Natural rubber (30 g carbon black on 100 g rubber)			
without antioxidant	0·13	0·22	0·05
5 g ceresin (ozocerite)	0·91	0·78	0·25
SKS–30 (30 g carbon black on 100 g rubber			
without antioxidant	0·30	0·33	0·13
2 g Topanol	0·70	0·67	0·28
5 g Topanol	0·98	0·97	0·73
Nairit			
without antioxidant	–	0·50	0·15
2 g Neozone D	–	0·15	0·04

will be approximately inverse to the change of rate of crack propagation on the steady section. This can be seen clearly on Fig. 173, where the greatest rate of crack propagation coincides with the smallest values of τ. One can, therefore, by maintaining the

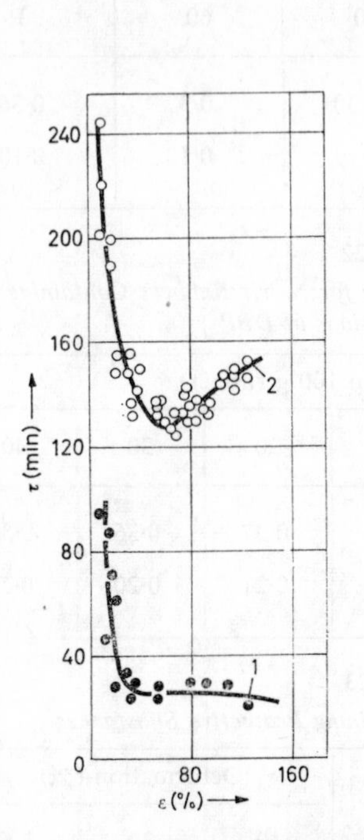

FIG. 175. Influence of thickness of SKS-30 samples with 30 g channel black and 5 g ceresine per 100 g rubber on the dependence of the durability on strain during ozone cracking (ozone concentr. 0·0032%). 1. Sample thickness 0·3 mm. 2. The same, 2 mm.

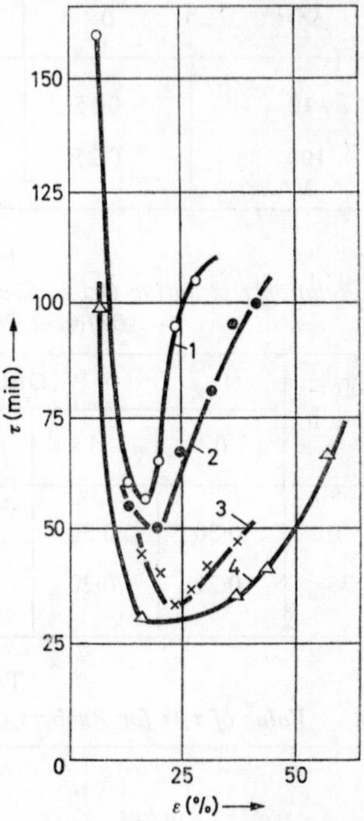

FIG. 176. Dependence of durability on strain of different rubbers under the attack of ozone (test at ε = const.). 1. SKS-10 (ozone concentr. 0·001%). 2. SKB (ozone concentr. 0·0007%). 3. SKS-50 (ozone concentr. 0·0016%). 4. SKN-18 (ozone concentr. 0·003%).

condition $\tau_i/\tau \ll 1$, observe and investigate the principles which relate to ε_{cr} by using a simple objective factor, namely the time to failure and not the rate of crack growth; for the determination of the latter a special device would be necessary.

The critical deformation has been discovered on all investigated rubbers which were non-resistant to ozone (Fig. 176), on films of chlorinated natural rubber,[40] and also on vulcanized SKS-30-1 under the attack by hydrochloric acid and on Nairit rubber under the attack by nitric acid (Figs. 177 and 178). No doubt whatsoever can be entertained at present concerning the existence of a region of critical deformation. How can this phenomenon be explained? Attempts to explain it by the rise of the reaction capability of the macromolecules of rubbers in the area of ε_{cr}[41] or by the change of the gas permeability of the rubber during deformation[42] proved to be unsuccessful (see pp. 384–5). The most widely accepted explanation is that by Newton[12] who established a definite relation of the maximum crack dimensions and their number to the size of the strain. Experiments show that in natu-

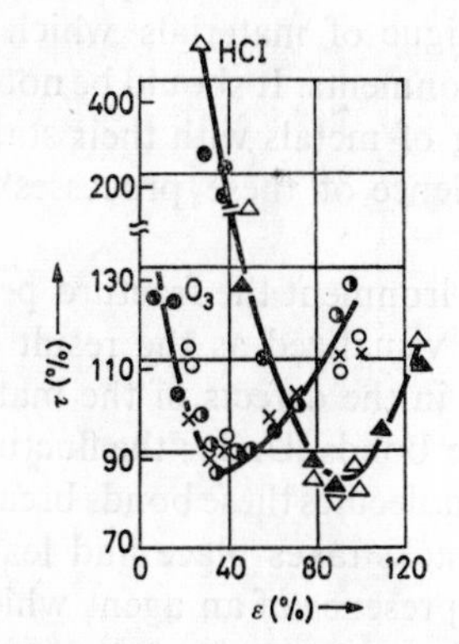

FIG. 177. Dependence of durability on strain of SKS-30 rubbers under the attack of O_3 and HCl (durability expressed in % of the time to fracture of rubber stretched by 70%). Different points on one curve: data of parallel tests.

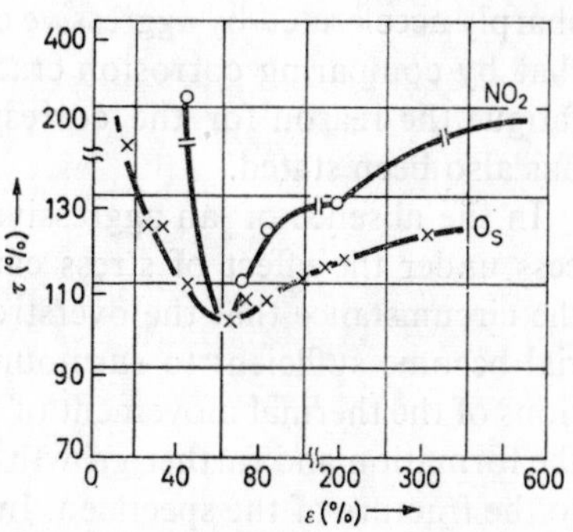

FIG. 178. Dependence of durability on strain of Nairit rubbers under the attack of O_3 and NO_2 (durability expressed in % of the time to fracture of rubber stretched by 60%).

ral rubber with a strain of less than 15–25% the number of cracks increases comparatively slowly. Here, the cracks which are relatively far from each other do not mutually influence the size of the stress concentration at their mouths, and the rate of their growth increases all the time in proportion to the increase of the strain. Under great deformations the number of the forming cracks is great but their dimensions are small. The cracks are placed very near to each other, so that during the growth of one of them a decrease of the stress concentration in the tips of the adjoining cracks takes place, and the cracking slows down. In the interstitial regions (at ε_{cr}) the dimensions of the cracks are already sufficiently large, and their number is not too great to inhibit their growth by their mutual influence; as a result the failure proceeds with maximum speed. In this way Newton established only the predetermined correlation between the size of the deformation and the maximum measurements of the cracks, without a deeper explanation of the essence of the phenomenon.

Experimental data which confirm the numerous features of similarity between static fatigue and corrosion failure permit to treat it as a special case of static fatigue of materials which is sharply accelerated by aggressive environments. It should be noted that by comparing corrosion cracking of metals with their static fatigue the reason for the correspondence of these processes[43] has also been stated.

In the absence of an aggressive environment the fracture process under the effect of stress can be visualized as the result of the circumstance that the overstresses in the defects of the material become sufficient to surmount the bonds. Under the fluctuations of the thermal movement of the molecules these bonds break, the formation and further growth of cracks takes place and leads to the fracture of the specimen. In the presence of an agent which reacts chemically with the polymer, the process of crack propagation may take place under considerably smaller stresses, as the bonds break in the defects under the influence of this agent. In both cases the failure process is accelerated by increasing stress.

The molecular orientation may impede the growth of both static fatigue[44, 45] and corrosion fatigue cracks. It has to be borne in mind that in this case we are not talking about the orientation of the chain molecules in the mass of the specimen but of their additional orientation at the roots of the cracks, where the deformation is considerably greater than the average.[46, 47] For instance, in filled SKB rubbers the maximum additional orientation at the fracture[46] exceeds the average orientation by a factor of

1·5–2, and in nitrile rubbers approximately by a factor of 3. With decreasing depth of the cut[47] the additional orientation decreases. The appearance of ε_{cr} at such comparatively small average strains as 10–20% is explained just by the presence of the additional orientation at the roots of the cracks. With the development of embryonic microscopic cracks in which the overstresses must be smaller than in the macroscopic ones, the effect of strengthening under small and medium strains is lacking. Consequently, the time τ_i until cracks do appear changes monotonously with an increase of ε, i.e. there is no ε_{cr}.

The process of corrosion cracking consists therefore in alternating stages of a chemical fracture of the molecules caused by their reaction with the aggressive medium and the opening of cracks under the influence of stress. The accepted connection between static fatigue and corrosion cracking makes it possible to explain the decrease of the rate of crack propagation with an increase of the strain above ε_{cr} by the development of the molecular orientation at the tips of the cracks.†

The change of the number and dimensions of cracks with growing deformation, although it is not the cause of the presence of a critical deformation, makes it easier to demonstrate its existence. In fact, in regions of small strains, individual cracks which do not influence each other develop with a greater coefficient of stress concentration than in regions of great strains. In the latter case the mutual influence of the cracks is very great due to their great number, and leads to a decrease of the overstresses at the edges of the cracks.‡ In essence the explanation for ε_{cr} from Newton's point of view leads also to the fact that while the cracks develop individually (in the region below ε_{cr}), an increase of strain leads to an increase of the rate of their growth. However, when (in the region after ε_{cr}) the cracks begin to influence each

† The tendency to corrosion cracking of metal specimens which are cut out parallel to the direction in which the material had been stretched is smaller than in specimens which are cut out across it. In particular steel samples which are bent in the direction of the rolling process have a smaller tendency to corrosion cracking in $NaNO_3$ and $CaCl_2$ solutions than bent samples which are cut out in the transversal direction.[48]

‡ The decrease of the overstresses in the depth of the cuts with an increase of their number has been observed by Patrikeev and Mel'nikov[18] when they investigated the tearing process.

other, their growth slows down, and the strain increases. From this point of view the absolute size of the stress at the tips of the cracks at ε_{cr} should be greater than at strains which exceed ε_{cr}. However, experiment does not confirm this.

Let us estimate the size of the stresses in natural rubbers at a strain of 5–10% (ε_{cr}) and at a strain of 200% under which τ is greater than under ε_{cr}. For the stress at a deformation of 5–10% to become equal to the nominal stress in a rubber which is stretched to 200%, the coefficient of stress concentration β at ε_{cr} must be 20–40. β does usually not reach such values.† This coefficient must, however, be even considerably greater, since under a strain of 200% there exists also an overstress at the tips of the cracks and, besides, the rate of crack propagation is considerably smaller at 200% than at ε_{cr}. Further, if the position of ε_{cr} would depend only on the number of cracks, ε_{cr} should shift with the introduction of active fillers towards smaller strains; since more cracks form in filled rubbers at all strains their number which corresponds to ε_{cr} must consequently be reached under smaller strains in filled than in unfilled rubber. However, as will be seen later on, with the introduction of active fillers ε_{cr} shifts towards greater deformations.[39]

In order to explain finally what connection exists between the number of cracks and the phenomenon of critical deformation, the following experiment was made.[50] Three types of natural rubber samples were ozonized under different strains. Samples of the first type had only one puncture in the middle, the remainder of the surface was either covered with a film of poly(isobutelene), or was oiled with silicone for the protection from ozone. Samples of the second type had the same number of cracks which had formed during their previous ozonization under the same conditions of deformation. The specimens of the third type (without cracks) were used as reference. If Newton's explanation were correct, the critical deformation ought to be found only on the control samples, and in the two other cases the speed of failure should grow monotonously with the growth of strain. Experiments disprove this. Critical deformations can be observed in all cases; if the data obtained on the relation τ to ε is drawn on

† The study of incised rubber samples by photo-elastic methods has shown[49] that at the apex of the cut β = 2–3, inside the specimen, and β = 6–8 on the edge.

one graph, with $\tau/\tau' \cdot 100$ instead of τ as ordinate (where τ' is the time to break at critical deformation), all points for the specimens of the second and third types in the region below ε_{cr} lie practically on one curve (Fig. 179).

The change of the number of cracks makes itself felt in the rate of their growth; as a consequence a shift of ε_{cr} may arise and a greater or smaller clarity of the phenomenon itself. The change

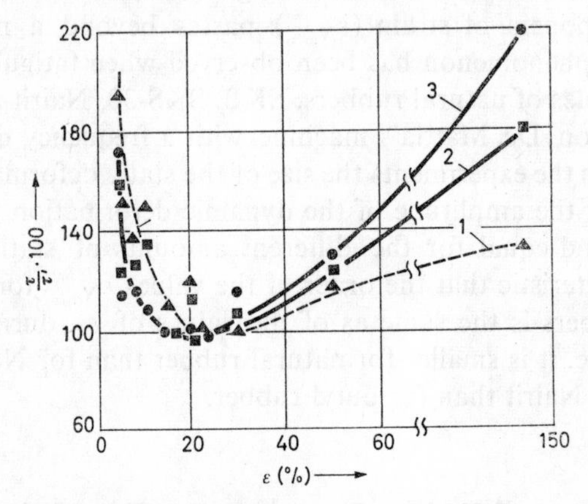

FIG. 179. Dependence of durability on deformation of natural rubbers with a different number of cracks. 1. perforated samples. 2. Samples previously ozonized at $\varepsilon = 50\%$. 3. Control samples. On the axis of the ordinate τ/τ' in %, where τ' is the time to fracture at ε_{cr}.

of speed of crack propagation is especially noticeable under great deformations. This can be seen clearly from the distribution of the points on Fig. 179 at $\varepsilon \approx 150\%$. The greatest relative speed of crack growth (i.e. the smallest durability τ) can be observed in the specimen with a puncture; a lower one in the specimen with a great number of cracks which has previously been ozonized at 50% strain; the smallest in the control sample which has the greatest number of cracks. In agreement with this ε_{cr} is most clearly observed in the control specimens.

These data show that the change of the number of cracks which is noticed with an increase of ε is not a necessary condition of the existence of ε_{cr}. The existence of ε_{cr} is connected with the change of the degree of orientation of the polymer under strain and with its strengthening. This is confirmed by the observation of a similar phenomenon by Patrikeev and Mel'nikov[18] during their study of the tearing of rubber samples with one cut in the absence of ozone. Evidently, the same kind of structural change during deformation lies at the basis of the extrema-possessing dependence of the strength of non-incised rubber specimens on the static component of deformation,[52–54] which has been observed under cyclic deformations in air: the strength at a certain value of the static component of strain ($\varepsilon_{N_{\min}}$) passes beyond a minimum. The same phenomenon has been observed when fatiguing punctured samples of natural rubbers, SKB, SKS-30, Nairit and butyl rubber[23] on De Mattia's machine with a frequency of 250 cycles/min. In the experiments the size of the static deformation was altered but the amplitude of the dynamic deformation remained constant and equal for the different amounts of static strain.† It is characteristic that the order of the values $\varepsilon_{N_{\min}}$ for a number of rubbers is the same as of the values of ε_{cr} during ozone cracking, i.e. it is smaller for natural rubber than for Nairit, and smaller for Nairit than for butyl rubber.

12.5. Influence of Various Factors on the Size of the Critical Deformation (Strain)

The position of the region of critical deformation is determined by two values:

(1) the extent of the increase of stress on the whole and of the overstress at the tips of the cracks with increasing strain, and

(2) the extent of strengthening of the rubber due to the additional orientation with deformation.

Both factors may in their turn greatly depend on the components of the rubber (the latter influences the intermolecular forces, the static modulus, the strength properties of the rubber, and the number of cracks developing), and also on the type of strain and the external environment (for instance, if the environment causes swelling).

The change of the intermolecular forces, as it influences the growth of stress and the strengthening with increasing strain, may

† The amplitude of the dynamic deformation was 120% for natural rubber, 100% for Nairit, 80% for butyl rubber and SKS-30, and 60% for SKB.

lead to a shift of ε_{cr} towards regions of greater or smaller strains. One can roughly estimate the role of this factor by examining on the graph the appearance and displacement of ε_{cr}, as shown on Fig. 180. On this diagram the curve 1 presents the hypothetical dependence of τ on ε under conditions of constancy of the rubber structure in regions of different strains. With the growth of ε the value of τ decreases (due to the simultaneous increase of σ).

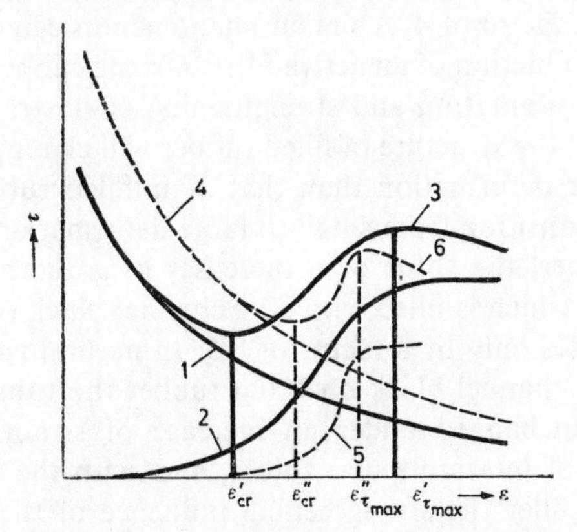

FIG. 180. Appearance and displacement of the critical deformation.

As an orientation and strengthening of the polymer takes place, τ must grow with increasing of strain. The influence of the strengthening is shown by the hypothetical curve 2. By adding the values of τ on curves 1 and 2 with the corresponding deformations, a really observable dependence of τ on ε with a minimum of ε_{cr} and maximum of $\varepsilon_{\tau\,max}$ (curve 3) is obtained. With the increase of the intermolecular forces the resistance to static fatigue will grow, and curve 1 will turn into curve 4; the orientation with the growth of deformation will be impeded, and curve 2 will turn into curve 5. Both these must lead to a shifting of ε_{cr} towards greater strains (curve 3 passes into curve 6). And inversely, with a weakening of the intermolecular forces ε_{cr} must shift towards regions of smaller strains.

The impeding of the orientation during deformation with an increase of the intermolecular forces is related to the existence of a certain structural order in the unstrained specimen. At room temperature this order will be greater in polychloroprene than in natural rubber, as polychloroprene does, and natural rubber does not crystallize under these conditions. It is natural to expect that under tension the structural change due to orientation and crystallization becomes more apparent in natural rubber than in polychloroprene. This can in fact be seen when comparing the values of the coefficient B in regions of small and great strains during ozone cracking (see p. 312). In natural rubber the coefficient B changes by a factor of 10,000; in polychloroprene approximately by a factor of 4. A similar phenomenon can be observed with the introduction of an active filler into the rubber. The active filler causes orientation and strengthening of unstrained rubber, and therefore the structure of filled rubber will change to a lesser degree under deformation than that of unfilled rubber. In fact, during the transition from small to large deformations of unfilled SKS-30 rubber, the value of B increases by a factor of 24, and for a rubber which is filled with 30 g channel black on 100 g rubber it increases only by a factor of 8·5. In natural rubbers which contain 60 g channel black on 100 g rubber the value B remains practically unchanged under an increase of strain. Both with the increase of intermolecular forces, and with the introduction of an active filler the strengthening influence of the orientation will cease with reduced strains, but with an increase of strain (and stress) the durability will decrease. In agreement with this the region of the maximum ($\varepsilon_{\tau\,max}$) on the curve τ–ε will in both cases shift towards smaller strains (see Fig. 180). If two rubbers with a different size of intermolecular forces are compared (for instance, natural rubber and Nairit), ε_{cr} lies usually in the region of a strain of 5–16% for unfilled, non-polar natural rubber, whereas for the polar Nairit it shifts to 65–100%.[23, 39] The introduction of carboxylic groups into non-polar rubbers exposed to ozone cracking leads also to a shift of ε_{cr} towards greater strains.

Type of rubber	ε_{cr} (%)	Literature
SKI	10	–
SKI–1	40	2
SKS–30	20	23
SKS–30–1	40	1

FILLERS

The introduction of active fillers into rubbers has the same effect. In non-polar rubbers (SKB, natural rubber) the interaction is greatly increased, and this impedes their orientation under

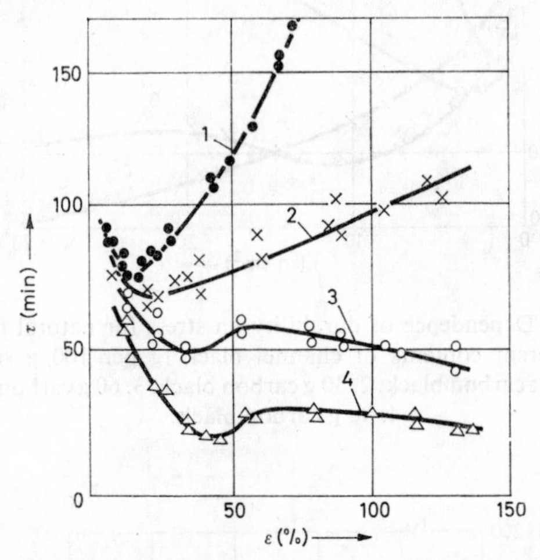

FIG. 181. Dependence of durability on the size of strain for natural rubbers with different contents of channel black (g per 100 g rubber). 1. Without filler. 2. 30 g carbon black. 3. 60 g carbon black. 4. 90 g carbon black. Ozone concentration 0·0036%.

deformation. As a consequence a shift of ε_{cr} towards greater strains (Fig. 181) and greater stresses (Fig. 182) can be observed when carbon black is added to SKB and natural rubbers. In such polar rubbers as SKN-40 and Nairit the intermolecular forces are sufficiently strong and the active filler affects them but little. Therefore ε_{cr} does not shift when SKN-40 rubbers or Nairit are filled with carbon black.

"White carbon black" (powdered Silicagel) affects ε_{cr} less than channel black which is obviously connected with the smaller reactivity of "white carbon black" with natural rubber. Non-reactive fillers which react only slightly with polymers have no influence on ε_{cr} [39] in SKB, SKS-30 and natural rubbers.

As we pointed out, the maximum on the curve τ–ε is usually shifted towards smaller deformations when a filler is added (see,

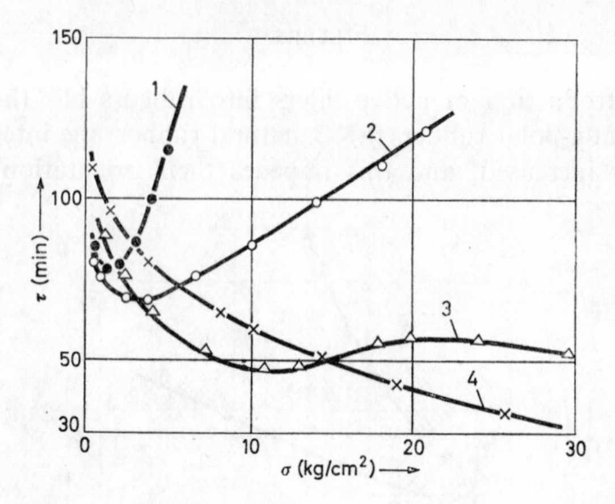

FIG. 182. Dependence of durability on stress for natural rubbers with different contents of channel black (g per 100 g rubber). 1. Without carbon black. 2. 30 g carbon black. 3. 60 g carbon black. 4. 90 g carbon black.

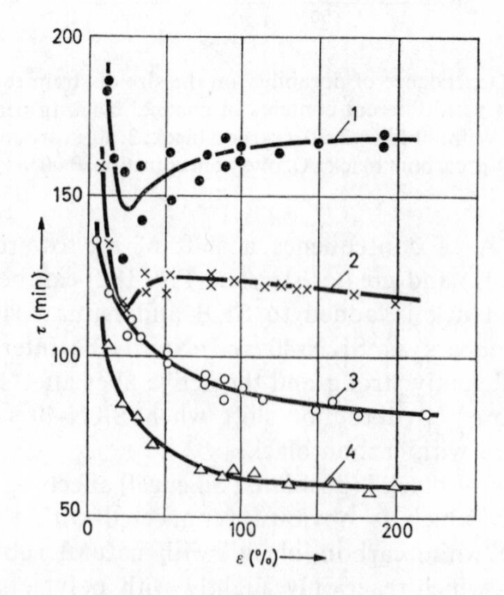

FIG. 183. Dependence of durability on deformation for SKN-40 rubbers with different quantities of powdered silica gel (in g per 100 g rubber). 1. Without filler. 2. 30 g SiO_2. 3. 60 g SiO_2. 4. 90 g SiO_2. Ozone concentration 0·0019%.

for instance, Fig. 181) and the curve itself lies lower than that for an unfilled rubber. Besides, the introduction of a filler causes an increase in the number of cracks, and due to the small difference in their number under different amounts of strain the region of

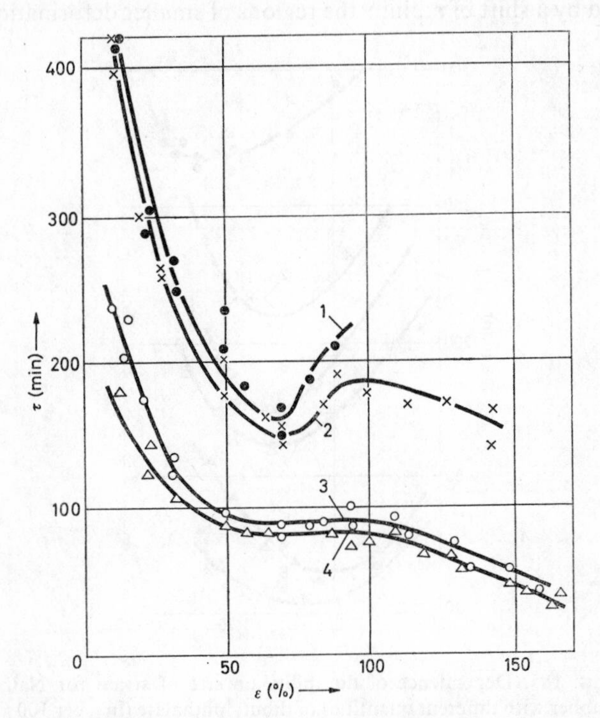

FIG. 184. Dependence of durability on size of strain for Nairit rubber with different quantities of channel black (in g per 100 g rubber). 1. Without filler. 2. 30 g carbon black. 3. 60 g carbon black. 4. 90 g carbon black. Ozone concentration 0·0041%.

critical deformation becomes less distinct. With great proportions of fillers this leads to the disappearance of ε_{cr} and to a degeneration of the curve either into a monotonously decreasing one (Fig. 183), or into a curve with a practically horizontal section in a wide region of strain (Fig. 184). It is possibly due to this phenomenon that indications are found in the literature of a non-existence of critical deformations in SK rubbers, since these are normally used with fillers.

PLASTICIZERS

The introduction of plasticizers, for instance dibutylphthalate which causes a weakening of the intermolecular forces, is accompanied by a shift of ε_{cr} into the regions of smaller deformations[39]

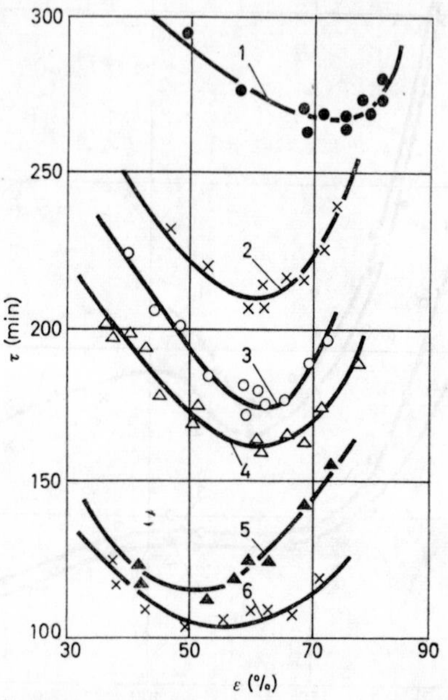

FIG. 185. Dependence of durability on size of strain for Nairit rubber with different quantities of dibutylphthalate (in g per 100 g rubber). 1. Without plasticizer. 2. 5 g DBP. 3. 10 g DBP. 4. 20 g DBP. 5. 30 g DBP. 6. 40 g DBP. Ozone concentration 0·0046%.

(Fig. 185). In SKN-40 and natural rubbers the corresponding shift cannot be found, because there ε_{cr} lies in the region of small deformations.

TEMPERATURE

The weakening of the intermolecular forces with rising temperature leads initially to a considerable shift of ε_{cr} towards smaller deformations;[6] this can be observed during the failure of Nairit (Fig. 186) and SKS-30 rubbers (Fig. 187) in the presence of ozone,

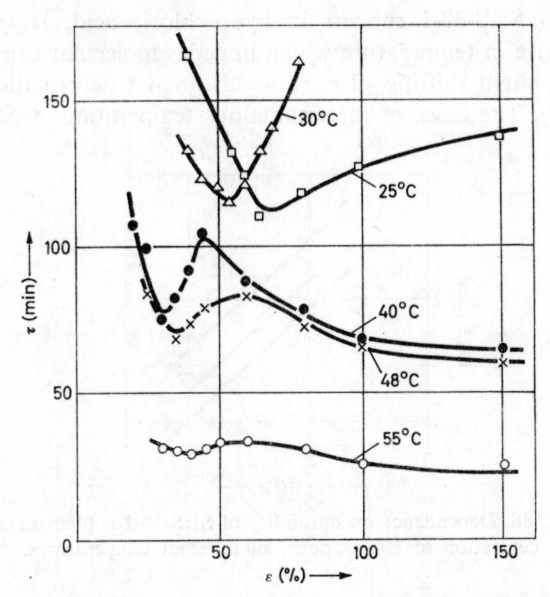

FIG. 186. Dependence of durability on deformation for Nairit rubbers under the attack of ozone at different temperatures.

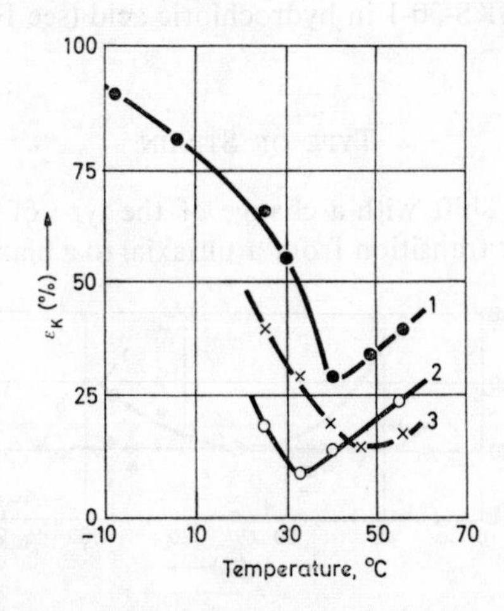

FIG. 187. Dependence of the value of ε_{cr} on temperature during the attack of ozone on different rubbers. 1. Nairit. 2. SKS-30. 3. SKS-30. 50 g of channel black per 100 g rubber.

and also SKS-30-1 rubbers in hydrochloric acid. However, a further rise in temperature which impedes molecular orientation causes a slight shifting of ε_{cr} towards greater deformations (see Fig. 187). The cases of the anomalous temperature dependence

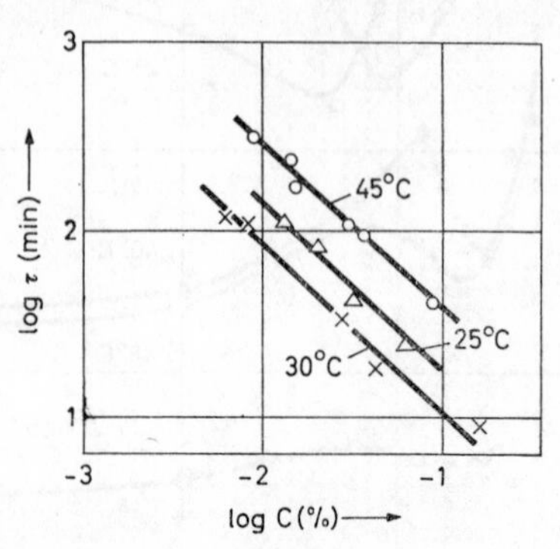

FIG. 188. Dependence on durability of SKS-30-1 rubber on concentration of HCl vapours at different temperatures.

(i.e. of the growth of τ with rising temperature) are a direct consequence of the shift of ε_{cr} with the temperature. This can be seen on the example of Nairit rubbers in ozone (see Fig. 186), and also of vulcanized SKS-30-1 in hydrochloric acid (see Fig. 188).

TYPE OF STRAIN

ε_{cr} may also shift with a change of the type of strain, for instance with the transition from a uniaxial to a biaxial strain. The

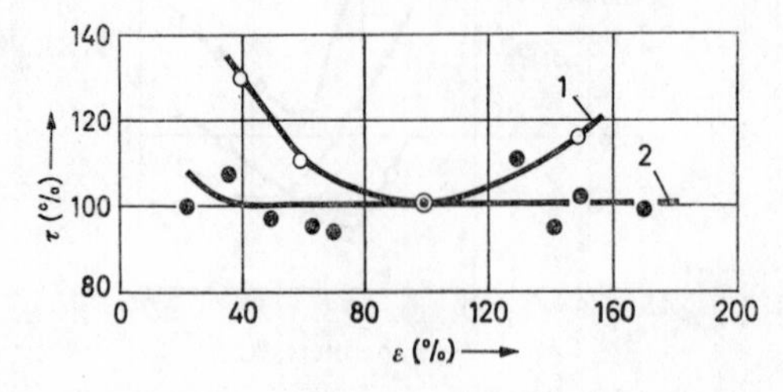

FIG. 189. Dependence of durability on deformation for Nairit rubbers under uniaxial (1) and biaxial (2) tension (τ expressed in % of its value at $\varepsilon = 100\%$).

orientation develops in the latter case to a smaller extent than with uniaxial strain, and the region of ε_{cr} practically disappears (Fig. 189).

Swelling

Due to the decrease of stress concentration at the roots of the cracks which, for instance, occurs during the swelling of the surface layer of the rubber, the strengthening begins to appear under

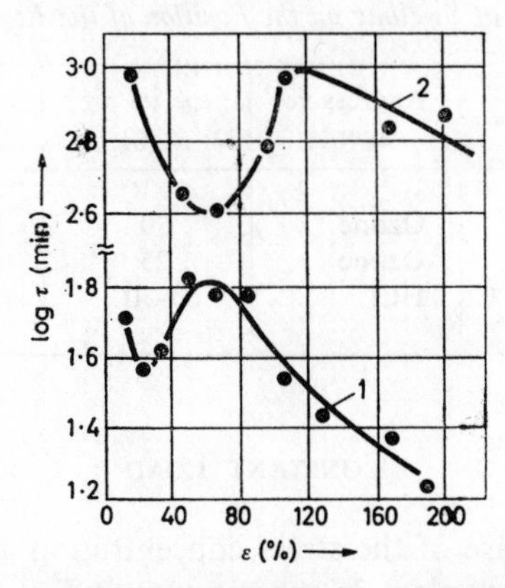

FIG. 190. Dependence of durability on strain during the ozonization of SKB rubbers in water and in air. 1. In air (ozone concentr. 0·01 mmol/mol). 2. In water (ozone concentr. 0·0012 mmol/mol).

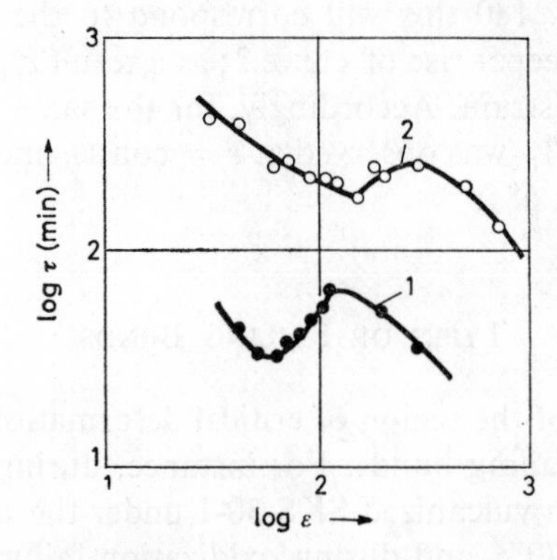

FIG. 191. Swelling and dependence of durability on deformation for vulcanized SKS-30-1 rubbers under the attack by HCl at 25°C. 1. Gaseous HCl (0·09 mmol/mol). 2. Hydrochloric acid ln (18 mmol/mol).

greater deformation, i.e. ε_{cr} shifts considerably towards greater strains. This can be clearly seen when comparing the action of ozone on SKB[55] rubbers in water and in air (Fig. 190), or of the effect of hydrochloric acid and gaseous HCl[6] on vulcanized SKS-30-1 rubber (Fig. 191). The same data have been obtained for SKI rubbers[55] (Table 24).

TABLE 24

Influence of Swelling on the Position of the Region of ε_{cr}

Type of rubber	Aggressive agent	ε_{cr} in air (%)	ε_{cr} in water (%)
SKI	Ozone	30	70
SKB	Ozone	25	70
SKS-30-1	HCl	60–70	150

CONSTANT LOAD

With an increase of the stress concentration at the tips of the cracks which takes place in experiments at $F = \text{const.}$, the equilibrium stress has no time to get established and will be greater than the stress under the same strain, but at $\varepsilon = \text{const.}$ On the diagram on Fig. 180 this will correspond to the steeper fall of curve 1, and a steeper rise of curve 2; as a result ε_{cr} shifts towards a smaller initial strain. Accordingly, for the same natural rubber sample $\varepsilon_{cr} = 20\%$ was observed at $\varepsilon = \text{const.}$, and at $F = \text{const.}$ it shifts to $\varepsilon_{cr} = 6\%$.

TYPES OF FAILING BONDS

The position of the region of critical deformation depends also on the type of failing bonds. For instance, during the failure of the cross-links in vulcanized SKS-30-1 under the action of gaseous HCl, $\varepsilon_{cr} = 70\%$ and during oxidization failure under ozone of the double bonds of the molecular chain $\varepsilon_{cr} \approx 40\%$.[1] The reason for the shifting of ε_{cr} is that the strengthening which is caused by the orientation of the polymer chains at the tips of the cracks during the failure of the cross-links decreases rather more than during the failure of double bonds. Consequently, curve 2 on Fig. 180 lies lower for failure of the cross-links, and ε_{cr} will shift to the right.

It should be noted that the changing type of the cross-links in unsaturated rubbers causes no changes whatsoever in the curve $\tau = f(\varepsilon)$ under the effect of ozone. This has been observed[39] on vulcanized SKB rubbers with the same equilibrium modulus (7·4 kg/cm²) and having mainly monosulphide cross-links (vulcanization with thiuram), with polysulphide (diphenylguanidine with sulphur) and with C—C bonds (thermal vulcanization). The data obtained confirm that ozone cracking is due to the reaction of ozone with the double C=C bonds.

Ozone Protective Substances (Antioxidants)

With the introduction of waxy substances into the rubber and, according to some data[20] of antioxidants, the number of cracks which form decreases sharply. ε_{cr} moves towards greater deformations, since the mutual influence of the cracks will appear under great strains as their great number is accompanied by a reduction of the overstresses. Such a displacement of ε_{cr} has been observed[12] as to be in proportion to the extent of cracking, the rate of crack growth,[20] and the change of τ.[56] In those cases when

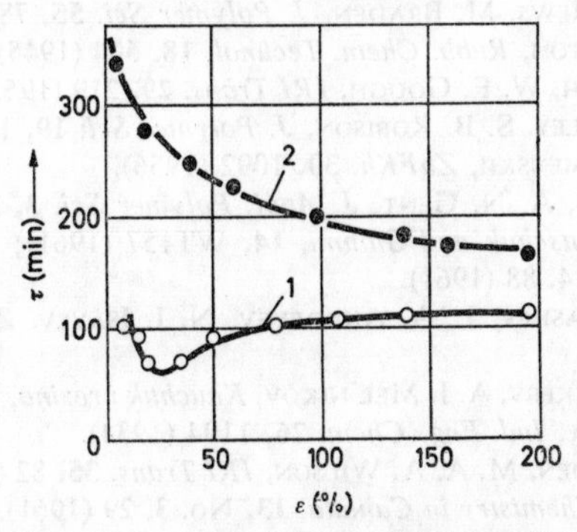

Fig. 192. Dependence of durability on strain under the attack of ozone (0·0017%) on vulcanized SKS-30 rubber, containing 30 g channel black per 100 g rubber. 1. Without antioxidant. 2. 5 g antioxidant UOP-88.

antioxidants are added which basically influence the duration of the induction period but have no real influence on the rate of crack propagation, i.e. when $\tau_i/\tau \rightarrow 1$, the region of ε_{cr} degenerates, and the relation becomes monotonous. This is seen, for instance, when ceresine (5 g on 100 g rubber) or N_1N-dioctyl-n-phenylenediamine is added to SKS-30 rubber (Fig. 192). If the antioxidants both lengthen the induction period and decrease the rate of crack propagation (for instance, 2–3 g ceresine or Topanole on 100 g rubber) ε_{cr} shifts towards greater strains.[56]

Literature

1. Yu. S. Zuyev, A. Z. Borshchevskaya, *DAN SSSR*, **124**, 613 (1959).
2. Yu. S. Zuyev, S. I. Pravednikova, *ZhFKh*, **32**, 1457 (1958); **31**, 2586 (1957).
3. Yu. S. Zuyev, *DAN SSSR*, **74**, 967 (1950).
4. F. N. Kendall, J. Mann, *J. Polymer Sci.* **93**, 503 (1956).
5. Yu. S. Zuyev, A. S. Kuz'minskii, *DAN SSSR*, **89**, 325 (1953).
6. Yu. S. Zuyev, A. Z. Borshchevskaya, S. I. Pravednikova, U. Yue-tsin', *Viyskomol. soyed.* **3**, 164 (1961).
7. A. S. Kuz'minskii, N. N. Lezhnev, Yu. S. Zuyev, *Okisleniye kauchukov i rezin*, ch. VII, Goskhimizdat, 1957.
8. E. R. Erickson, R. A. Bernsten, E. L. Hill, P. Kusy, *Rubb. Chem. Technol.* **32**, 1062 (1959).
9. H. Tucker, *Rubb. Chem. Technol.* **32**, 269 (1959).
10. J. Kishimoto, H. Fujita, *Koll. – Z.* **150**, No. 1, 24 (1957).
11. E. H. Andrews, M. Braden, *J. Polymer Sci.* **55**, 787 (1961).
12. R. G. Newton, *Rubb. Chem. Technol.* **18**, 504 (1945).
13. D. M. Smith, W. E. Gough, *IRI Trans.* **29**, 219 (1953).
14. D. J. Buckley, S. B. Robison, *J. Polymer Sci.* **19**, 145 (1956).
15. N. N. Znamenskii, *ZhFKh*, **30**, 1092 (1956).
16. M. Braden, A. N. Gent, *J. Appl. Polymer Sci.* **3**, 90 (1960); **3**, 100 (1960); *Kautschuk und Gummi*, **14**, WT157 (1961); *Proc. Inst. Rubb. Ind.* **8**, No. 4, 88 (1961).
17. N. D. Tomashev, L. A. Andreyev, N. I. Isayev, *Zav. lab.* **25**, 1200 (1959).
18. G. A. Patrikeev, A. I. Mel'nikov, *Kauchuk i rezina*, No. 12, 12 (1940).
19. W. F. Busse, *Ind. Eng. Chem.* **26**, 1194 (1934).
20. H. A. Vodden, M. A. A. Wilson, *IRI Trans.* **35**, 82 (1959).
21. J. Payne, *Chemistry in Canada*, **13**, No. 3, 29 (1961).
22. A. G. Veith, *Rubb. Chem. Technol.* **32**, 346 (1959).
23. Yu. S. Zuyev, S. I. Pravednikova, *DAN SSSR*, **116**, 813 (1957); *Kauchuk i rezina*, No. 1, 30 (1961).
24. C. E. Inglis, *Trans. Inst. Nav. Archit.* **55**, No. 1, 219 (1913).

25. K. V. Bunn, *Khimiya bol'shikh molekul*, No. 2. Izdatinlit, 1948, p. 137.
26. G. R. Guthbertson, D. D. Dunnom, *Rubb. Chem. Technol* **25**, 878 (1952).
27. G. R. Crabtree, A. R. Kemp, *Rubb. Chem. Technol.* **19**, 712 (1946).
28. L. A. Smirnova, *v sb. "Stareniye kauchukov i rezin i poviysheniye ikh stoikosti"*, Goskhimizdat, 1952, p. 28.
29. E. P. Kearsley, *Rubb. Age*, **27**, 649 (1930).
30. J. M. Ball, R. A. Joumans, A. F. Raussell, *Rubb. Age*, **55**, 481 (1944).
31. E. W. Ford, L. V. Cooper, *India Rubb. World*, **125**, 55 (1951).
32. J. Williams, *Ind. Eng. Chem.* **18**, 367 (1926).
33. C. H. Leigh-Dugmore, *Rubb. Age Synt.* **33**, 398, 444 (1952).
34. A. Hartman, F. Glander, *Kautschuk und Gummi*, **8**, 35 (1955).
35. E. Powell, V. Gough, *Rubb. Chem. Technol.* **19**, 406 (1946).
36. D. C. Thompson, R. H. Baker, *Rubb. Chem. Technol.* **25**, 928 (1952).
37. I. S. Rugg, *Anal. Chem.* **24**, 818 (1952).
38. L. S. Carlson, R. S. Havenhill, *Rubb. World*, **138**, 883 (1958).
39. Yu. S. Zuyev, S. I. Pravednikova, G. V. Kotel'nikova, *Kauchuk i rezina*, No. 11, 15 (1961); No. 3, 21 (1962).
40. G. Salomon, F. van Bloois, *Proceedings of Fourth Rubber Technology Conference*, London, May 1962, Preprint 58.
41. F. Colton, *IRI Trans.* **6**, 165 (1930).
42. Matsuda, Tanaka, *Hikhou gomu kekaisi*, **28**, No. 3, 131 (1955); **25**, No. 7, 208, 214 (1952).
43. K. Matthaes, *Werkstoffe und Korrosion*, **8**, No. 5, 261 (1957).
44. J. A. Sauer, S. S. Hsiao, *India Rubb. World*, **128**, 355 (1953).
45. A. V. Stepanov, *ZhETF*, **19**, No. 11, 973 (1949).
46. V. Ye. Gul', B. A. Dogadkin, Van Man'-sia, *Kauchuk i rezina*, No. 2, 17 (1959).
47. V. Ye. Gul', V. V. Kovriga, A. N. Kamenskii, *DAN SSSR*, **133**, 1364 (1960).
48. R. L. McGlasson, W. D. Greathouse, C. M. Hudgins, *Corrosion*, **16**, No. 11, 113 (1960).
49. A. Angioletti, *Rubb. Chem. Technol.* **29**, 753 (1956); *Khimiya i khimicheskaya tekhnologia polimerov*, No. 4, Izdatinlit, 1957, p. 34.
50. Yu. S. Zuyev, *Viysokomol. soyed.* **5**, 1479 (1963).
51. G. Sh. Izrayelit, *Mekhanicheskiye ispiytaniya reziniy i kauchuka*, *Goskhimizdat*, 1949.
52. J. H. Fielding, *Ind. Eng. Chem.* **35**, 1259 (1943).
53. M. K. Khromov, L. S. Priss, M. M. Rezinkovskii, *v sb. "Fiziko-mekhanicheskiye ispiytaniya Kauchuka i reziniy"*, Goskhimizdat, 1960 (*Trudiy NIIShP*, No. 7).
54. I. Kh. Dillon, *vsb. "Ustalost' viysokopolimerov"*, Goskhimizdat, 1957.
55. Yu. S. Zuyev, S. I. Pravednikova, *Kauchuk i rezina*, No. 6, 26 (1961).
56. Yu. S. Zuyev, S. I. Pravednikova, *Kauchuk i rezina*, No. 1, 30 (1961).

CHAPTER 13

SPECIAL FEATURES OF FAILURE OF RUBBER IN AGGRESSIVE ENVIRONMENTS

13.1. Influence of Concentration of Aggressive Media on the Failure of Rubber

THE investigation of ozone cracking at constant strain shows that a certain relation exists between the intensity of the cracking (which can be defined by the time before the appearance of the first cracks) and the ozone concentration. An analysis of all papers in this field which were known up to 1945 made it possible for Newton[(1)] to draw the conclusion that the product of ozone concentration C and the time τ_i to the appearance of the first cracks is approximately constant for natural rubbers. Later[(2)] on this relation was more precisely determined for the concentration range of 10^{-2}–10^{-4}%:

$$\tau_i C^{n_1} = K_1 \tag{13.1}$$

where the coefficient n_1 takes the values from 0·9 to 1·5 for different rubbers (for natural rubber $n_1 = 1$).

This relation has also been confirmed[(3)] for the concentration ranges 6×10^{-1}–3×10^{-2} and 2×10^{-5}–2×10^{-4}%. Equation (13.1) underlies the accelerated method for the determination of the resistance of rubber to ozone cracking.

With the growth of strain the sensitivity of rubber to a change of ozone concentration decreases somewhat; to a greater extent in SKS-30 rubbers than in Nairit. This shows up in the corresponding value of n_1 which can be explained by the improvement of the conditions of adsorption under tension.

The dependence of the growth rate of the visible cracks in the steady stage (v_r) and of the durability of the rubber (τ) on the ozone concentration is described by equations which are analogous to eqn. (13.1):

$$v_r C^{-n_2} = K_2 \tag{13.2}$$

$$\tau C^{n_3} = K_3 \tag{13.3}$$

The coefficients n_2 and n_3 depend only little on the deformation. In agreement with this the dependence of both the speeds of crack propagation and the durability on ozone concentrations for different strains is expressed in logarithmic coordinates by a bundle of practically parallel straight lines (Fig. 193).

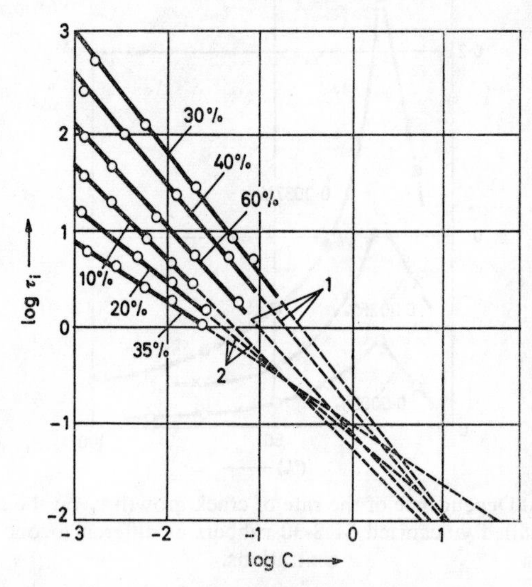

FIG. 193. Dependence of the time to the appearance of cracks τ_i on the ozone concentration C for unfilled vulcanized rubbers. 1. Nairit. 2. SKS-30.

In corrosion cracking the position of ε_{cr} does not depend on the concentration of the aggressive medium. The rate of crack growth, however, grows[4] with increasing ozone concentration under all strains, and the maxima on the curves become more pronounced (Fig. 194).

However, if the dependence of the rate of crack propagation is described in relative coordinates, assuming as 100% the rate of growth at the critical deformation, the curves for different ozone concentrations merge into one curve with a maximum in the region of critical deformation. The opposite can be observed for the relation of the durability to the deformation at different concentrations of the aggressive medium. In this case the durabilities (in relative units) change their dependence on the concentration of the aggressive medium, and have a minimum at ε_{cr}.

The relation $\tau = f(C)$ for rubbers[5] which are non-resistant to acids has the same character in the region of comparatively great acid concentrations as in the case of their cracking under

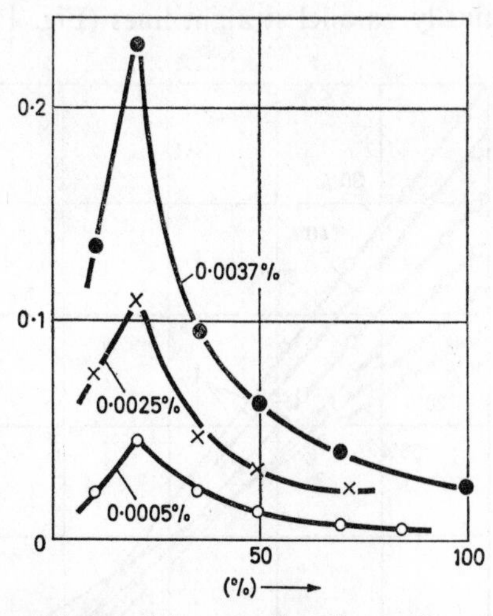

FIG. 194. Dependence of the rate of crack growth v_r on the strain ε for unfilled vulcanized SKS-30 rubbers at different ozone concentrations.

ozone (Fig. 195). In the same way as the number of cracks changes with the concentration of the aggressive agent, the region of critical strain is independent from the concentration of the aggressive substance and this correlates with the independence of ε_{cr} from the number of cracks (if there is a sufficiency of them) as was shown earlier.

Ozone cracking of vulcanized butyl rubber at constant stress was studied by Buckley and Robison.[6] They considered the relative work of the formation of cracks $\Delta W/W_1$ as the criterium of the extent of the action of ozone, where W is the store of energy in the originally stretched specimen, and ΔW is the work during the additional deformation of the specimen caused by its cracking. If, according to Buckley's and Robison's[6] data one determines the time τ' to reach the same values $\Delta W/W'$ for different ozone concentrations (in the range of 0·2–0·03%), then it appears that $\tau'C$ = const., i.e. the same relation can be observed as for the time to the appearance of cracks (at $n = 1$). If the test is carried out at constant load[7] when both strain and stress change, it is found that this product is not constant in the range of ozone concentration of 6×10^{-4}–4×10^{-3}.

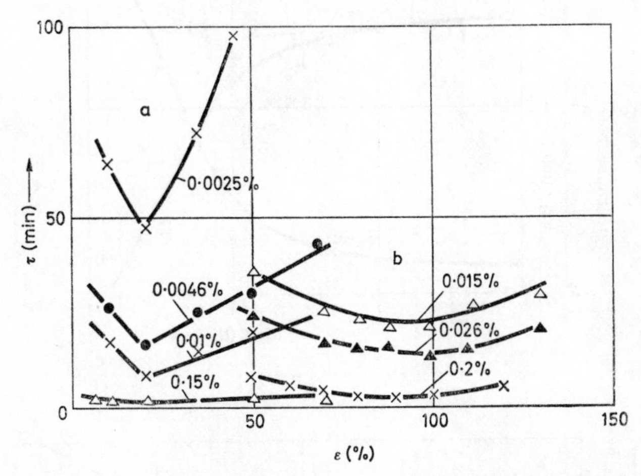

FIG. 195. Dependence of durability on strain at different concentrations of the aggressive agent. (a) Unfilled vulcanized SKS-30, attack by ozone. (b) Unfilled vulcanized SKS-30-1 (MgO) rubber, attack by gaseous HCl.

In Braden's and Gent's[8] tests which were carried out on incised specimens under constant load, a straight line relation of the mean rate $\bar{v}$ of the cut growth on the ozone concentration (in a range of ozone concentration of 0·125–0·005%) was discovered for natural rubbers, and a curvilinear one for BSK and SKN rubbers. A reconsideration of the obtained data showed, however, that they lie on a straight line in the coordinates $\log \bar{v}$–$\log C$.

Considering that corrosion cracking is a peculiar type of static fatigue of rubbers, one might expect that in the absence of an aggressive medium a continuous transition exists of the rate of the failure process (and, consequently, of the durability) to the values connected with the increase of the concentration.

In connection with this investigations were carried out into the dependence of the rate of creep $\dot{\varepsilon}$ and of durability τ on a wide range of concentrations of the aggressive medium, beginning with O, at a constant stress[9] (σ = const.) on a helical type apparatus.[10]

It has been established that the curves $\dot{\varepsilon} = \dot{\varepsilon}(C)$ and $\tau = \tau(C)$ consist of three sections (Fig. 196a):

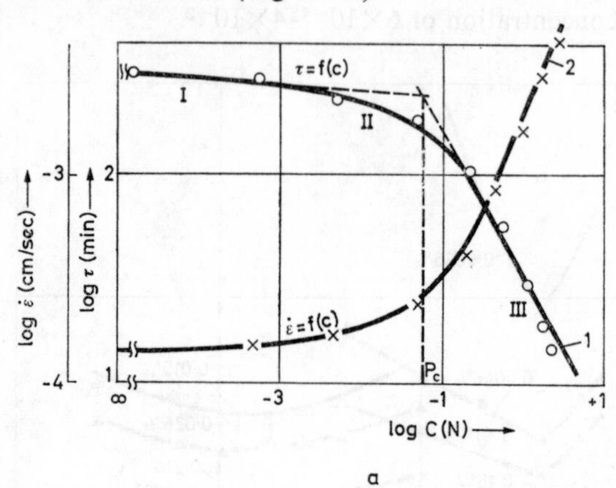

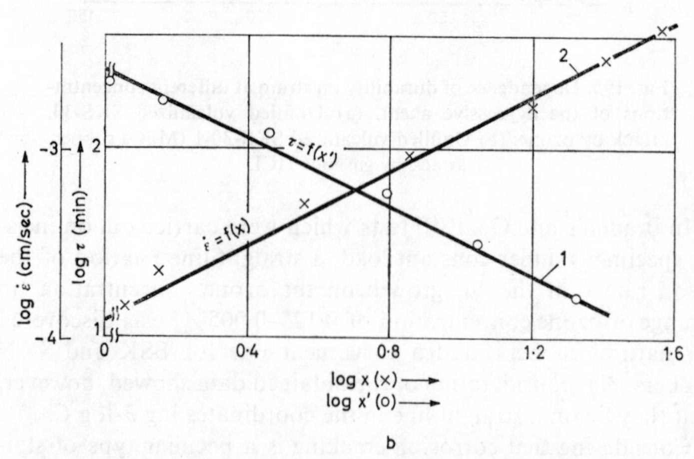

FIG. 196. Dependence of durability and creep of SKS-30-1 rubber on concentration of hydrochloric acid at constant stress (σ = 83 kg/cm^2) and a temperature of 35°C. (a) In coordinates $\log \tau - \log C$ and $\log \dot{\varepsilon} - \log C$; (b) In coordinates $\log \tau - \log x'$ and $\log \dot{\varepsilon} - \log x$.

I. The section on which, beginning with 0 and up to a certain concentration of the aggressive substance, τ and $\dot{\varepsilon}$ are practically constants;

II. The transitional section where the relations $\log \tau$–$\log C$ and $\log \dot{\varepsilon}$–$\log C$ are non-linear;

III. The section where the formulae

$$\tau C^{n_3} = K_3$$

and

$$\dot{\varepsilon}_x = K_x C^k \tag{13.4}$$

are fulfilled.

If the speed of creep $\dot{\varepsilon}$ is written as

$$\dot{\varepsilon} = \dot{\varepsilon}_{st} + \dot{\varepsilon}_x$$

where $\dot{\varepsilon}_{st}$ is the rate of creep during static fatigue under the effect of stress only (this process develops on section I), and $\dot{\varepsilon}_x$ is the rate of creep during corrosion failure under the influence of stress and environment (section III), and if it is assumed that on section III the value $\dot{\varepsilon}_{st}$ can be disregarded, i.e. that $\dot{\varepsilon} = \dot{\varepsilon}_x = K_x C^k$, then

$$\dot{\varepsilon} = \dot{\varepsilon}_{st} + K_x C^k \quad \text{and} \quad \dot{\varepsilon} = \dot{\varepsilon}_{st}\left(1 + C^k \frac{K_x}{\dot{\varepsilon}_{st}}\right).$$

Experiment shows[9] that between $\dot{\varepsilon}$ and τ, which are determined at different concentrations of the aggressive medium, there exists the relation

$$\tau \dot{\varepsilon}^m = \text{const} \tag{13.5}$$

where m is a constant. It is similar to the relation which can be observed during the creep of metals[11] and plastics[12] under different stresses in the absence of an aggressive medium.

As follows from relations (13.3), (13.4) and (13.5) $n_3 = mk$. The relation which has been found is, however, not only observed for concentrations of the environment where the relations (13.3) and (13.4) are fulfilled but in almost all investigated cases, right to the zero concentration of the aggressive agent (Fig. 197), i.e. the value m does not depend on its concentration.

Starting from the additivity of the rates of static fatigue and corrosion failure, the durability can be expressed by the equation

$$\tau = \frac{\tau_{st}}{1 + (\tau_{st}/K_x')C^{mk}} \tag{13.6}$$

where τ_{st} is the time to failure under static fatigue.

In the absence of an aggressive medium ($C = 0$) $\tau = \tau_{st}$. At great values of C

$$\frac{\tau_{st}}{K'_x} C^{mk} \gg 1$$

If the values $\dot{\varepsilon}$ and τ are plotted in coordinates $\log \dot{\varepsilon}$–$\log X$ and $\log \tau$–$\log X$, where

$$X = 1 + \frac{K_x}{\dot{\varepsilon}_{st}} C^k \quad \text{and} \quad X' = 1 + \frac{\tau_{st}}{K'_x} C^{n_2}$$

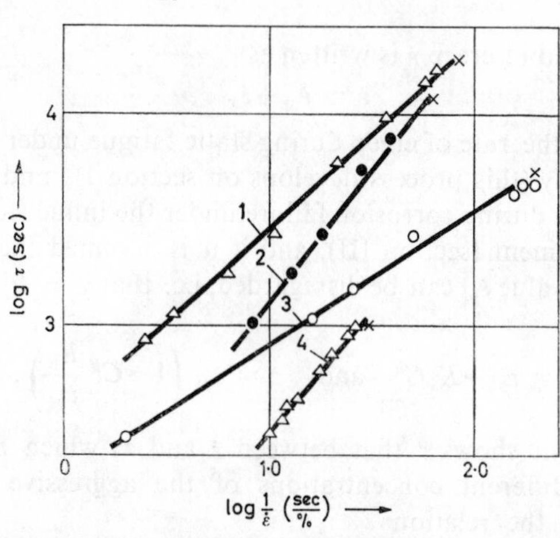

FIG. 197. Connection between durability and creep of rubber at different concentrations of the aggressive agent. 1. SKS-30-1 in hydrochloric acid (according to data of Fig. 196). 2. SKS-30-1 in ozone (in gaseous phase). 3. Butyl rubber in nitric acid. 4. Fluorine rubber in nitric acid.
The crosses mark the values of τ and $1/\dot{\varepsilon}$ in the absence of an aggressive medium.

straight lines are obtained (see Fig. 196b). This relation can be used for a quick determination of τ_{st}, if one measures τ both in the transitional region and in the region of great ozone concentrations.

The studies of corrosion failure of rubber show that the greater the part played by chemical interaction during failure, the smaller the concentrations of the aggressive agent at which it will appear. Now we can follow the influence of the concentration of the aggressive agent on the relative durability $D = \tau_{st}/\tau$ or the relative creep

$E = \dot{\varepsilon}/\dot{\varepsilon}_{st}$ (τ_{st} and $\dot{\varepsilon}_{st}$ are the durability and creep in the absence of an aggressive medium, i.e. in water or in air).† The smallest concentration of aggressive medium at which the speed of failure begins to increase sharply is suitably called "the threshold of concentration" (P_c). It can be defined as the point of intersection of the sections I and III of the curve log τ–log C (see Fig. 196). At $C < P_c$ the values D and E equal 1. At $C > P_c$ these values begin to increase, i.e. P_c is the smallest concentration at which D

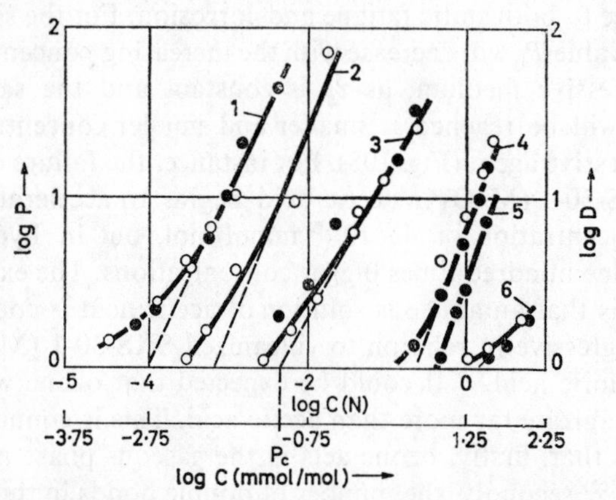

FIG. 198. Dependence of relative durability (black circles) and relative creep (white circles) on the concentration of the aggressive agent at 35°C for different rubbers. 1. SKS-30-1 (ozone, σ = 434 kg/cm²). 2. Nairit (ozone, σ = 434 kg/cm²). 3. SKS-30-1, vulcanized with MgO (acetic acid, σ = 91 kg/cm²). 4. SKS-30-1, vulcanized with MgO (hydrochloric acid, σ = 84 kg/cm²). 5. SKS-30-1 vulcanized with MgO (hydrochloric acid, σ = 15 kg/cm²). 6. SKS-30-1, vulcanized with sulphur (hydrochloric acid, σ = 50 kg/cm²).

and E assume values greater than 1. If the relation of log D to log C is presented graphically (Fig. 198), P_c will be defined in these coordinates as the point of interception of the straight line section of the curve with the abscissa. From eqn. (13.5) it follows that

$$\tau\dot{\varepsilon}^m = \tau_{st}\dot{\varepsilon}_{st}^m,$$

i.e.

$$\frac{\tau}{\tau_{st}} = \left(\frac{\dot{\varepsilon}_{st}}{\dot{\varepsilon}}\right)^m \quad \text{or} \quad \log D = m \log E.$$

† The aggressive effect of water or air is assumed as negligibly small, excepting specially mentioned cases.

In those cases when m differs only little from 1, the values D and E are usually similar, and they may be used equally, as can be seen on Fig. 198. When comparing different polymers a greater value of D corresponds normally, at a given concentration of the aggressive medium, to a smaller P_c.

The value of the threshold of concentration depends on the relation of the "mechanical" to the "chemical-mechanical" (proportional to the chemical) resistance of the rubber, i.e. on the resistance to both static fatigue and corrosion. For the same polymer the value P_c will decrease with the increasing concentration of the aggressive medium, as τ_{st} is constant and the same value $\tau > \tau_{st}$ will be reached at smaller and smaller concentrations of the aggressive agent (Fig. 198). For instance, the failure of vulcanized SKS-30-1 (MgO) in acetic acid begins to accelerate sharply at a concentration of 4×10^{-2} mmol/mol, but in hydrochloric acid at one hundred times bigger concentrations. The explanation for this is that an aqueous solution of acetic acid is considerably more aggressive in relation to vulcanized SKS-30-1 (MgO) than hydrochloric acid.[5] It could be expected that ozone will attack this vulcanizate far more than acetic acid. This is connected with the facts that, firstly, ozone acts in the gaseous phase and not in solution,[2] secondly, the number of double bonds in this polymer which can react with ozone is considerably greater than the number of cross-links which react with acid, and thirdly, the activation energy of the reaction of ozone with the polymer[13] is considerably smaller (2–4 kcal/mol), than that of acid (19–26 kcal/mol). In fact, as can be seen on Fig. 198, under the attack of ozone the process accelerates sharply at concentrations which are one hundred times smaller than the concentrations of acetic acid.

A comparison of the values P_c for different polymers shows that the increase of the chemical resistance and the decrease of the durability lead to an increase of P_c when D decreases, and conversely an opposite change of these parameters causes a decrease of P_c. As an example one may examine the behaviour of two SKS-30-1 rubbers in hydrochloric acid—one of them vulcanized with MgO, and another with sulphur (see Fig. 198). The failure of the rubber which is vulcanized with sulphur, the acid resistance of which is greater than that of the MgO vulcanized one, but the strength of which is smaller, is sharply accelerated at a concentration of the aggressive agent which is ten times greater than that of the stronger but less acid resistant type. If the mechanical strength and chemical resistance change in the same way (for instance, with their simultaneous increase) P_c, depending on their relation, may shift in different directions. It has been found when comparing the relative creep of different rubbers in ozone, that P_c in Nairit rubbers is ten times greater than in SKS-30-1 rubbers (see Fig. 198). This can be explained by the great difference in the chemical resistance between Nairit and SKS-30-1 rubber, whereas the strength properties of Nairit and SKS-30-1 rubbers differ but little.

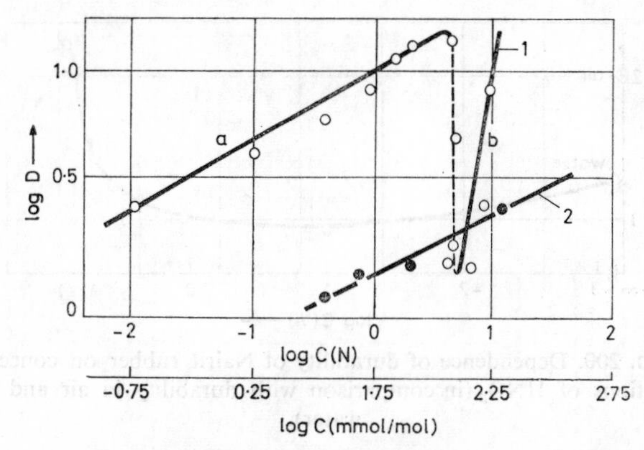

FIG. 199. Dependence of relative durability of rubber D on concentrations of HNO_3 at 60°C and $\sigma = 97$ kg/cm². 1. (a and b) Butyl rubber. 2. Fluorine rubber.

Very interesting results have been obtained when comparing stressed fluorine rubbers of the Kel-F type and butyl rubbers in nitric acid (Fig. 199). It is well known that unstressed fluorine containing polymers are incomparably more acid resistant than butyl rubbers. Correspondingly, for the same failure type and especially in the absence of cracking, $P_c = 4{\cdot}5$ mmol/mol for fluorine rubbers of the Kel-F type, and for butyl rubbers $P_c = 5{\cdot}6 \times 10^{-3}$ mmol/mol, i.e. it is a thousand times smaller. If, however, the HNO_3 concentration is increased to $4n$ and higher, the nature of the failure of butyl rubber changes sharply; its surface oxidizes to a considerable extent which obviously impedes the diffusion of the acid—whereas the nature of the failure of fluorine polymer does not change. This leads to a considerable increase of the durability of the butyl rubber (curve 1b, Fig. 199), in spite of the increase of the HNO_3 concentration. As a result the life (durability) of butyl rubbers exceeds at certain concentrations even the life of fluorine rubbers. With a further increase of the HNO_3 concentration a great number of cracks appear which leads to a great stress concentration and to a more pronounced dependence of the durability on the concentration of the aggressive agent.

The increase of the durability of the rubber with increased acid concentration can be observed even more during the attack

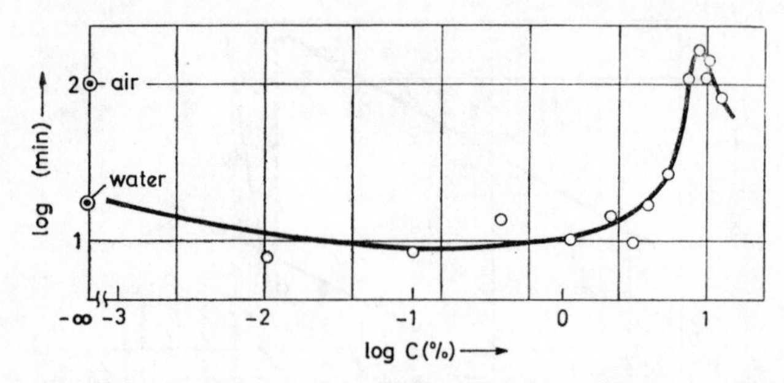

FIG. 200. Dependence of durability of Nairit rubber on concentrations of HNO_3 (in comparison with durability in air and in water).

of HNO_3 on polychloroprene rubber:[14] its durability in concentrated (4–9*n*) nitric acid exceeds its durability in air and in water (Fig. 200), in spite of the fact that in HNO_3 the thickness of the specimen under the chosen experimental conditions decreases approximately by one-half due to the degradation and cracking of the surface layer.

13.2. Influence of Stress on the Failure of Rubber in Aggressive Environments

As has been shown earlier, the influence of stress (or deformation) on the durability of rubber in an aggressive environment is described by a curve which has extremes. In the regions of small strains and of strains near the failure values, τ decreases with the growth of σ (i.e. the "normal" relation of τ to σ is observed which is characteristic for solids), and in the intermediate region τ increases. Accordingly, one might also expect a modified influence of the stress at the threshold of concentration P_c.

For these stress regions where a "normal" dependence of the durability on stress exists, i.e. $\tau = B\sigma^{-b}$, the value of P_c must increase with increasing stress, as the relative role of stress in the failure of rubber grows and D (or E) decreases and tends towards 1. This was shown on the example of filled SKS-30 rubber in ozone.[34] If at $\sigma = 400$ kg/cm^2 (Fig. 201) the threshold of concentration of ozone is approximately $3{\cdot}6 \times 10^{-5}$%, then it is at

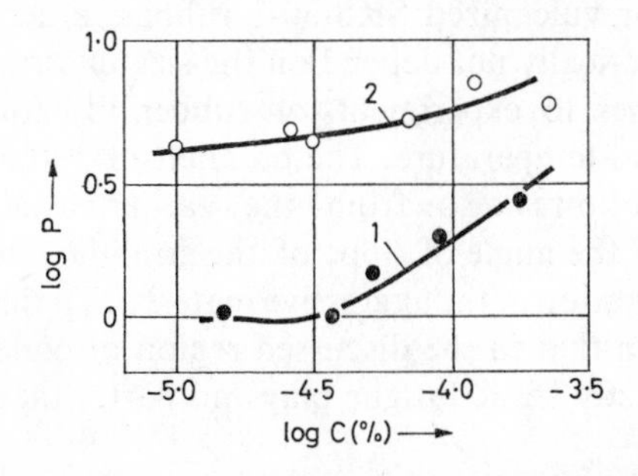

FIG. 201. Dependence of relative creep P of SKS-30 rubber, filled with channel black, on ozone concentration at 35°C and different stresses. 1. $\sigma = 400$ kg/cm^2. 2. $\sigma = 200$ kg/cm^2.

$\sigma = 200$ kg/cm^2 considerably below 10^{-6}% (i.e. smaller than the atmospheric ozone concentrations). It is clearly correct in the case of the small stresses, which rubber articles experience during their use, to extrapolate the data which have been obtained with great ozone concentrations to atmospheric concentrations, using eqns. (13.2) and (13.3), but not the more complicated (13.6).

In regions of stress where, due to the increase of molecular orientation, a strengthening of the polymer takes place, the value of P_c shifts with increasing stress towards smaller concentrations of the aggressive agent. This can, for instance, be observed when testing SKS-30-1 rubbers in HCl solution. At $\sigma = 15$ kg/cm^2 (average deformation about 100%) $P_c \approx 0{\cdot}4n$, whilst for $\sigma \approx 84$ kg/cm^2 (average deformation approx. 400%) $P_c \approx 0{\cdot}2n$, i.e. one-half less (see curves 4 and 5 on Fig. 198). ε_{cr} for this rubber in hydrochloric acid is about 120%.

In the stress range mentioned a strengthening of the rubber takes place, the relative role played by corrosion failure as compared with static fatigue grows, and P_c shifts towards smaller concentrations.

13.3. Study of the Quantitative Dependence of the Durability of Rubber on the Concentration of the Aggressive Medium

At concentrations of the aggressive medium where the relation (13.3) is fulfilled, the angle of the slope of the straight lines $\log \tau$–$\log C$ for vulcanized SKS-30-1 rubber in acetic acid solutions does practically not depend on the size of stress† (Fig. 202); the same applies to experiments on rubber in ozone under constant strain and temperature. The parallelism of the straight lines $\log \tau$–$\log C$ follows also from the earlier established, non-dependence of the angle of slope of the straight lines $\log \tau$–$\log \sigma$ on the concentration of the aggressive material. To this corresponds the proposition that in the discussed region of concentration the process of strictly static fatigue plays no part. Since the mechan-

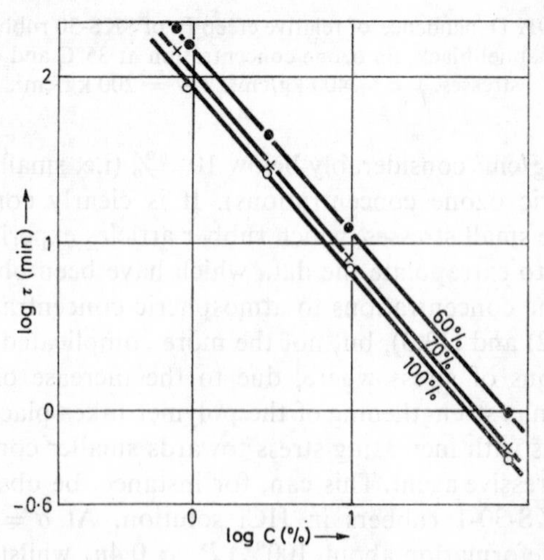

FIG. 202. Dependence of durability of vulcanized SKS-30-1 on concentration of acetic acid at strains from 60 to 100%.

† Very great stresses constitute an exception.

ism of the failure process does not change, its activation energy which in this case characterizes the chemical interaction of the polymer with the environment, does not depend on C.

The interaction of the polymer and the aggressive medium takes place after its adsorption, and therefore the relation $\tau = K_3 C^{-n_3}$ which is valid at comparatively great concentrations, may be connected with Freindlich's adsorption formula

$$C_0 = pC^n \tag{13.7}$$

where C is the concentration of the aggressive agent per volume of the medium; C_0 the same on the rubber surface; p and n are constants ($n \leqslant 1$).

In simplified form the chemical interaction of the aggressive agent with the rubber can, in general, be presented by the formula

$$E + qA \rightarrow EA_q \tag{13.8}$$

$$K + rA \rightarrow KA_r \tag{13.9}$$

Equation (13.8) reflects the interaction of the aggressive agent A with the active sections of the rubber, but eqn. (13.9) is the interaction of the aggressive agent with the rubber components K which is not directly connected with its cracking (q and r are stoichiometric coefficients).

Under usual experimental conditions, particularly during ozone cracking of rubber, the second reaction does not affect the first, since the process proceeds under an excess of the aggressive agent, and its rate does not depend on the feed of ozone. Taking into account the surface nature of the interaction of ozone with the polymer, the rate of the chemical reaction can be written in the form:

$$v_1 = k_1[E]C_0^q \tag{13.10}$$

We may assume in the first approximation that the rate of crack propagation in the absence of diffusion impediments is proportional to the rate of the chemical reaction, i.e.

$$v_r = k_2 v_1 \tag{13.11}$$

and that the concentration of the active parts of the rubber E remains constant. If we take into account that the concentration of the aggressive medium is also constant, in each experiment we obtain for those cases at $\varepsilon =$ const. in which τ is determined primarily by the rate of growth of the visible cracks (v_r), the expression

$$\tau C^{n_3} = \text{const.}$$

where $n_3 = qn$.

If a reaction of the first order takes place (as, for instance, under the action of ozone on unsaturated rubber or under the attack of acids on vulcanized carboxylate rubber), $q = 1$ and n_3 must be $\leqslant 1$; this can be observed in the majority of cases of corrosion failure of Nairit rubber and of filled and unfilled SKS-30 rubbers at room temperature and above. At temperatures of $+6°C$ and $-8°C$, however, a sharp increase of n_3 to 2·5–5·8 is found in Nairit rubbers.[4]

The explanation for the growth of the factor n_3 during a transition to lower temperatures has evidently to be looked for in the lacking constancy of the concentration of the active sections $[E]$. Indeed, $[E]$ is not simply the concentration of the active sections of the polymer but a concentration of the stressed active sections, as the crack propagation occurs only by the interaction of the aggressive medium with the stressed polymer chains at the tips of the cracks. Generally, $[E]$ depends on the temperature, ozone concentration and on the size of the surface of the polymer at the tips of the cracks which actually interact with the ozone and, consequently, also on the number and dimensions of the cracks. Owing to the overstresses which appear during the development of the cracks, the number of molecules which are stressed to a certain degree will depend on the rate of their relaxation and on the rate of their interaction with the aggressive medium. The value $[E]$ will increase with lowering temperature, as the rate of stress relaxation decreases, and also with increasing ozone concentration, as the probability of an interaction of the lower stressed sections grows. The latter is confirmed by the increase in the number of forming cracks with the growth of ozone concentration which must be accompanied by an increase of n_3. When examining eqn. (13.10) it can be seen that the increase of $[E]$ with the growth of C_0 is formally equivalent to an increase of the order of magnitude of the speed of reaction, i.e. to the increase of the exponent of C_0, we assume k_1 $[E]$ as invariable.

Testing of rubber at $\sigma =$ const. (or at $F =$ const.) occurs under conditions when, due to the small durability (at great ozone concentrations) the stress relaxation has no time to make itself felt; the stress distribution is then less even than in the case of

ε = const. under the same average stress, when the samples are specially subject to stress relaxation. One can, therefore, expect that in experiments at σ = const. or F = const. the factor n_3 will be larger than at ε = const., as in the first case more molecules are stressed to the necessary level. Experiment confirms this assumption. For instance, $n_3 = 1{\cdot}8$ for vulcanized SKB in the area of comparatively great ozone concentrations, and at smaller ozone concentrations $n_3 = 0{\cdot}4$.

If Braden's and Gent's[8] data, which were obtained at $F =$ = const., are plotted in the coordinates log τ–log C, then $n_3 = 1$ for natural rubbers, for BSK $n_3 = 1{\cdot}1$, whereas in experiments at ε = const. the value of $n_3 = 0{\cdot}6$–1 for natural rubbers and for BSK $n_3 = 0{\cdot}45$-$0{\cdot}9$. Braden and Gent determined in their experiments the rate of growth of a lateral incision in a rubber specimen the surface of which was covered with silicone grease. In these conditions the polymer molecules on the flat surface of the specimen in front of the apex of the crack which undergo complicated stresses do not take part in the process of crack propagation,† which ought to decrease the value of n_3. In spite of this circumstance n_3, as can be seen from the quoted data, is of the same order as in experiments under σ = const. The comparison of results[34] for unfilled SKN-40 rubber shows that at ε = const., $n_3 = 0{\cdot}75$ and at σ = const., $n_3 = 1{\cdot}2$. On the other hand, parallel experiments on vulcanized carboxylate SKS-30-1 rubber in solutions of acetic and hydrochloric acid at ε = const. and at σ = const. gave equal values for n_3, which is connected with the swelling of their surface layer, the resorption of stresses and their more even distribution.

The transition from failure accompanied by cracking to failure in an aggressive environment without cracking[34] is of basic and practical interest from the point of view of the dependence of the durability of rubber on the concentration of the aggressive medium. In the latter case n_3 is always smaller than 1, as in failure in the absence of macrocracks the value of $[E]$ can be considered a constant. Thus, $n_3 = 0{\cdot}63$ for sulphur vulcanized SKS-30-1 rubber in hydrochloric acid, $n_3 = 0{\cdot}31$ for rubbers of the fluorine rubber type Kel-F in nitric acid, and $n_3 = 0{\cdot}3$ for butyl rubber in

† In the absence of a protective coating the value of $[E]$ grows due just to these molecules.

diluted nitric acid, when no cracks form, whereas at great concentrations of HNO_3 in the presence of cracks $n_3 = 5{\cdot}7$.

The change of the order of speed of reaction of the aggressive medium with the polymer under the circumstances which were indicated on p. 373 will be reflected in the values of K_3 and n_3 in eqn. (13.3) but cannot be reflected on the nature of the dependence itself (13.3).

13.4. Influence of Temperature on the Failure of Rubber in Aggressive Environments

When establishing the temperature dependence of any process it is most interesting to determine the activation energy U, the size of which makes it possible in many cases to form an opinion on the mechanism of the process. An analysis of the experimental data shows that one has to take into account in such determinations the complicated nature of corrosion cracking which is a physical process of crack propagation as a result of the chemical reaction of the polymer with the aggressive medium. In connection with this it is expedient to examine separately the behaviour of rubbers at room and higher temperatures at which they are elastomers (i. e. the chain molecules are sufficiently mobile not to impede the crack propagation), and at low temperatures where the rubbers begin to lose their elastomeric properties. In regions of low temperatures, the lowering of temperature brings about not only, and not so much, the slowing down of the chemical interaction, as especially the reduction of the mobility of the macromolecules.

It has been shown[13] in the study of corrosion cracking at normal temperatures of stretched vulcanized carboxylate SKS--30-1 rubber in hydrochloric acid and in ozone that the temperature dependences of all chosen characteristics of the process (rate of build-up of Mg^{2+} ions in HCl solution, durability τ at σ = const. and at ε = const., rate of creep $\dot{\varepsilon}$ at σ = const.) is well described by equations of an exponential type (Figs. 203 and 204). Values approaching an order of 20 kcal/mol are obtained independent of the method of determination for the activation energy of failure of vulcanized SKS-30-1 (MgO), stretched by 30–250% in hydrochloric acid (Table 25, lines 1, 3, 4, 6).

The data in the literature on the activation energy of viscous flow of different salts of low-molecular copolymers of butadiene with methacrylic acid reflect partly the failure energy of the Me–O bonds, since the author[15] considers that the flow may be achieved by the successive failure and reforming of the Me–O bonds.

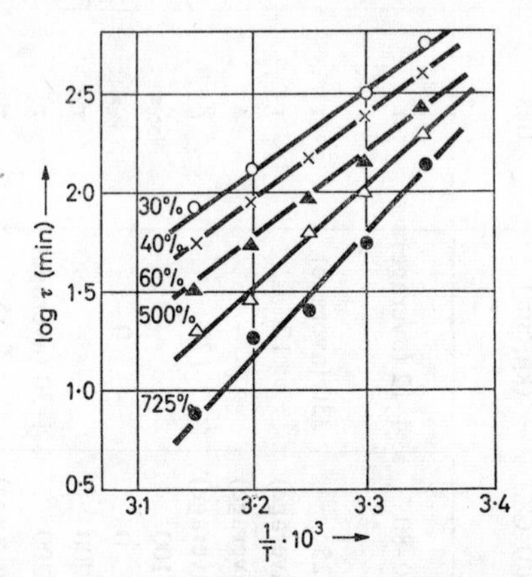

FIG. 203. Temperature dependence of durability of vulcanized SKS-30-1 (ε = const., 1 n solution of HCl) at different strains.

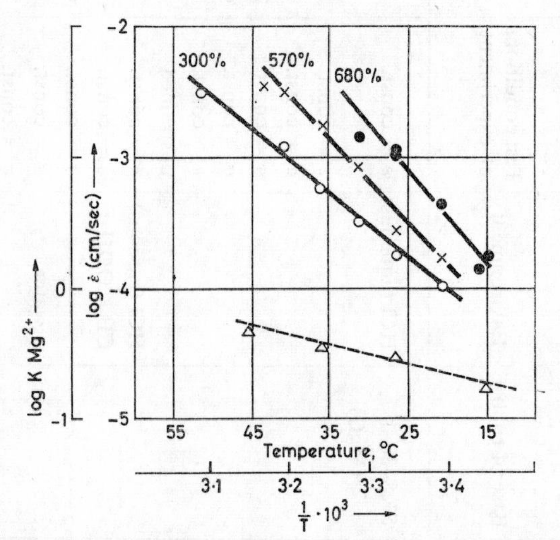

FIG. 204. Temperature dependence of the failure speed of vulcanized SKS-30-1 in 1 n HCl: (on the rate of creep at different deformations – solid lines; on the speed of dissolving of Mg^{2+} ions for unstressed rubbers – dotted line).

TABLE 25

Activation Energy of Failure of Rubber in Aggressive Environments Determined by Various Characteristics

No.	Base of rubber	Environment	Test conditions of experiment	Deformation (strain) (%)	Stress (kg/cm²)	Characteristics	Activation energy U (kcal/mol)
1	SKS–30–1 (MgO)	HCl (aqueous)	ε = const.	30–80	4–12 (average)	τ	19–21·6
2			ε = const.	725	330 (average)	τ	29
3			σ = const.	240 (average)	17	τ	18
4			σ = const.	240 (average)	–	$\dot{\varepsilon}$	18·4
5			σ = const.	680 (average)	178	$\dot{\varepsilon}$	31
6			ε = const.	100	–	v_{Mg^2}	21
7			unstrained	0	0	v_{Mg^2}	6·4
8		HCl (gaseous)	ε = const.	200	–	τ	9·5
9		CH_3COOH	ε = const.	30–200	4–30 (average)	τ	19
10			σ = const.	660 (average)	91·5	τ	27
11		water	σ = const.	277 (average)	41·2	τ	10–12
12			σ = const.	277 (average)	60	$\dot{\varepsilon}$	13·3
13			σ = const.	680 (average)	88	τ	28

14	SKS–30–1 (MgO)	air	σ = const.	303 (average)	51·4	$\dot{\varepsilon}$	8·2
15			σ = const.	740 (average)	432	τ	18
16		O_3 (gas)	ε = const.	25–125	4–25 (average)	τ	2·3
17			ε = const.	725	375 (average)	τ	6·4
18	SKS–3–1 (MgO thiuram)	HCl (aqueous)	σ = const.	710 (average)	166	τ	26·7
19	SKS–30	O_3 (gas)	ε = const.	20–80	3–14 (average)	τ	3·4
20	Nairit	O_3 (gas)	ε = const.	40–150	6–24 (average)	τ	8
21			ε = const.	600	160 (average)	τ	10
22	Natural rubber	O_3 (gas)	ε = const.	50	6 (average)	τ	2·3
23			ε = const.	650	220 (average)	τ	4·4
24	Fluorine rubber	HNO_3 (54%)	σ = const.	118 (average)	41	τ	22
25	Type Kel-F (benzoyl peroxide, powdered silica gel)		σ = const.	183 (average)	109	τ	30
26	Fluorine rubber type Kel-F (benzoyl peroxide)	HNO_3 (54%)	σ = const.	–	48	τ	35

The value U lies in this case in the range from 12·7 kcal/mol (zincate) to 30 kcal/mol (magnesium salt), increasing with the proportion of $-COOH$ groups, due to the increase of the fraction of breaking Me–O bonds. During the chemical interaction of the Me–O bonds with the acid the activation energy of the process is, naturally, smaller than with the mechanical fracture of these bonds.

Values for U from 2 to 4 kcal/mol are obtained (see Table 25) under the attack of ozone on vulcanized SKS-30-1, SKS-30 and natural rubbers (unfilled, and filled with channel black) which did not contain antioxidants, in conditions when the durability was determined by the steady rate of crack propagation with ε of an order of 20–100%. These values are in agreement with the low values of the activation energy of the interaction of ozone with olefines.[16]

In Matsuda's and Tanaka's paper[17] the following values of the activation energy of ozonized rubbers have been published, which were determined by the fall of viscosity:

	U (kcal/mol)
Natural rubber	2·3
Neoprene W	2·7
GR-S	1·5
Hycar	1·8

This is comparable with the values obtained with ozone cracking of rubber (with the exception of Neoprene W).

It is interesting that the activation energies of ozonized natural rubber and of polychloroprene are almost identical; this tallies with the practically identical speed of ozone absorption (samples in sheet form),[18] and also with the data on the speed of the accumulation of the products of ozonization in the solution which are obtained by infrared spectroscopy.[19] In connection with this the proposition appears more probable that the increase of the resistance of polychloroprene rubber to ozone cracking is not connected with its lesser reactivity with ozone but with its physical structure which is more favourable than that of natural rubber. Such a conclusion is confirmed by the recently detected fact[20] that the tips of ozone cracks in polychloroprene rubber are rounded and are acute in natural rubber, i.e. the stress concentration is considerably greater in natural rubber than in polychloroprene. The presence of a great number of polar groups in polychloroprene which impede the mobility of its chains hinders the growth of cracks. With the formation of supramolecular structures this effect must increase even more and it is known that the tendency towards the formation of such structures (particularly to crystallization) is more pronounced in polychloroprene than in natural rubber. The high value of the activation energy of fracture of vulcanized polychloroprene in ozone (8 kcal/mol) compared with the activation energy of its ozonization in solution (2·6 kcal/mol) can be explained by the intensified degradation of the supramolecular structure if during its determination the temperature rises. The degradation of the supramolecular structure must facilitate crack propagation and is therefore accompanied by a considerable decrease of strength. The hypothesis of the failure of the supramolecular structure of polychloroprene was also used as an explanation of the temperature dependence of its durability in the absence of an aggressive medium (see p. 258). Therefore, the activation energy of fracture of vulcanized polychloroprene in ozone does evidently not tally with the activation energy of the chemical interaction of ozone with polychloroprene but is a fictitious value.

As can be seen from Table 25, under the attack of concentrated HNO_3 on fluorine rubber of the Kel-F type the activation energy of failure has the value of 22–35 kcal/mol. An estimate of the activation energy of the chemical relaxation of such rubbers gives a value[21] of an order of 30–40 kcal/mol but the activation energy of the oxidation of the polymer itself is[22] 30 kcal/mol on glass, and 36 kcal/mol on platinum. Allowing that HNO_3 is a strong oxidizer, one can assume that during its attack on fluorine rubber an oxidizing reaction develops which causes in the first place the failure of those chemical bonds which are weaker† than of those of the fluorine rubber. The activation energy of the failure process of rubber in HNO_3 must, on account of this, also be somewhat smaller than that of the oxidation of fluorine rubber.

† The C–C cross-links and those which appear with the vulcanization of the C = C double bonds are weaker than the CF-CF links.

All results obtained permit to draw the following conclusions: (1) the rate of crack propagation in rubbers in the presence of an aggressive medium is determined by the rate of the chemical interaction of the medium with the polymer; (2) the test conditions (ε = const. or σ = const.) have no noticeable influence on the temperature dependence of the process. The activation energy of the failure process of a polymer in an aggressive environment depends to a great extent not only on the nature of the chemical interaction with the environment but also on adsorption effects, since this reaction is heterogeneous. Data on the influence of the aggressive media on vulcanized SKS-30-1 show that the size of the activation energy in gaseous HCl which attacks the O–Me cross-links is greater than in ozone, and is 9·5 kcal/mol (ε = 200%). The apparent activation energy of the chemical reaction of HCl with a polymer in an aqueous solution must be higher than that of the reaction of the polymer with gaseous HCl, as it consists of the activation energy of the dehydration of HCl (according to existing data[23] it is 8·6 kcal/mol), the activation energy of the dehydration of the active centres of the polymer, and the activation energy of the interaction of the dehydrated HCl with the polymer. The apparent activation energy of the failure process of the rubber in CH_3COOH solutions is, both under small and large strains, somewhat lower (see Table 25) than in HCl solutions; this is obviously connected with the smaller energy of dehydration of acetic acid molecules and with its better adsorption on the polymer.

Data can be found in a number of papers[3, 17, 24] on the temperature dependence of the time to the appearance of cracks under the action of ozone on rubber. Matsuda and Tanaka[17], using this as an index, carried out a comprehensive research into the failure of natural rubber containing different fillers.

The values of the apparent activation energy U (Table 26) which were found by the use of this index are in general considerably higher than those of the activation energy of crack propagation. This proves the greater temperature dependence of the process of crack formation than of the process of their growth.

This is, however, not so much connected with the different mechanism of the process, as with the fact that the time to the appearance of visible cracks is basically not determined by the rate of the reaction of ozone with the polymer but with the rate of the reaction of ozone with the substances which are found on the rubber surface (stearic acid, antioxidants, resins, waxes and foreign bodies and impurities of all kinds). Their quantity on the surface depends to a great extent on the composition and conditions of preparation of the rubber, and also on the temperature of the experiment which all have an influence on the solubility of these impurities in the rubber.

TABLE 26

Value of the Apparent Activation Energy (U_k) of the Crack Formation Under the Attack of Ozone on Rubber

Rubber	Ingredients	U_k (kcal/mol)	Reference
Natural rubber	–	5–8	17
	–	11·5	24
	–	7·4	3
	Wax (2%) w/w	52·4	24
	Wax (5%) w/w	72·6	24
	$CaCO_3$ (10–54%) w/w	8–13	17
	Aragonite (18·5%) v/v	19	17
	Various carbon blacks	5–15	17
	Various carbon blacks	7·4	3
	Various carbon blacks	27·1	24
Polychloroprene	–	17	3
butyl rubber	–	30	3

The discrepancy of the data obtained by different authors and the sharp dependence of U on the presence of a surface protective layer is a confirmation of the reasons expressed concerning the fictitiousness of the value of U which was determined by this method. In natural rubbers which contain, for instance wax, U reaches 72 kcal/mol which is improbable for ozonization.

When estimating the size of the activation energy of corrosion failure of stressed rubbers one has to examine the question of the activation by the stress caused by the action of the aggressive agent and of the role of the diffusion of the aggressive medium during cracking. The opinions which have been expressed in a number of papers, namely that in the presence of stress the process of interaction of the aggressive medium with the polymer is activated, lack explanation. With the application of stress the speed of the process increases due to the transition from the region of diffusion kinetics to the region of chemical kinetics. Consequently, the role of stress leads to elimination of the impediments to diffusion, but not to an activation of the reaction itself. A formal examination of the conceptions of activation as a lowering of the activation energy under the influence of the aggressive medium may lead to misunderstandings. Thus, under the attack of hydrochloric acid on SKS-30-1 rubber the activation energy of diffusion is smaller than the activation energy of the chemical interaction. U does, therefore, not only not decrease but it even grows with the application of stress. However, under the effect of ozone on rubber which contains double bonds, the activation energy of the chemical reaction (2–3 kcal/mol) is smaller than the activation energy of the diffusion of ozone into the rubber which must exceed the activation energy of diffusion of oxygen (7–12 kcal/mol) in the rubber.[25] Consequently, in this case the application of stress must be accompanied by a decrease of the activation energy of the failure process. The question of the role of diffusion in the processes of corrosion cracking may be solved in connection with this. In Japanese papers[17] the speed of ozone cracking is put into direct dependence on the gas permeability of the rubber. Buckley and Robison[6] consider that the cracking of rubber proceeds both by surface interaction with ozone, and by the penetration of ozone into the depth of the rubber (the latter causes the formation of internal cracks which then merge with the external ones). From this point of view the resistance to ozone of SKN and of Neoprene rubbers must be the same as their permeability by oxygen[26] is the same (there exist no data for ozone). However, experiment shows that SKN rubbers resist ozone less than Neoprene rubbers or natural rubbers though these have a 6 times greater permeability.[26]

The values of the activation energy of corrosion cracking of different rubbers (see Table 25) can in no way be attributed to the diffusion process, since in the one case they are too small (ozone), in the other too great (HCl). Finally, a straightforward calculation shows[27] that, for instance, ozone at a concentration of 1 mg/l ($5\times10^{-2}\%$) diffuses in 24 hr in natural rubbers to a depth of about 6 μ, whereas cracks at this same ozone concentration and a strain of 10–15% penetrate to a depth of 1 mm in

10 min. Electron-microscopic investigations have also shown that the depth of the layer which is destroyed through the diffusion of ozone does not depend on the size of the strain,[27] in contrast to the rate of crack growth which depends very much on the strain. There can, therefore, be no doubt that the process of corrosion cracking cannot be connected with the rate of diffusion of the aggressive agent.

In regions of strains up to 100–150% and corresponding stresses the activation energy of the failure process does in an aggressive environment not depend on the magnitude of the stress. Under great strains and stresses a growth of the activation energy can be found in all investigated cases (see Table 25). It is known that in the failure of solids in the absence of aggressive media the activation energy decreases with the growth of stress.

However, in the failure of SKS-30-1 rubber in air an increase of the activation energy with the growth of stress has been observed (see Table 25). Taking into account that these experiments must not be considered as experiments carried out in the absence of an aggressive medium (the part of the aggressive agent may be played by the humidity of the air), numerous determinations of the durability of radiation vulcanized SKN-18, SKN-26, SKN-40 and SKT with various densities of cross-links have been made in air[28] in order to define more exactly the relation of U to τ. In many cases at temperatures between 25 and 50°C an increase of the apparent activation energy has been observed with a growth of stress, whilst at temperatures of 60–150°C it does not depend on the stress. This can be seen clearly from the following data:

Rubber	U_k (kcal/mol)		
	60–150° C	25–60° C	
		$\delta = 50$ kg/cm²	$\sigma = 100$ kg/cm²
SKN–18	10	17	22
SKN–40	22	40	63

The growth of the apparent activation energy observed in the absence of aggressive agents may be connected both with the failure of the supramolecular structures under stress (similar to their failure in Nairit rubbers at elevated temperatures and with the introduction of plasticizers), and with an increase of the fraction of the breaking chemical bonds. In the presence of aggressive media an analogous principle is observed which confirms again the similarity between corrosion failure and static fatigue.

In general the activation energy of failure in the presence of an aggressive medium may consist of the following elements:

1. Activation energy of fracture of the Van der Waals bonds (U_1).
2. Activation energy of failure of the chemical bonds with chemical reaction (U_2).
3. Activation energy of failure of the chemical bonds due to mechanical action (U_3).

The total activation energy (U) of fracture calculated per mol of broken bonds is equal to the sum of the products of the activation energies (U_1, U_2, U_3) per corresponding molar fraction $a+b+c = 1$:

$$U = aU_1+bU_2+cU_3.$$

If the failure takes place in a solution, the chemical interaction may be preceded by the falling out of solution of the aggressive agent and the active groups of polymer molecule, on which also the corresponding quantities of energy U_2' and U_2'' are spent. In this case

$$U = aU_1+b(U_2+U_2'+U_2'')+cU_3.$$

With small strains and an active medium $a = 0$ and $c = 0$. With great prefailure strains of highly orientated polymers the fracture of the chemical bonds begins to play an essential part, c acquires appreciable values and, as U_3 is very great, U grows. The greatest increase of U can be observed under deformations which exceed $\varepsilon_{\tau_{max}}$, i.e. in that region where the speed of the failure process is to a considerable extent determined by the mechanical failure of the rubber, namely by its static fatigue. This happens both for natural rubbers and Nairit in ozone; $\varepsilon_{\tau_{max}}$ lies for natural rubber and Nairit in the region $\varepsilon > 400$–500%, and for vulcanized SKS-30-1 (MgO) in HCl solution $\varepsilon_{\tau_{max}}$ is in the region $\varepsilon > 300\%$. In vulcanized SKS-30-1 (MgO) under great strains the C–C bonds must break besides the Me–O bonds, and in a rubber vulcanized with MgO and thiuram the weaker sulphur bonds break. The activation energy of failure must, therefore, be smaller in the second case than in the first. Experiment confirms this (see Table 25, lines 5 and 18).

Passing to media which are less active than acid (water and water vapour in air), the coefficient *a* acquires an appreciable value, due to which a decrease of the activation energy of failure takes place (see Table 25, lines 3, 11 and 14, 13 and 15).

Taking into account the established principles of the dependence of τ on stress σ, the ozone concentration C and the temperature T, one can calculate τ quantitatively if the rubbers contain no antioxidants and their deformation is smaller than ε_{cr}. This is most frequently met in the practical working conditions of technical rubber articles, when the operating temperature is between 20 and 50°C. In this case the following empirical formula is valid:

$$\tau = \tau_0 \sigma^{-b} C^{-\alpha} e^{U/RT}$$

In this formula τ is in min, σ in kg/cm², C in vol %, U in kcal/mol; the order of magnitude of τ_0, b, α and U for some unfilled rubbers under the attack of ozone is as follows:

	Natural rubber	SKS-30	Nairit
b	0·35	0·90	0·76
α	0·6–1·0	0·5–0·9	0·5–1·0
U	2000–3000	2000–3000	8000
τ_0	$1{\cdot}2\times10^{-3}$- 7×10^{-2}	$1{\cdot}9\times10^{-3}$- $0{\cdot}11\times10^{-2}$	$2{\cdot}5\times10^{-5}$- $1{\cdot}4\times10^{-4}$

With a decrease of temperature a considerable retardation of the ozone cracking of rubber which is under constant strain takes place.

Thus, for instance,[29] ozone cracking practically ceases at temperatures which lie 15–20°C above the temperature of the glass transition point (Table 27). The latter was defined as the temperature at which the residual deformation of rubber, which was stretched by 20% for 10 min, was 98–99%.

The cracking of rubber was determined with an elongation of 20%, followed by cooling, with an ozone concentration of 0·4% and observed for 1 hr.

Depending on the test conditions (size of strain, ozone concentration, duration of cooling and observation), different temperatures have been mentioned in different papers below which practically no cracking occurs. For instance, −18°C[30] and −40°C %[24, 29] are mentioned for natural rubbers. With increasing deformation the temperature threshold of cracking decreases:[31] for instance,[32] filled, and plasticizer containing butadiene-styrene rubber with $T_2 < -60$°C, stretched by 50%, does not crack at a temperature of −40°C, but stretched by 100% at a temperature of −60°C.

The sharp deceleration of the process cannot be related to the decrease of the speed of the chemical reaction of ozone with the double bonds, as the activation energy of this reaction is very small. It has been shown experimentally[29] that at −50°C an instant reaction of ozone with SKB, SKN, SKS and natural rubbers, containing carbon black, takes place. This can be judged by the rapid formation of a suspension of carbon black in the chloroform in which the ozonization was carried out.

The possible reasons for the deceleration and ceasing of the cracking may be:

(1) the sharp decrease of the mobility of the molecules, the "freezing-in" of the stress which acts in the specimen, since in all experiments the rubber was stretched before being cooled. Consequently, the cracks do not open, and do not grow;

(2) increase of strength of the rubber;

(3) increase of stability of the ozonides with decreasing temperature.[32]

In one of the papers[33] data were obtained showing that glassy natural rubber cracks at a temperature of −80°C, and breaks at once in an atmosphere of 0·5% ozone at a stress of about 60 kg/cm², produced by a load which was applied after the cooling of the rubber. This proves that the decrease of the mobility of the molecules is one of the fundamental factors with which the sharp increase of the resistance of the rubber to cracking at ε = const. and at decreasing temperatures is connected.

Literature

1. R. G. Newton, *Rubb. Chem. Technol.* **18** 504 (1945).
2. Yu. S. Zuyev, *Khim. prom.* No. 9, 272 (1950).
3. W. L. Dunkel, R. R. Phelan, *Rubb. Age*, **83,** 281 (1958).

4. Yu. S. Zuyev, S. I. Pravednikova, *v sb.* "*Stareniye i zashchita rezin*", Goskhimizdat, 1960, p. 3 (*Trudiy NIIRP*, No. 6).
5. Yu. S. Zuyev, A. Z. Borshchevskaya, *DAN SSSR*, **124**, 613 (1959).
6. D. J. Buckley, B. B. Robison, *Rubb. Chem. Technol.* **32**, 257 (1959).
7. N. N. Znamenskii, *ZhFKh*, **30**, 1092 (1956).
8. M. Braden, A. N. Gent, *J. Appl. Polymer Sci.* **3**, 90, 100 (1960); *Kautschuk und Gummi*, **14**, WT157 (1961).
9. Yu. S. Zuyev, A. Z. Borshchevskaya, *DAN SSSR*, **144**, 849 (1962).
10. Yu. S. Zuyev, N. N. Bukhanova, T. I. Dorfman, *Kauchuk i rezina*, No. 10, 44 (1960).
11. S. N. Zhurkov, T. P. Sanfirova, *ZhTF*, **28**, 1719 (1958).
12. M. I. Bessonov, Ye. V. Kuvshinskii, *Viysokomol. soyed.* **2**, 397 (1960).
13. Yu. S. Zuyev, A. Z. Borshchevskaya, S. I. Pravednikova, U. Yue-tsin′, *Viysokomol. soyed.* **3**, 164 (1961).
14. Yu. S. Zuyev, A. Z. Borshchevskaya, *Kauchuk i rezina*, No. 10, 23 (1963).
15. W. Cooper, *J. Polymer Sci.* **28**, 1951 (1953).
16. B. D. Cadle, S. Schadt, *J. Am. Chem. Soc.* **74**, 6002 (1952); *J. Chem. Phys.* **21**, 163 (1953).
17. Matsuda, Tanaka, *Nikhon gomu kokaisi*, **28**, No. 3, 131 (1955); **25**, No. 7, 208, 214 (1952).
18. J. H. Gilbert, *Proceedings of the Fourth Rubber Technology Conference*, London, May, 1962. Preprint 56.
19. A. R. Allison, I. I. Stanley, *Anal. Chem.* **24**, 630 (1952).
20. G. Salamon, F. van Bloois, *Proceedings of the Fourth Rubber Technology Conference*, London, May 1962. Preprint 58.
21. A. S. Novikov, V. A. Kargin, F. A., Galil-Ogly, *Kauchuk i rezina*, No. 1, 33 (1959).
22. T. G. Degteva, A. S. Kuz'minskii, *Viysokomol. soyed (v pechati)*.
23. A. V. Rakovskii, *Vvedeniye v fizicheskuyu khimiyu, GONTI, Red. Khim. lit.* (1938).
24. J. Crabtree, A. R. Kemp, *Ind. Eng. Chem., Anal. Ed.* **18**, 769 (1946).
25. S. A. Reitlinger, *sm.* A. S. Kuz'minskii, N. N. Lezhnev, Yu. S. Zuyev, *Okisleniye kauchukov i rezin*, Goskhimizdat, 1957, p. 308.
26. G. Amerongen, *J. Appl. Phys.* **17**, 972 (1946); *J. Polymer Sci.* **5**, 307 (1950).
27. E. H. Andrews, M. Braden, *J. Polymer Sci.* **55**, 787 (1961).
28. Yu. S. Zuyev, G. M. Bartenev, N. I. Kirchenshtein, *Viysokomol. soyed.* **6**, No. 9, 16 (1964).
29. Yu. S. Zuyev, A. S. Kuz'minskii, *DAN SSSR*, **89**, 325 (1953).
30. J. M. Ball, R. A. Joumans, A. F. Rausell, *Rubb. Age*, **55**, 481 (1944).
31. G. R. Guthbertson, D. D. Dunnom, *Ind. Eng. Chem.* **44**, 834 (1952).
32. R. F. Grossman, A. C. Bluestein, *Rubb. Age*, **85**, 96 (1959).
33. Yu. S. Zuyev, S. I. Pravednikova, T. V. Likhtman, *Viysokomol. soyed.* **5**, 262 (1963).
34. Yu. S. Zuyev, A. Z. Borshchevskaya, *Viysokomol. soyed.* **6**, 323 (1964).

CHAPTER 14

METHODS OF INCREASING THE DURABILITY OF RUBBERS IN AGGRESSIVE ENVIRONMENTS

IN THE literature there exist data which refer only to methods of protecting the rubber from ozone cracking.[1–4]

A part of these methods is of general interest, another part describes specific methods of protection from ozone, the acquaintance with which, however, makes it possible to consider ways of protection also from other aggressive media.

First of all it is necessary to have a clear conception of how resistant the various rubbers are to different chemically aggressive media, how most important ingredients influence their resistance, to know something of the conditions of their use, and then to examine the general and specific methods of protection.

The following methods of protection are general, independent from the nature of the medium;

(1) decreasing the tensile stress and creating a compressive stress in the surface layer of the material;

(2) establishing a physical barrier between the surface of the rubber and the environment in the form of a layer of a plastic or elastic coating (waxy substances† and elastomers which are resistant to the medium). This method is to a certain degree specific, as the coating must be inert to the actual aggressive medium.

† Under waxy substances (waxes) we understand here and further on not only the narrow class of waxes proper, but blends of isomeric or similar compounds of different chemical structure which have the consistency of ordinary waxes.

14.1. Resistance of Rubber to Ozone and Other Aggressive Media

Rubbers which are used at present can be divided into three groups[1] according to their resistance to ozone cracking.

(1) *Highly resistant.* To this group belong the rubbers on the base of polymers which have no double bonds; they do not fail for a long time (years) neither under atmospheric ozone concentrations, nor under significant concentrations (0·1–1%). To such polymers belong different fluorine-containing rubbers of the Kel-F type, Viton A, poly-FBA,[5] sulphochlorinated polyethylene,[6] copolymers of ethylene with propylene,[7] and to a lesser degree silicone rubbers.[8]

(2) *Moderately resistant.* To this group belong the vulcanizates on the base of butyl rubber, bromebutyl rubber, polychloroprene, and also of saturated polyester rubbers and Thiokols[9] (depending on the composition, polyester rubbers can have a resistance which approaches that of polychloroprene or natural rubbers[10]); all these stand up well to atmospheric ageing. Vulcanized rubbers of this group which contain more than 50% filler are already covered with cracks after a few months. At ozone concentrations of an order of 10^{-2}–10^{-1}% the vulcanized rubbers crack at room temperature within some tens of minutes.

(3) *Non-resistant.* To this group belong vulcanizates on the base of such unsaturated rubbers as natural rubber, SKI, SKB, SKD, SKN, SKS, etc., which fail extensively (within several days) in atmospheric ozone concentrations.

In experiments on (or in practical use of) rubbers under the condition ε = const. the addition of fillers, active and non-active, and also of plasticizers, leads to an increase of the time to the appearance of ozone cracks and the time to failure (durability)[11]. This is connected with the fact that in the presence of fillers at a given deformation a higher stress is produced, and that in the presence of plasticizers the strength properties are lowered. However, in experiments under the condition σ = const. the addition of fillers, especially of active ones, has a favourable influence, and increases the durability in regions of small stresses (up to 5 kg/cm^2)[11]. Under these conditions the deformation of the specimens decreases and a greater number of cracks form but they are smaller; thanks to this a more favourable stress distribution is established, and the strength properties of the rubber are improved. The type of the vulcanization network, assuming an equal density, has practically no influence on the resistance of the rubber to ozone cracking at static strains,[11] as the failure proceeds along the double C = C bonds which are in the main chains of the macromolecules.

No information can be found in the literature concerning the effect of most aggressive media on stressed rubbers (with the exception of ozone), in view of the lack of suitable experimental methods. Very limited information exists only on the effect of chemically active media on unstressed vulcanized rubbers.

Chemically active environments with which the rubbers are in contact, either in the form of anticorrosion coverings or in the shape of various articles vary widely. They can, however, be divided into the following groups:

(1) Strong oxidizers (nitric, chromic, concentrated sulphuric acids, hydrogen peroxide, calcium hypochlorite and others).

(2) Mineral acids (hydrochloric, sulphuric, phosphoric, hydrofluoric).

(3) Organic acids (formic, acetic, oxalic, citric, tartaric, and others).

(4) Alkalis (caustic potash, sodium and others).

(5) Salts (mixture of salts in sea water, copper sulphate, and others).

The resistance of various rubbers to these media is determined by two factors:

(1) By their capacity to be wetted by the given medium and to swell in it.

(2) By the intensity of the chemical reaction and the nature of the material which is formed as a result.

A good wetting is an indispensible condition for the process of chemical reaction between the polymer and the environment; the less good the wetting, the more resistant is the polymer with the same tendency to react. Pronounced swelling in the mass leads even in the absence of a chemical interaction to the weakening and failure of the polymer. However, though the swelling of the surface layer leads to a sharp increase of the durability of the stressed rubber in this case, the reduction of the wettability decreases the durability; this can, for instance, be observed during the attack of ozone in water on rubber which contains paraffin wax. Therefore, we have to evaluate in each actual case the relation between the favourable and the unfavourable influence of the wetting on the durability of the rubber in the medium.

At present it is hardly possible to give a valid classification of rubbers according to their resistance to specific environments. One can only formulate some conclusions. Rubbers which are manufactured at present can be conveniently examined from the point of view of their relation to the aggressive media according to the generally accepted chemical classification of polymers.

(1) Carbon chain unsaturated polymers (natural rubber, BSK, BNK, SKB, polychloroprene).

(2) Carbon chain saturated polymers, or polymers with little saturation (polyisobutylene, butyl rubber, fluorine rubber, sulphochlorinated polyethylene);

(3) Hetero-chain polymers (polysiloxanes, Thiokols, polyester rubbers).

Rubbers of the first group are under suitable conditions capable of all reactions which are characteristic for the monomeric unsaturated compounds (they oxidize easily, halogenate, incorporate hydrogen halide acids, sulphonate, acetylate). Rubbers of the second group are little reactive and rubbers of the third group break up easily under the attack of acids, alkalis and even of hot water (polyurethanes).[12–14] Of the different types of cross links in vulcanized rubbers the monosulphide bonds are the most resistant ones to the attack of aggressive media at elevated temperature, and the polysulphide bonds are the least resistant.

Under the attack of the aggressive media rubbers display, however, a number of features in contrast to the corresponding low-molecular compounds. For instance, in contrast to the low-molecular substances, the chemical interaction with the aggressive medium is slowed down due to impeded diffusion. Besides, under the attack of such media as H_2SO_4 and HCl on natural rubber there proceeds, besides the usual reactions, a process of cyclization of the rubber[15] which leads to a consolidation of the rubber surface and to a sharp deceleration of the diffusion of the acid into the rubber. The hydrochlorination of natural rubber leads also to the formation of a compact layer on the surface (as distinguished from the effect of HCl on BSK and butyl rubber).[16]

Only rubbers of the first two groups have found a practical use as anticorrosion materials (in some cases Thiokol, for instance, is used in sea water). In the literature exist references to the use of specific rubbers.[17, 18] So, for instance, is it mentioned that chemically inert polymers are resistant to strong oxidants, for example, fluorine containing rubbers of the Kel-F type[5] and others,[19] to a lesser degree sulphochlorinated polyethylene, poly-(isobutylene) combined with polyethylene,[20] butyl rubber.[21] As far as unsaturated polymers are concerned, their resistance to such oxidants as HNO_3 decreases[22] with an increase of the content of double bonds in their main chain (a structural change of the rubbers under the action of HNO_3 takes place not only as a result of oxidation but also due to nitration[23]). A combination of unsaturated polymers with poly(isobutylene) increases their resistance to oxidizing agents.[24] Resistant to hydrochloric acid are natural rubbers,[16, 20] SKB,[25] polychloroprene,[20] poly-(isobutylene),[20] divinyl-styrene rubbers;[14] to sulphuric acid: butyl rubbers;[26] poly(chloroprenes)[27] and SKB rubbers[25] to a lesser extent. Comparatively resistant to acetic acid are sulphochlorinated polyethylenes and butyl rubbers.[28] Alkalis are well resisted by almost all rubbers, apart from fluorine rubbers.

The resistance of rubbers to aggressive media can to a considerable extent be adjusted by the addition of different ingredients into the composition of the rubber blends. This question has, however, not been sufficiently investigated. It is, for instance, known that the adding of fillers introduces factors which influence the resistance of the rubbers both favourably and unfavourably. Particularly, if the filler does not react chemically with the aggressive medium and is wetted by it like rubber, or less, then its addition must make itself favourably felt as the volume content of the rubber decreases and the diffusion path of the aggressive agent is lengthened. If the filler is wetted and adsorbs the medium better than the rubber, and also if it promotes the chemical interaction of the rubber with the medium, then its presence makes itself felt in an unfavourable way. There exists some information on the favourable influence of powdered silica gel[30] and of litharge in fluorine rubbers,[29] and also on the unfavourable influence of carbon black on the resistance of rubber in an oxidizing environment[24] and of butyl rubber in sulphuric acid.[26] Recently, more data have been obtained on the influence of fillers (lamp black, powdered silica gel, talc, barytes) on the resistance of SKS-30 rubbers, Nairit, SKN-26, fluorine rubber, butyl rubber, sulphochlorinated polyethylene,[31] and also SKI-3 and natural rubber[32] to different aggressive media. It has been shown that lamp black increases the resistance of rubber to 30% nitric acid, powdered silica gel increases the resistance of rubber to acetic and hydrochloric acids. These data, however, do not, for the time being, make it possible to infer general principles, the more so as fillers have sometimes a sharply different influence on rubbers of the same type. For instance, SKI rubbers which contain powdered silica gel resist HCl considerably better than rubbers with lamp black, but natural rubbers with both fillers are practically resistant to HCl. These fillers have the opposite effect on the behaviour of these rubbers in H_2SO_4. Powdered silica gel improves the resistance of SKI rubbers far more than lamp black but the latter is much more effective in natural rubbers than in SKI.[32]

It must not be forgotten that all the mentioned data refer to the behaviour of unstrained rubbers. Their resistance in the aggressive environment may be quite different when strained, like under the action of ozone on unsaturated rubbers.

14.2. Protection of Stressed Rubbers by the Change of the Size of the Tensile Stresses

In view of the fact that the intense failure which is accompanied by cracking occurs only in the presence of tensile stresses, one of the methods of retarding the failure is the reduction of the amount of stretching or the creation of compressive stresses in the surface layer. This method is employed in the manufacture of glass objects[33] (toughened glass) and metal articles[34, 35] (rolling, treatment by shot peening) to increase their resistance to static fatigue.

Some increase of strength can also be found in plastics when using such methods. It has been shown,[36] for example, that the strength of chilled polystyrene on the surface of which compressive stresses were created, increased by 20%.

The reduction of tensile stresses can be achieved by a change in the method of production, during which a compressive surface stress is created. Similar stresses may also appear as the result of swelling of the surface layer.

For the creation of compressive stresses rubber tubes (diameter 44 mm, 1·4 mm thick in the case of SKS-30, and 46 mm and 2·6 mm in the case of SKB) were cut lengthwise after vulcanization and cemented together on their interior surfaces. Both external surfaces of the resultant flat laminate were under compression; approximately 3–4% in the SKS-30 and 5–6% in the SKB. The testing of these laminates, which were bent to a tube of a diameter of 30 mm (SKS-30) and 40 mm (SKB) showed that the type of the internal stresses in the laminate has a great influence on the resistance to ozone cracking and that in the presence of compressive stresses the resistance increases greatly[37] (Table 28).

TABLE 28

Time to the Appearance of Cracks (in min) under the Attack of Ozone on Laminated Specimens (N are Laminates Cemented Together by their Outsides, B by their Insides, P—Flat Laminates)

Specimens	Ozone (5×10^{-4}%) + light	Ozone (10^{-3}%)	Ozone ($3{\cdot}7\times10^{-3}$%
SKB rubbers			
N	54	11	–
P	200	22	–
B	–	132	–
SKS-30 rubbers			
N	32	40	10
P	–	–	20
B	200	70	85

Swelling produces an especially striking, though temporary, protective effect (Table 29).

In proportion to the diffusion of the agent into the depth of the specimen the protective effect decreases, and with its even distribution the resistance to ozone becomes smaller than the initial one.

In order to keep the extent of swelling of the surface constant, the polarity of the rubber was changed by a preliminary treatment of the surface; then the specimen was placed in a swelling agent which did not combine with the rubber itself. For this pur-

TABLE 29

Time to the Appearance of Cracks (in min) Under the Attack of Ozone (6×10^{-4}–3×10^{-3}%) on Rubber Samples which had been Treated with Swelling Agents

Swelling agents (action 20 min at 90°C)	Ozone; light static strain 20%			Ozone; dynamic strain 15–25%		
	SKB	natural rubber	SKN-20	SKB	SKB	SKB-30
Without swelling	10	30	6	12	4	14
Rosin antioxidant	500	87	120	320	300	200
Mineral oil	–	–	–	60	–	–
Castor oil	100	150	–	–	100	145

pose triethanolamine was used, in which only the surface layer of natural rubber swelled; this had been pretreated with sulphuric acid.

14.3. Protection of Rubbers by the Creation of an Inert Surface Film

To protect rubber from ozone its surface is made inert by a treatment which leads to a reduction of the unsaturation of the polymer(1) (hydrogenation, treatment with mercaptans) or by covering the surfaces with ozone resisting polymers or by the introduction of waxy substances into the rubber mix which migrate to the surface. The latter method(38) is the most widespread one; the waxes are employed either alone or in combination with antioxidants.

It is usually considered that the waxy substances have a protective effect only under static deformations. There are, however, indications that in individual cases they are also effective with dynamically strained rubbers.(1, 39–41)

The basic parameters which determine the protective properties of the waxy substances are not yet wholly understood.

An attempt to connect these properties with the "basic viscosity" (basic viscosity is the value which is obtained from the temperature dependence of the viscosity of the molten wax extrapolated to 25°C, i.e. to a temperature at which the wax is solid) was not successful, of which the diametrically opposed conclusions which were drawn in two papers, bear witness. According to one reference the protective wax should possess a small basic viscosity,[42] according to the other[43]—a large one.

It follows from the majority of papers that waxes which have good protective properties should be microcrystalline in order to produce a solid layer on the surface, and the temperature range

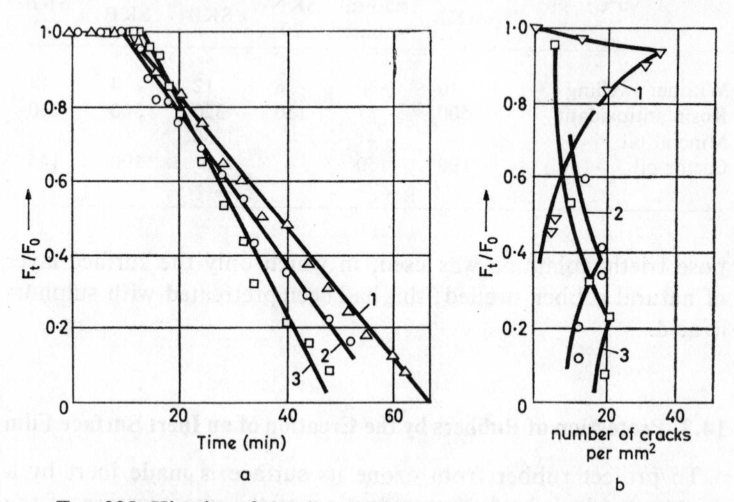

FIG. 205. Kinetics of the drop of force (stress) (a) and change of the number of cracks (b) during ozonization of rubber, stretched by 20% (SKS-30, unfilled). 1. Without ceresine. 2. 2 g ceresine per 100 g rubber. 3. 5 g ceresine per 100 g rubber. Ozone concentration 0·0007%.

of their softening points must agree with the intended temperature at which the rubber articles will be used. The study of the influence of waxes on the durability of rubber by objective methods (by the kinetics of the drop of the force sustained by the stretched sample and by the kinetics of the change of the number of cracks) has shown[44] that at great strains less cracks form on rubber which contains wax than on rubber without wax. Owing to the great overstress at the tips of the cracks the drop of the force which characterizes the rate of the crack propagation proceeds faster and the time to failure is shortened. The more wax is added the smaller is the durability of the rubber (i.e. the time in which F_t/F_0 becomes zero) at great deformations in ozone (Fig. 205), since the number of cracks decreases.† Under small strains (in investigations of SKS-30 rubber with ceresine at $\varepsilon =$ 10%) the wax shows a protective effect. This derives from the fact that we have on the surface of the rubber a solid layer (it fails at great strains) which causes the appearance of the induction period τ_i, owing to which the durability increases as well (Fig.

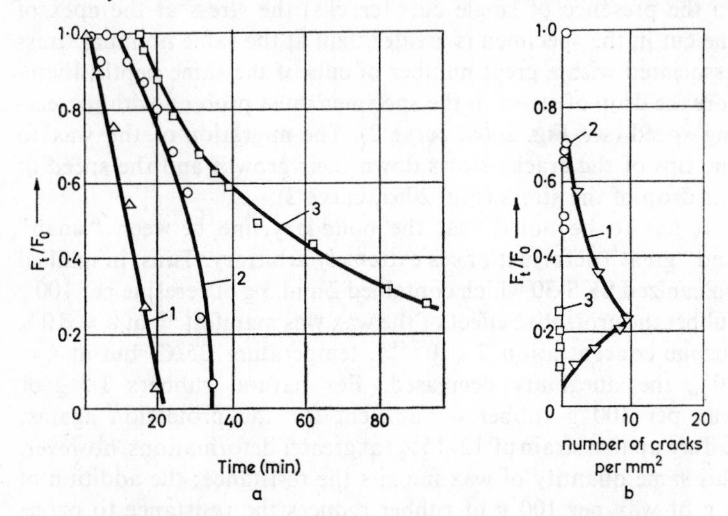

FIG. 206. Kinetics of drop of force (stress) (a) and change of number of cracks (b) during ozonization of rubber, stretched by 10% (SKS-30, unfilled). 1. Without ceresine. 2. 2 g ceresine per 100 g rubber. 3. 5 g ceresine per 100 g rubber.

206a, curve 2). When the wax forms a sufficiently great proportion of the rubber the growth of durability under small deformations is caused not only by the presence of the film of wax which has formed on the surface and which increases the induction period but also by the reduction in the rate of crack growth. This can be observed in those cases where the migration speed of the wax to the rubber surface is comparable to the rate of crack propagation (see, for instance, Fig. 206a, curve 3).

† For a clearer graphic description and convenience of comparison the kinetics of the change of the number of cracks is depicted on Figs. 205b and 206b not in direct relation to time but to the size of the drop of the force which decreases with time.

A characteristic peculiarity of the failure of rubbers which contain wax is the development of single cracks under small strains. The drop of the stress in the specimen must proceed in this case in a different manner than in the presence of a great number of cracks. This can be seen on p. 281, where data are shown on the drop of the stress in a stretched specimen on which a greater and greater number of pairs of incisions have been made. In the presence of single cuts (cracks) the stress at the apex of the cut in the specimen is greater than at the same nominal stress associated with a great number of cuts of the same depth; therefore the drop of stress in the specimen must proceed with increasing speed (see Fig. 206a, curve 2). The migration of the wax to the tips of the cracks slows down their growth and the speed of the drop of the stress (Fig. 206a, curve 3).

It has to be noted that the boundary line between "small" and "great" deformations is extremely arbitrary. Thus, in unfilled vulcanized SKS-30 which contained 2 and 5 g of ceresine per 100 g rubber the protective effect of the wax was manifest[44] at $\varepsilon = 10\%$ (ozone concentration $7 \times 10^{-4}\%$, temperature 25°C) but at $\varepsilon = 30\%$ the durability decreased. For natural rubbers 1·5 g of wax per 100 g rubber is sufficient for the protection against failure up to a strain of 12–15%; at greater deformations, however, this same quantity of wax impairs the resistance; the addition of 3 g of wax per 100 g of rubber reduces the resistance to ozone sharply under all deformations.[45] According to other data[46] the wax in proportions of 1–5 g per 100 g rubber protects natural rubber at deformations up to $\varepsilon = 12\%$, but at greater strains it reduces its resistance to the attack by ozone. There are indications that the deformation of rubbers at which one can count on the protective effect of the wax, does not exceed 30–50%.[1]

If we take into account that the protective effect of wax is connected with the continuity of its film (depending on the plasticity of the wax) and with the migration speed onto the rubber surface, it is clear that such factors as temperature, ozone concentration and strain must have a considerable influence on the efficiency of this method of protection. At temperatures below the softening temperature of the wax, the wax film loses its ductile properties and becomes brittle. This leads to an undesirable localization of the attack by the ozone along the sparse cracks which grow rapidly, as the migration of the wax to the surface of the growing cracks does under these conditions practically not take place. With rising temperature the solubility of the wax in the rubber mass increases and the protective effect is weakened as well.[1]

It is assumed that waxes which have a softening point of about 80°C lose their protective properties already at 50°C, those with a softening point of 53°C lose them at 30°C. A softening temperature of paraffin waxes of 60–65°C,[42] 65–72°C[47] is considered the most favourable one from the point of view of the possibility of formating a protective layer. However, in our opinion, the greatest protective effect of wax must be found when the temperature at which the rubber articles are used approaches the softening point of the wax, i.e. when the wax film is plastic. Waxes which are used at present for the protection of technical rubber articles at room, or lower, and also at somewhat higher temperatures, do obviously not possess the most favourable properties. This explains also the harmful effect of adding wax to rubber on the resistance to ozone, which is extremely frequently observed.

A smaller proportion of paraffin wax (1%) is needed for the protection of the polar rubbers (polychloroprenes, butadiene-nitriles) than for the protection of non-polar rubbers (3%), which is obviously connected with the lesser solubility of wax in the former. This may also be the cause of the good protection from ozone of the polar synthetic wax Agawax 290 in non-polar butyl rubber.[48]

With an increase of the ozone concentration the speed of its diffusion through the surface film of wax increases which leads to the decrease and disappearance of the induction period and, consequently, of the protective effect of the wax. The rate of crack propagation grows, and at that ozone concentration at which it exceeds the diffusion speed of the wax, the latter's protective effect, which is caused by it, ceases. With the addition of a great quantity of wax the speed of its diffusion to the tips of the cracks increases, and their growth slows down. The addition of 1% wax to rubber which is stretched by 20% does not have anymore a noticeable effect with an ozone concentration exceeding $0{\cdot}5\times 10^{-4}\%$; the addition of 2% wax protects rubber up to an ozone concentration of $1{\cdot}5\times 10^{-4}\%$, and 4% wax up to a concentration of $5\times 10^{-4}\%$.[1]

The resistance of rubber to ozone cracking immediately after vulcanization is smaller than after ageing, since as a result of ageing both the thickness of the protective film and its evenness increase.[1] An extremely essential role in the protective properties of wax is attributed to the dimensions of the wax crystals which are on the rubber surface, and it is assumed that microcrystalline waxes are more effective. The microcrystalline structure is characteristic for the naphtenic hydrocarbons which are contained in ozocerite, the macrocrystalline structure for the

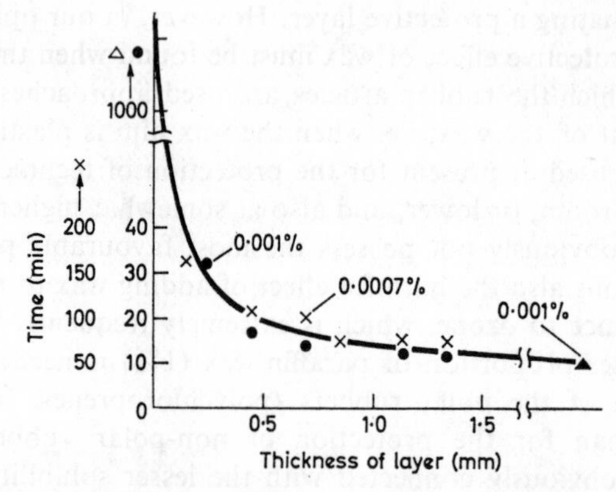

FIG. 207. Dependence of time to appearance of cracks on the thickness of the rubber layer underneath the fabric layer, at different ozone concentrations. △ — specimen without fabric.

normal paraffins.[49] The paraffins possess also a greater migration speed than the ceresines (ozocerites) (meaning fractions with melting temperatures which are usually chosen for protective waxes).[50] This leads to the opinions expressed that a mixture of paraffins and ceresines must possess the best protective properties.

Hardly any research has been made into the use of waxy substances for the protection of rubber from other aggressive media, apart from ozone. It is only known that in some cases the addition of paraffin wax to poly(isobutylene) improves its acid resistance[20] considerably. The increasing softening of Nairit rubber with the addition of mineral oil also improves its acid resistance.[51]

The use of coverings from reinforcing fabric or metal is also a method which partially assists with the acid resistance by decreasing the tensile stresses and assures the separation of the rubber from the aggressive medium. The resistance to ozone cracking in this case is increased more than one hundred times, and a sharp increase of resistance can be observed if the fabric is vulcanized on to the rubber surface (Fig. 207).

14.4. Chemical Means of Protecting Rubber from the Attack by Ozone

The protective effect of antiozonizers (antioxidants) is most effective on poly(chloroprene) rubbers. The addition of the usual inhibitors into them—phenyl β-naphtylamine,[3] its blends with *N*-phenyl-N′-cyclohexyl-n-phenylenediamine (1:1),[52] and also of such antiozonizers as UOP-88, UOP-288, Santoflex AW, Flectol H and others,[53] increases the resistance to ozone cracking more than ten times. For BSK rubbers the most effective ones are *N*-isopropyl-*N*′-phenyl-n-phenylenediamine (4010 Na),[39] *N*-phenyl-*N*′-cyclohexyl-n-phenylenediamine (4010), *N*,*N*′-di-*sec.*-octyl-n-phenylenediamine, 6-ethoxy-2,2,4-trimethyl-1,2-dihydroquinoline,[54] Vingstey 100.[55] These antioxidants increase the time to the appearance of cracks by a factor of 4–5.

Natural rubbers are far less easy to protect. For them are effective: 6-ethoxy-2,2,4-trimethyl-2,2,4-dihydroquinoline,[56] *N*-isopropyl-*N*′-phenyl-phenylenediamine,[57] a blend of *N*,*N*′-diphenyl-n-phenylenediamine with *N*-phenyl-*N*′-cyclohexyl-n-phenylenediamine,[47, 58] and also aromatic sulphides and aromatic derivatives of carbamides.[59]

BNK rubbers require even greater proportions than natural rubbers.[60]

Stereoregular *cis*-polyisoprene differs little in its resistance to ozone from natural rubber but *cis*-poly(butadiene) is practically not protected by antioxidants. For methylvinylpyridine rubber (MVP) derivatives of n-phenylenediamine are not effective, as they react with organic halide compounds which are employed with rubbers of this type.[9] The use of the antioxidant 4010 NA for the protection of butyl rubber at great ozone concentrations (0·02%) is shown by a qualitative evaluation to reduce the resistance of these rubbers to ozone.[89]

Under great strains the efficiency of antioxidants is smaller, evidently, because the size of the surface grows and sharply increases the number of cracks.

Intensive research is carried out at present into new antioxidants. According to patent data, the following are recommended as antioxidants: dioxyalkyl substituted derivatives of n-phenylenediamine[61] [for instance, *N,N'*-di-(oxyhexyl)-n-phenylenediamine], nitrosoamines[62] (for instance, *N*-cyclohexyl-*N'*-nitroso-n-phenetidene), derivatives of aniline[63, 64] (4,6-di-methoxy-m-toluidene), carbazoles[65] (3-cyclohexylaminocarbazole or 3-benzylaminocarbazole), etc.[66] For the protection of light natural rubbers is suggested 2,6-dialkoxy-4-alkylphenol in the proportion of 0·25–10 g per 100 g rubber (for instance, 2,6-dimethoxy-4-propylphenol).[67]

The protective effect of antioxidants ceases after 1½ to 2 years' storing due to its gradual deterioration (oxidation, volatility).[68] The type of carbon black which is added to the rubber has an essential influence on the protective effect of the antioxidants. The more the carbon black surface is oxidized, the greater is the quantity of the antioxidant which is bound, and the weaker is its protective effect.[69]

At present one can postulate several mechanisms of the protective action of the antioxidants.

1. Antioxidants like the derivatives of n-phenylenediamine which migrate to the rubber surface, form a barrier on it between the ozone and the rubber which acts as a rival to the rubber in the reaction with ozone.[10] This is confirmed by the fact that the viscosity of solutions of rubber which contain antioxidants does not increase until all the antioxidants have been used up.[70, 71] It has been proposed that, as the antioxidant is used up, it is replenished on the surface by diffusion from the mass of the specimen. However, a calculation of the migration speed of one of the antioxidants (on the basis of the experimentally found coefficient of its diffusion) showed[72] that under conditions, under which a strong ozone protective action of this substance is displayed, its surface concentration is smaller by an order of 6, than would be necessary for a full interaction with ozone. Besides, the migration speed of the antioxidant onto the rubber surface passes in the course of time over a sharp maximum, and the efficiency of the antioxidant does not change during this time. These

data prove that the mechanism of the action of the antioxidants is more complicated than has been supposed.

2. The addition of phenyl-β-naphthylamine to Nairit rubber does not increase the induction period τ_i but sharply impedes the rate of crack propagation.[11] From this the conclusion can be drawn that the ozone-protective action is caused in this case by a different mechanism than in the first case. It is possible that phenyl-β-naphthylamine acts as an inhibitor of the oxidation, and that its influence on the process is connected with its very good solubility (the solubility of phenyl-β-naphthylamine in poly(chloroprene) exceeds its solubility in other rubbers by a factor of 3–4[73]). However, the first mechanism of the action of the antioxidant is connected with its bad solubility and its migration to the surface.

There can be no doubt that the action of ozone is accompanied by oxidizing processes of a chain character, proceeding, as is known, with the action of ozone on hydrocarbons, including aliphatic,[74–77] and also on SKB.[78] The oxidizing processes which are accompanied by degradation must accelerate the cracking; they also play an essential part under the action of ozone on technical rubbers.[79, 80] Sometimes the oxidizing processes during ozonization are inhibited to a considerable extent by those antioxidants with which the rubber is stabilized, and they can, consequently, be disregarded. This is proved by the very low value of the activation energy of the ozonization of natural rubber and SKS-30. Evidently, under the effect of ozone on poly(chloroprene) rubbers the oxidizing processes play also some part; the comparatively great value of the activation energy (8–10 kcal per mol) of the ozone cracking of these rubbers may also be related to it. The circumstance that the processes which develop with chemical relaxation, fatigue and, possibly during ozone cracking, are noticeably slowed down by derivatives of n-phenylenediamine, particularly by the antioxidants 4010 and 4010 NA[81] has also an influence on the undisputed role played by the oxidizing processes during ozone cracking, and on their characteristic performance in stressed rubbers. A third point of view on the mechanism of the effect of antioxidants must here be stated.

3. It is assumed that derivatives of n-phenylenediamine interact with the peroxides which form under the reaction of rubber and ozones; this is accompanied by vulcanization which prevents the

development of cracks.[82] This suggestion has recently been developed experimentally. In the ozonization products of a sample of *cis*-poly(butadiene) (2-butene) and of a sample of poly(isoprene) (2,6-dimethyl-2,6-octadiene[83]) with and without certain antioxidants, more than eighteen different products have been discovered by means of chromatographic analysis. During ozonization of vulcanized *cis*-poly(butadiene) the kinetics of the consumption of the antioxidant and the kinetics of its combining with the polymer have been determined. The present authors showed that the antioxidants act at least in three ways: (a) they react quickly with ozone; (b) they interact actively with ozonides and slowly with polymer peroxides; (c) they react actively with the aldehydes which form during ozonization, giving polymeric compounds. As a result of these reactions a protective film is formed on the polymer surface and molecules which have been degraded during ozonization cross-link. This point of view is also held by Andrews and Braden[84] who, however, consider that the prevalence of the one or the other of these reactions depends on the ozone concentration. Reaction (a) predominates under great concentrations. In the case of natural rubbers the reaction (c) leads to the appearance of a brittle, tough surface layer which is not capable of a further interaction with ozone and which impedes its diffusion into the specimen.

Antioxidants are often employed together with waxes; they improve their solubility and their migration to the surface. In some cases a reduction of the protective effect of the antioxidants under static deformations[56, 84] can be observed in the presence of wax, in others an improvement.[57, 71, 85–87] In the majority of papers it is shown that the presence of wax does not improve the protective effect of the antioxidants at dynamic deformation, and at a concentration of 4% even reduced it. On the other hand, there are indications that a combination of antioxidants with wax has a favourable influence also at dynamic strains.[86, 89, 90]

In order to increase the concentration on the surface antioxidants have recently been directly applied to the article itself, if the antioxidant is liquid; but if it is solid then it is applied in the form of concentrated solutions or aqueous dispersions.[88] Such a method gives a considerable increase of the resistance of rubber to ozone cracking (by more than ten times). Dispersions which also contain microcrystalline wax, besides antioxidants, are also applied. In this case the wax is not meant to play the part of the solvent and carrier of the antioxidant but is a protective means both for the rubber and for the antioxidant, and protects it from oxidation.

With a view to increasing their concentration as carriers for the antioxidants on the rubber surface it has been suggested[91] to use liquid soaps of polyvalent metals, liquid organic polysulphides, amino alcohols and other compounds. At present, as recommended by Monsanto, the application of blends of microcrystalline wax, *N*-phenyl-*N*'-isopropyl-n-phenylenediamine and β-ethoxy-2,2,4-trimethyl-1,2-dihydroxy-quinoline in the relation 3:0·5:1 has been accepted to protect rubber from ozone under static strain, and also under dynamic, alternating with static strains.

Literature

1. Yu. S. Zuyev, A. Z. Borshchevskaya, *Metodiy zashchitiy rezin of ozonogo rastreskivaniya*, izd. NIIRP, 1957.
2. L. A. Smirnova, *Khimiya i tekhnologiya polimerov*, No. 11, Izdatinlit, 1960, p. 33.
3. Yu. S. Zuyev, V. F. Malafeyevskaya, *v sb. "Stareniye i zashchita rezin"*, Goskhimizdat, 1960, p. 27 (*Trudiy NIIRP*, No. 6).
4. J. Szurrat, *Gummi und Asbest*, **13,** No. 2, 74 (1960).
5. V. F. Malafeyevskaya, Yu. S. Zuyev, *Kauchuk i rezina*, No. 2, 35 (1958).
6. Yu. S. Zuyev, *Khim. prom.* No. 8, 55 (1954).
7. G. Natta, G. Cresp, E. D. Giulio, G. Ballini, M. Bruzzone, *Rubb. Plast. Age*, **42,** No. 1, 53 (1961).
8. A. S. Novikov, K. F. Kaluzhenina, *Khim. prom.* No. 1, 21 (1954); *Rubb. Plast. Age*, **41,** No. 7, 771, 829 (1960).
9. Z. T. Ossefort, *Rubb. Chem. Technol.* **32,** 1088 (1959).
10. B. S. Biggs, *Rubb. Chem. Technol.* **31,** 1015 (1958).
11. Yu. S. Zuyev, S. I. Pravednikova, G. V. Kotel'nikova, *Kauchuk i rezina*, No. 11, 15 (1961); No. 3, 21 (1962).
12. *Kauchuki spetsial'nogo naznacheniya*, edited by I. V. Garmonova, Izd. AN SSSR, VINITI, 1961.
13. R. Moakes, C. W. Payne, *J. Rubb. Res.* **19,** 77 (1950).
14. A. L. Labutin, *Khim. nauka i prom.* **2,** No. 3, 359 (1957).
15. B. A. Dogadkin, *Fizika i khimiya Kauchuka*, Goskhimizdat, 1947.
16. F. D. McNamee, *Chem. Eng.* **61,** No. 8, 234; No. 9, 230; No. 10, 233 (1954).
17. A. L. Labutin, *Kauchuki v antikorrozionnoi tekhnike*, Goskhimizdat, 1962.

18. *RTM 22–61, Pokriytiya zashchitniyye gummirovaiyem*, Standartgiz, 1961.
19. *Rubb. World*, **141**, 770 (1960); **142**, No. 1, 40, 127 (1960); *Rubb. Plast. Age*, **41**, No. 7, 829 (1960).
20. I. Ya. Klinov, M. A. Vorob'yeva, *Trudiy Moskovskogo instituta khimicheskogo mashinostroyeniya*, vol. 22, Mashgiz, 1960, p. 159.
21. W. A. Heath, *Corr. Pres. Const.* **5**, No. 1, 49 (1958).
22. A. S. Kuz'minskii, Ye. V. Shemastina, *ZhPKh*, **30**, 433 (1957).
23. A. F. Postovskaya, A. S. Kuz'minskii, *ZhFKh*, **33**, 447 (1959); A. F. Postovskaya, M. A. Salimov, *Vestnik Moskovskogo universiteta*, No. 3, 24 (1960).
24. A. N. Kornev, *Avtoreferat dissertatsii, MITKhT im. M. V. Lomonosova* 1958; F. F. Koshelev, A. Ye. Kornev, *Kauchuk i rezina*, No. 3, 16 (1958).
25. I. Ya. Klinov, D. I. Siychev, *Trudiy Moskovskogo instituta Khimicheskogo mashinostroyeniya*, vol. 12, Mashgiz, 1957, p. 95.
26. G. S. Uitbi, *Sinteticheskii kauchuk*, Goskhimizdat, 1957, p. 848.
27. N. L. Ketton, *Neopreniy*, Goskhimizdat, 1958.
28. A. L. Labutin, *Korroziya i sposobiy zachchitiy oborudovaniya v proizvodstve organicheskikh kislot*, Goskhimizdat, 1959.
29. A. S. Moran, *Gummi und Asbest*, **13**, No. 12, 1006 (1960).
30. N. V. Biryukov, *Zashchita Khimicheskogo oborudovaniya ot Korrozin*, byull. NIITEKhIM, 1960, p. 22.
31. A. P. Bogayevskii, S. K. Zherebkov, Ye. M. Grozhan, A. D. Chelmodeyev, *Kauchuk i rezina*, No. 12, 11 (1962).
32. A. P. Bogayevskii, S. K. Zherebkov, Ye. M. Grozhan, L. M. Polyakova, A. D. Chelmodeyev, *Kauchuk i rezina (v pechati)*.
33. G. M. Bartenev, *Mekhanicheskiye svoistva i teplovaya obrabotka stekla*, Gosstroiizdat, 1960.
34. *Korroziya metallov pod napriazhiyem i sposobiy zashchitiy pod red.* G. V. Akimova, kn. 31, Izd. AN SSSR, 1950.
35. N. A. Karasev, *Vestnik mashinostroyeniya*, No. 8, 70 (1953).
36. M. I. Bessonov, Ye. V. Kuvshinskii, *Viysokomol. soyed.* **1**, 1561 (1959).
37. Yu. S. Zuyev, *Khim. prom.* No. 9, 21 (1953).
38. A. I. Marei, M. P. Divova, *Stareniye i utomleniye kauchukov i rezin i noviysheniye ikh stoikosti*, Goskimizdat (1955), p. 185.
39. C. H. Avons, Firi, *Rubb. Plast. Weekly*, **141**, No. 17, 671 (1961).
40. A. E. Hoffman, W. L. Cox, *The Report on American Chemical Society Polymer Group North Jersey*, New York, May 25, 1959, Orange (Preprint).
41. G. T. Hodgkinson, C. E. Kendall, *Proceedings of the Fourth Rubber Technology Conference*, London, May, 1962, Preprint 57.
42. R. Musch, L. Prolss, *Kautschuk und Gummi*, **13**, WT34 (1960).
43. I. A. Levitin, Ye. N. Poloskin, V. D. Petrova, E. D. Marchenko, *Kauchuk i rezina*, No. 4, 14 (1963).
44. Yu. S. Zuyev, V. D. Zaitseva, *Kauchuk i rezina*, No. 2, 22 (1963).
45. H. A. Vodden, M. A. A. Wilson, *IRI Trans.* **35**, 82 (1959).
46. A. G. Veith, *Rubb. Chem. Technol.* **32**, 346 (1959).

47. B. van Pul, *IRI Trans.* **34,** 37 (1958).
48. *Rev. Gén. Caout.* **38,** No. 11, 1617 (1961).
49. N. I. Chernozhukov, *Izv. viysshikh uchebniykh zavedenii. Neft' i gaz,* No. 8, (1961); No. 11 (1962).
50. F. U. Wilcocs, *Rubb. Age,* **90,** 646 (1962).
51. V. L. Stezhenskii, *Trudiy Moskovskogo instituta khimicheskogo mashinostroyeniya,* vol. 22, Mashgiz, 1960, p. 179.
52. A. Hartmann, F. Glander, *Rubb. Chem. Technol.* **29,** No. 1 (1956).
53. R. M. Murray, *Rubb. Age,* **85,** 623 (1959).
54. H. W. Kielbourne, G. R. Wilder, J. E. van Verth, J. O. Harris, C. C. Tung, *Rubb. Chem. Technol.* **32,** 1155 (1959).
55. R. B. Spacht, W. S. Hollingshead, J. C. Lichty, *Kautschuk und Gummi,* **12,** 166 WT (1959).
56. *Rubb. World,* **135,** 723 (1957).
57. F. B. Smith, *Rubb. Age,* **85,** 619 (1959).
58. H. E. Albert, Z. O. Beutz, pat. U.S.A. 2822414, 4/II/1958; *Rubb. Abs.* **36,** 387, 3664 (1958); pat. U.S.A. 2822415, 4/II/1958; *Rubb. Abs.* **36,** 381, (1958).
59. H. E. Albert, pat U.S.A. 2822413 4/II/1957; pat. U.S.A. 2802892 13/VIII/1957; *C.A.* **52,** 777, (1958).
60. A. G. Buswell, J. T. Watts, *IRI Trans.* **37,** 175 (1961).
61. J. C. Ambelang, pat. U.S.A. 2929796, 22/III/1960.
62. D. J. Beaver, P. J. Stoffel, pat. U.S.A. 2931785, 5/IV/1960.
63. A. J. Green, pat. U.S.A. 2926155, 23/II/1960.
64. B. E. Wilde, M. J. H. Wilde, Brit. pat. 857628, 4/I/1961.
65. J. O. Harris, pat. U.S.A. 2921922, 19/I/1960.
66. H. E. Albert, pat. U.S.A. 2956981, 18/X/1960.
67. Brit. pat. 810529, 18/III/1959.
68. J. M. Willis, G. Allinger, R. S. McFadden, *Chemistry in Canada,* **11,** No. 11, 39 (1959).
69. *Rubb. Age,* **87,** 954 (1960).
70. A. D. Dellman, B. B. Simms, A. K. Allison, *Anal. Chem.* **26,** 10 (1954).
71. F. B. Smith, W. F. Tuley, *Rubb. World,* **140,** 243 (1959).
72. M. Braden, *J. Appl. Polymer Sci.* **6,** 56 (1962).
73. S. A. Reitlinger, A. S. Kuz'minskii, *ZhPKh,* **27,** 214 (1954).
74. K. Zugler, D. Wagner, H. Wilms, *Ann.* **567,** 99 (1950).
75. I. Durland, K. Adkins, *J. Am. Chem. Soc.* **61,** 429 (1939).
76. C. Clarence, S. I. Schubert, R. N. Pease, *J. Am. Chem. Soc.* **78,** 5553 (1956).
77. S. I. Schubert, R. N. Pease, *J. Am. Chem. Soc.* **78,** 2044 (1956).
78. A. S. Kuz'minskii, N. N. Lezhnev, Yu. S. Zuyev, *Okisleniye kauchukov i rezin,* Goskhimizdat, 1957.
79. Yu. S. Zuyev, A. F. Postovskaya, G. M. Podchufarova, *Kauchuk i rezina,* No. 7, 14 (1963).
80. Yu. S. Zuyev, A. F. Postovskaya, *Kauchuk i rezina v pechati.*
81. L. G. Angert, A. S. Kuz'minskii, *Rol' i primeneniye antioksidantov v Kauchakakh i rezinakh,* Goskhimizdat, 1957.
82. R. W. Murrey, D. R. Story, *Rubb. Age,* **89,** 115 (1961).
83. O. Lorenz, C. R. Parks, *Rubb. Chem. Technol.* **36,** 194, 201 (1963).

84. E. H. ANDREWS, M. BRADEN, *J. Appl. Polymer Sci.* **7**, 1003 (1963).
85. W. L. COX, *Rubb. Chem. Technol.* **32**, 364 (1959).
86. T. H. NEWBY, *Chemistry in Canada*, **11**, No. 11, 52 (1959).
87. D. C. EDWARDS, E. B. STOREY, *Rubb. Age*, **79**, 787 (1956).
88. J. M. BUIST, G. E. WILLIAMS, Brit. pat. 815263, 24/IV/1959.
89. W. A. WILSON, *Rubb. Age*, **90**, 85 (1961).
90. R. KING, J. PAYNE, K. TAYLOR, *Proceedings of the Fourth Rubber Technology Conference*, London, May 1962, Preprint 54.
91. *Rubb. Plast Age*, **44**, No. 2, 159 (1963).

INDEX